Deltas of the World

Series Editor

Abdelazim Negm, Zagazig, Egypt

Worldwide, over 600 million people live in deltas. Extensive human activities in deltas, their supplying rivers, and related maritime currents increase the pressure on deltas' natural resources and habitat equilibrium, leading to degradation of the available resources and jeopardizing inhabitants' livelihoods – and in some cases even their lives.

This book series is intended to promote a sustainable future for deltas around the globe and the communities that live in them. Accordingly, we publish studies focusing on various aspects of deltas, including hydrogeology, soil pollution, water pollution, environmental hydrology, environmental hydraulics, groundwater engineering and management, morphology, anthropology, socioeconomics, wastewater management, water resources engineering and management, agriculture, greenhouse agriculture, irrigation and drainage engineering, hydraulic structures, geophysics and geology, sedimentology, mineralogy, remote sensing and GIS studies, climate change variabilities and impacts, marine-delta interactions and sustainability, and solid waste management. The findings presented here support the SDGs and Agenda 2030 with regard to helping countries with deltas and related environments to preserve their natural resources for future generations.

The editors invite scientists, researchers and scientific communities to contribute to the series by submitting delta atlases, edited books, monographs, and high-quality conference proceedings focused on deltas around the world.

Chiranjit Singha · Satiprasad Sahoo ·
Ajit Govind
Editors

Sustainable Management of Delta Ecosystems Resilience

Case Studies on Delta Adaptation Pathways

Springer

Editors
Chiranjit Singha
Department of Agricultural Engineering,
Institute of Agriculture
Visva-Bharati (A Central University)
Sriniketan, West Bengal, India

Satiprasad Sahoo
International Center for Agricultural Research
in the Dry Areas (ICARDA)
Maadi, Cairo, Egypt

Ajit Govind
International Center for Agricultural Research
in the Dry Areas (ICARDA)
Maadi, Cairo, Egypt

ISSN 2731-832X ISSN 2731-8338 (electronic)
Deltas of the World
ISBN 978-3-032-16260-1 ISBN 978-3-032-16261-8 (eBook)
https://doi.org/10.1007/978-3-032-16261-8

This Springer imprint is published by the registered company Springer Nature Switzerland AG
The registered company address is: Gewerbestrasse 11, 6330 Cham, Switzerland

Preface

Deltas stand among the most vibrant yet vulnerable landscapes on Earth. These dynamic environments where the flow of rivers meets the pulse of the sea sustain rich biodiversity, fertile agricultural lands, and millions of people. Yet, they now face unprecedented pressures. Rapid urban expansion, unsustainable resource extraction, pollution, and the intensifying impacts of climate change have begun to erode the ecological and social foundations of these regions. The urgent need to safeguard these intricate systems forms the core motivation for this book.

Delta Ecosystems Resilience brings together a diverse body of research that examines the multifaceted challenges confronting deltaic ecosystems, with particular emphasis on regions such as the Indian Sundarbans. By weaving insights drawn from remote sensing, GIS applications, ecological assessments, and socio-economic analyses, the volume showcases how scientific innovation can illuminate the pathways toward environmental stability. Across the chapters, satellite-based monitoring reveals hidden patterns of landscape transformation, field-based ecological studies document the stresses facing mangroves and wetlands, and socio-economic investigations highlight the everyday vulnerabilities of delta communities.

A central message emerges from this compilation: resilience cannot be achieved through isolated interventions. Protecting deltas requires an integrated and inclusive framework that balances nature-based solutions with sustainable agriculture, efficient wastewater management, and participatory governance. Mangroves and wetlands must be preserved not merely as ecological assets but as frontline defenses against climate extremes. Community livelihoods and institutional mechanisms must be strengthened to withstand shocks. And technological tools ranging from vegetation indices to precision agriculture and environmental modelling must be leveraged to support evidence-based policy and planning.

The chapters that follow reflect this holistic vision. They explore flood-induced agricultural losses, ecosystem quality fluctuations, wetland evolution, mangrove decline, livelihood challenges, pollution dynamics, wastewater management, agro-ecosystem monitoring, precision farming practices, bioremediation applications, urbanization impacts, industrial pressures, and climate-driven land-use transformations. Together, these contributions provide a comprehensive understanding of the environmental and human dimensions of delta sustainability.

This book is intended for researchers, environmental scientists, geographers, policy analysts, and students seeking to deepen their knowledge of delta systems and resilience-building strategies. It also aims to serve as a practical reference for planners and decision-makers charged with shaping the future of these fragile yet vital environments.

Ultimately, *Sustainable Management of Delta Ecosystems Resilience* underscores a simple but powerful truth: the survival of delta ecosystems and the millions who depend on them rests on our collective ability to integrate science, technology, and community-centered action. By bringing together diverse perspectives and approaches, this volume aspires to contribute to the global conversation on building resilient, equitable, and sustainable deltas for generations to come.

Chapter Summaries.

Chapter 1 uses remote sensing and GIS tools to examine how floods affect vegetation and agricultural land. By analyzing satellite images from different times and applying indices like NDFI and DVDI, it maps where and how severely crops have been damaged. The findings offer practical information that can help guide flood management strategies and support sustainable farming in vulnerable delta regions.

Chapter 2 examines how weather and environmental conditions shape the resilience of delta ecosystems. By analyzing remote sensing data from different seasons and applying ecological indicators, it tracks changes in vegetation health over time and space. The results show that mangrove and coastal areas become particularly stressed during the pre-monsoon dry period. These insights highlight the need for adaptive management, mangrove rehabilitation, and improved water conservation to maintain ecological stability in a changing climate.

Chapter 3 explores how a developing wetland in West Bengal has formed and evolved by analyzing its landforms, water flow patterns, and biodiversity. Using combined geological mapping and ecological surveys, it shows how rock types, water movement, and surrounding land use shape the long-term health of the wetland. The chapter highlights the importance of wetlands as natural shields against climate-related hazards and as key elements supporting resilience and productivity in delta environments.

Chapter 4 assesses the overall condition and species richness of mangrove ecosystems using both field surveys and satellite data. It pinpoints zones where vegetation cover is decreasing and salinity levels are rising. The chapter calls for community-led mangrove protection and climate adaptation measures, emphasizing that restoring these ecosystems also strengthens local livelihoods and enhances resilience to natural hazards.

Chapter 5 explores how environmental pressures affect the social and economic well-being of delta communities. It shows that rising sea levels, increasing soil salinity, and repeated cyclones are severely impacting traditional livelihoods like farming and fishing. By reviewing governance practices and institutional actions, the study stresses the need for coordinated policies, community participation, and diversified livelihood options to strengthen both social and ecological resilience in delta regions.

Chapter 6 focuses on the serious issue of pollution and declining water quality in delta regions. It reviews how much wastewater is produced, how much is treated, and how it is disposed of across different delta areas, showing the harmful impacts of untreated waste on soil and water. The study recommends sustainable reuse practices, nature-based treatment methods, and decentralized management systems to improve environmental health and long-term sustainability in deltas.

Chapter 7 extends the discussion on pollution by pinpointing major contamination hotspots and assessing strategies to reduce environmental damage. Using geospatial tools and environmental models, it analyzes how farm runoff, industrial waste, and growing urban areas affect delta ecosystems. The findings support the use of nature-based solutions, integrated water management, and cooperative governance to protect ecological health.

Chapter 8 shows how satellite observations can be used to evaluate crop conditions and environmental pressures in delta regions. By combining vegetation indices with geospatial modeling, it maps crop health, soil moisture levels, and patterns of land-use change. These insights support precision farming and help promote sustainable resource management in sensitive delta agro-ecosystems.

Chapter 9 explores how modern technologies can enhance agricultural efficiency. It shows how satellite images, GIS tools, and spatial analysis can help plan irrigation, assess soil fertility, and estimate crop yields more accurately. The findings demonstrate how geospatial intelligence can boost farm productivity while reducing environmental pressures.

Chapter 10 presents an eco-biotechnological approach by examining how bioremediation can help repair polluted soils and water bodies. It shows that microbial and plant-based remediation methods can reduce contaminants, improve soil health, and support sustainable farming in delta areas impacted by industrial and chemical pollutants.

Chapter 11 examines fast-paced urban growth and its environmental impacts using satellite images from different years. It shows a sharp rise in built-up areas and growing thermal hotspots, which intensify the urban heat island effect and increase ecological stress. The findings highlight the need for sustainable urban planning and stronger land-use controls to protect the resilience of delta ecosystems.

Chapter 12 offers a broader global view by examining environmental pressures in Southeast Asian deltas. It shows how fast industrial growth and extensive land reclamation have disrupted natural water systems, damaged wetlands, and intensified pollution. The study emphasizes the need for integrated coastal zone management and sustainable development strategies to balance economic expansion with environmental protection.

Chapter 13 brings together the central themes of the book. Using satellite data over time, it shows how changes in land cover are influenced by climate shifts and human activities. The study provides a broad understanding of how climate, land use, and ecosystems interact, emphasizing the importance of adaptive management, innovative policies, and advanced technologies for supporting sustainable delta development.

Together, the chapters provide an in-depth examination of the environmental, technological, and socio-economic aspects of managing delta ecosystems. By combining tools such as remote sensing, GIS, machine learning, and community-based governance, the volume advances data-informed strategies for building resilience and promoting sustainable delta management in a changing climate.

Sriniketan, West Bengal, India Chiranjit Singha
Cairo, Egypt Satiprasad Sahoo
Cairo, Egypt Ajit Govind

Contents

Flood-Induced Crop Damage Assessment Using DVDI and Geospatial Methods in Hugli District of West Bengal in India

1

Rajib Patra, Tanmoy Basu, and Biraj Kanti Mondal

Abstract

The study investigates the impact of flooding on vegetation health and agricultural productivity in Hugli district, West Bengal, between 2022 and 2024 using remote sensing and statistical modeling techniques. Flooding is a recurrent hazard in the region, significantly affecting cropland, soil health, and ecosystem stability, necessitating a systematic assessment of flood-induced vegetation stress. The primary objective of this study is to analyze the relationship between flood intensity and vegetation damage using the Normalized Difference Flood Index (NDFI), Normalized Difference Vegetation Index (NDVI), Vegetation Condition Index (VCI), and Disaster Vegetation Damage Index (DVDI). The study utilizes Bhuvan data to derive spatio-temporal flood and vegetation indices, combined with GIS-based (Geographic Information System-based) buffer analysis and regression modeling to examine the extent of crop damage. The results indicate that flood severity increased in 2024, with the highest DVDI values (> 0.65) recorded in river-adjacent farmlands of Khanakul-II, Singur, and Chinsurah-Magra, while higher-elevation agricultural zones (e.g., Pandua, Tarakeswar) exhibited minimal damage (DVDI < 0.2). The regression model ($R^2 = 0.152$, N is 1981) confirms a statistically significant negative correlation (− 0.63) between NDFI and DVDI, validating flood impact predictions. The findings emphasize the need for adaptive flood mitigation strategies, improved drainage infrastructure, and sustainable land-use planning. This study provides a quantitative basis for disaster management policies and

R. Patra · B. K. Mondal (✉)
Department of Geography, Netaji Subhas Open University, Kolkata, West Bengal, India
e-mail: birajmondal.kolkata@gmail.com

T. Basu
Department of Geography, Katwa College, Katwa, Purba Bardhaman, West Bengal, India

C. Singha et al. (eds.), *Sustainable Management of Delta Ecosystems Resilience*,
Deltas of the World, https://doi.org/10.1007/978-3-032-16261-8_1

agricultural resilience planning, while future research should incorporate hydrological modeling and long-term trend analysis for improved predictive accuracy.

List of Abbreviations

AI	Artificial Intelligence
AWiFS	Advanced Wide Field Sensor
CNN	Convolutional Neural Network
COV	Coefficient of Variation
CWC	Central Water Commission
DEM	Digital Elevation Model
DOS1	Dark Object Subtraction (Method 1—Atmospheric Correction)
DN	Digital Number
DVDI	Disaster Vegetation Damage Index
DVC	Damodar Valley Corporation
FAO	Food and Agriculture Organization
FLAASH	Fast Line-of-sight Atmospheric Analysis of Spectral Hypercubes
GCP	Ground Control Point
GIS	Geographic Information System
GPS	Global Positioning System
ICAR	Indian Council of Agricultural Research
IMD	India Meteorological Department
IPCC	Intergovernmental Panel on Climate Change
ISRO	Indian Space Research Organisation
LISA	Local Indicators of Spatial Association
LMCI	Lower Marginal Confidence Interval
LULC	Land Use Land Cover
MAE	Mean Absolute Error
MLC	Maximum Likelihood Classifier
mVCI	Median Vegetation Condition Index
NDFI	Normalized Difference Flood Index
NDVI	Normalized Difference Vegetation Index
NIDM	National Institute of Disaster Management
NIR	Near-Infrared
NRSC	National Remote Sensing Centre
PMFBY	Pradhan Mantri Fasal Bima Yojana
R2	Correlation Coefficient
RMSE	Root Mean Square Error
SAPCC	State Action Plan on Climate Change
SDGs	Sustainable Development Goals

SRTM	Shuttle Radar Topography Mission
SWIR	Short-Wave Infrared
UMCI	Upper Marginal Confidence Interval
VCI	Vegetation Condition Index
VIF	Variance Inflation Factor
WBDMD	West Bengal Disaster Management Department
WBIWD	West Bengal Irrigation and Waterways Department
WBMIFMP	West Bengal Major Irrigation and Flood Management Project
ZPR	Standardized Predicted Residual
ZRE	Standardized Residual Error

1.1 Introduction

Climate change and increasing hydrological variability have exacerbated the frequency and intensity of extreme weather events, particularly floods, which severely impact agricultural productivity and food security worldwide (Intergovernmental Panel on Climate Change 2021; Food and Agriculture Organization 2022). Agriculture in flood-prone regions faces mounting challenges due to erratic rainfall, rising temperatures, and shifting hydrological patterns, necessitating the development of climate-resilient agricultural systems. In India, where agriculture contributes nearly 17% to the national GDP and supports over 50% of the workforce, climate-induced disasters have emerged as a critical concern (Ministry of Agriculture & Farmers Welfare 2022). Among various climate-induced risks, floods remain among the most devastating, particularly in low-lying, riverine districts such as Hugli in West Bengal (Sharma et al. 2020). Several studies emphasize the significance of geospatial technologies in mapping and monitoring flood-prone agricultural landscapes. Remote sensing-based indices such as the Normalized Difference Vegetation Index (NDVI) and the Disaster Vegetation Damage Index (DVDI) have been extensively applied in assessing vegetation health and identifying anomalies caused by extreme climatic events (Huete et al. 2002). While NDVI has been traditionally used to monitor crop stress, recent advancements indicate that DVDI provides a more robust assessment of vegetation loss, particularly in flood-affected regions (Rahman et al. 2021). Recent trends in climate-resilient agriculture emphasize integrating real-time geospatial analysis with adaptive farming practices such as crop diversification, flood-resistant varieties, and precision agriculture techniques (Verrelst et al. 2021; Wang et al. 2023). In India, initiatives like the Pradhan Mantri Fasal Bima Yojana (PMFBY) and the National Adaptation Fund for Climate Change (NAFCC) underscore the urgency of incorporating scientific methodologies into disaster risk reduction strategies (Mondal et al. 2020). This study leverages geospatial technology and multi-temporal satellite data to assess flood-induced crop damage in the Hugli district. By employing DVDI, this research provides a comprehensive framework to evaluate spatial variations in vegetation health, bridging the gap between remote sensing applications and climate-resilient agricultural planning. The findings of this study contribute to enhancing adaptive strategies in flood-prone regions, ensuring long-term agricultural sustainability and food security.

1.1.1 The Rationale of the Study

Floods significantly disrupt agricultural activities, causing yield losses, soil erosion, and long-term degradation of farmlands. Hugli district in West Bengal is highly vulnerable to recurrent flood events due to its proximity to the Hooghly River and its low-lying physiography. The impact of floods on agricultural productivity in this region is exacerbated by unpredictable monsoonal variations and inadequate drainage infrastructure. Despite existing flood management strategies, there remains a lack of precise, data-driven assessments of crop damage at the block level, which is essential for targeted interventions and adaptive agricultural planning. Geospatial techniques provide a scientific basis for monitoring, assessing, and mitigating flood-related agricultural damage. The use of satellite-derived vegetation indices such as DVDI allows for a more accurate assessment of flood-induced stress on crops, surpassing traditional NDVI-based methodologies. This study is motivated by the need to enhance the precision of flood impact assessments using advanced remote sensing techniques, thereby contributing to climate-resilient agricultural development in Hugli. Furthermore, the study aligns with global and national priorities, such as the UN Sustainable Development Goals (SDGs), particularly Goal 13 (Climate Action) and Goal 2 (Zero Hunger). Strengthening agricultural resilience through scientific methodologies can support policymakers, agronomists, and farmers in adopting sustainable and adaptive practices to mitigate the adverse impacts of floods. The findings of this study will contribute to designing region-specific mitigation strategies, improving early warning systems, and promoting sustainable land-use practices to ensure long-term agricultural productivity and food security.

1.1.2 Contemporary Studies on the Issue and Literature on Adopted Methodologies

Flood-induced agricultural damage has been widely studied across the globe using remote sensing and geospatial technologies. Globally, studies by Kogan et al. (2019) have demonstrated the effectiveness of vegetation indices, including NDVI, in detecting flood-induced stress in crops. Research in Bangladesh (Rahman et al. 2021) and Thailand (Sangwan et al. 2020) highlights the applicability of NDVI and DVDI for monitoring post-flood agricultural recovery. In China, Liu et al. (2022) employed Sentinel-2 imagery to assess flood damage in rice fields, proving the efficacy of geospatial techniques for crop health monitoring. Studies in the United States and Australia have also integrated remote sensing with hydrological models to predict flood-prone agricultural zones (Jones et al. 2020; Anderson et al. 2019). Global Perspective Flood-induced crop damage has been a significant concern globally, with various studies highlighting the impacts of extreme weather events on agricultural systems. Research by Smith et al. (2020) emphasizes that climate change has intensified the frequency and severity of floods, causing substantial losses in food production. Studies in the United States and China have shown that remote sensing techniques, including NDVI and DVDI, are effective in assessing vegetation stress caused by flood events (Jones et al. 2020; Wang et al. 2021). In Europe, particularly in the Netherlands and Germany, geospatial analysis using Sentinel-2 and Landsat-8 imagery has proven crucial for flood monitoring and mitigation strategies (Tarpanelli et al. 2022). Furthermore, African nations have increasingly adopted GIS-based flood risk assessment models to improve agricultural resilience (Ahmed and Kamara 2021). Indian Perspective India experiences frequent flooding, particularly in the Indo-Gangetic Plains, leading to significant agricultural losses. Studies by Sharma et al. (2018) and Gupta et al. (2020) indicate that flood-prone states such as Bihar, Assam, and West Bengal suffer extensive damage to paddy crops due to monsoon-related inundation. The use of geospatial techniques in flood damage assessment has gained traction, with the National Remote Sensing Centre (NRSC) and Indian Space Research Organisation (ISRO) deploying satellite-based indices like NDVI and land use land cover (LULC) mapping for flood impact analysis (Rao et al. 2021). Additionally, the India Meteorological Department (IMD) and Central Water Commission (CWC) provide crucial hydro-meteorological data that assist in real-time flood forecasting and damage estimation (Mishra et al. 2019). West Bengal Perspective West Bengal, particularly the Hugli district, faces recurrent flooding due to its geographical location and extensive river network. Studies by Chatterjee and Saha (2017) highlight that the Damodar and Hooghly rivers contribute significantly to seasonal flooding, affecting crop productivity in low-lying areas like Khanakul and Arambagh. Remote sensing-based research conducted by the West Bengal State Disaster Management Department (WBDMD) has demonstrated that DVDI and NDVI are effective indicators for flood-induced vegetation loss (Mukherjee et al. 2023). Moreover, regional studies emphasize the necessity of integrating geospatial analysis with field-based surveys to enhance flood mitigation strategies (National Remote Sensing Centrel (2017). In India, recent research has focused on using GIS and remote sensing for flood damage assessments. Mishra et al. (2020) used MODIS NDVI time-series data to evaluate flood impacts in Bihar and Assam. Furthermore, researchers employed machine learning models integrated with remote sensing data to predict flood-prone agricultural areas. Kumar et al. (2023) demonstrated how geospatial data fusion improves accuracy in flood-prone agricultural mapping in Odisha and West Bengal. These studies underscore the significance of vegetation indices in capturing flood-induced crop damage with high accuracy. In West Bengal, flood-related agricultural losses have been extensively documented in the lower Gangetic plains (Chakraborty and Das 2018). Using Landsat-derived NDVI data, Sarker et al. (2021) assessed flood-induced vegetation stress in Murshidabad and Bardhaman districts, emphasizing the necessity of early warning systems. Studies by Mukhopadhyay et al. (2022) incorporated geospatial analysis with hydrological models to predict flood susceptibility zones in South Bengal. However, limited research has applied DVDI in this region to evaluate crop damage due to floods. This study aims to fill this gap by employing DVDI and other geospatial indices to develop a comprehensive flood assessment framework specific to Hugli district. Existing flood damage assessments in India often rely on conventional vegetation indices, which may not fully capture vegetation stress and recovery patterns. The application of DVDI in flood-induced crop damage assessment remains limited, particularly at high-resolution spatial scales. Additionally, most studies focus on broad regional analyses, lacking localized, block-level insights essential for targeted intervention strategies. Moreover, there is inadequate integration of hydrological and socio-economic parameters with remote sensing data, which is crucial for comprehensive flood risk assessment. This study addresses these gaps by employing multi-source remote sensing data and DVDI to provide a detailed, block-level flood damage assessment framework for Hugli district, contributing to more effective disaster mitigation and climate-resilient agricultural planning. Despite significant advancements in flood damage assessment methodologies, certain gaps remain. While remote sensing techniques have proven effective, their integration with socio-economic factors remains underexplored. Most studies focus on NDVI-based assessments, whereas newer indices like DVDI require further validation in Indian contexts. Additionally, real-time flood monitoring and early warning systems are yet to be fully integrated with agricultural planning in Hugli district. A more comprehensive approach, combining geospatial analysis, hydrological data, and socio-economic assessments, is necessary to enhance the resilience of flood-affected communities. The main objectives of the study are to assess block-wise variations in flood-induced crop damage using DVDI and geospatial techniques, analyze the correlation between flood events and vegetation health using multi-temporal satellite imagery, and identify priority areas for mitigation strategies based on spatial crop damage patterns.

The methodologies employed in flood-induced crop damage assessment predominantly rely on geospatial and hydrological models. NDVI and DVDI are widely used to quantify vegetation stress, as demonstrated by previous studies in flood-prone regions (Tucker 1979). DEM-based flood modelling (Digital Elevation Model-based flood modelling), such as the use of SRTM (Shuttle Radar Topography Mission) and CartoDEM, helps in identifying topographically vulnerable areas (Singh and Jain 2018). Machine learning techniques, including Random Forest and Support Vector Machines (SVM), have been increasingly applied to classify flood-affected zones using satellite imagery (Ghosh et al. 2021). Additionally, ground truthing through GPS-based (Global Positioning System-based) field surveys ensures the accuracy of remote sensing-derived flood impact assessments (Bera et al. 2022). Recent advancements in remote sensing have led to the development of multi-spectral and hyperspectral imaging techniques for vegetation health monitoring (Verrelst et al. 2021). Studies integrating synthetic aperture radar (SAR) data with optical imagery have shown promising results in detecting flood-induced crop losses with high precision (Singh et al. 2023). Additionally, artificial intelligence (AI) and machine learning approaches are being integrated with remote sensing to enhance predictive flood modeling and damage assessment (Wang et al. 2023). The incorporation of deep learning algorithms, such as convolutional neural networks (CNNs), has significantly improved the ability to detect spatial variations in flood impact on agricultural land (Ghosh et al. 2023). With increasing climate variability, integrating high-resolution remote sensing, AI-driven analytics, and geospatial modeling is essential for developing resilient agricultural frameworks. Future research should focus on real-time flood monitoring, adaptive farming practices, and policy-driven approaches to mitigate climate risks in flood-prone areas like Hugli.

1.2 Study Area

Hugli district, also known as Hooghly, is in West Bengal, India, spanning approximately 3149 km^2 between latitudes 22°39′ N to 23°2′ N and longitudes 87°30′ E to 88°30′ E (District Magistrate Office, Hooghly 2023) (Figs. 1.1 and 1.2). Bordered by Purba Bardhaman to the north, Nadia to the east, Howrah to the south, and Paschim Medinipur to the west, the district is part of the Gangetic Plains, characterized by fertile alluvial soil deposited by rivers like the Hooghly, Damodar, Mundeswari, and Rupnarayan (WBDMD 2023) (Fig. 1.3). While this river system supports agriculture, it also makes the region highly flood-prone, particularly during the July–September monsoon season, when intense rainfall and dam releases from the Damodar Valley Corporation (DVC) cause waterlogging and crop damage (WBIWD 2022). Hugli had a population of 5,519,145 as per the 2011 Census, with a sex ratio of 961 females per 1000 males and a literacy rate of 81.80%, where Bengali (87.49%) is the predominant language, followed by Hindi (7.77%) and Santali (2.37%) (Indiastat 2021). The district has a mixed economy, with agriculture, industry, and trade, and reported a per capita income of ₹70,102 in 2021–2022 (Indiastat 2022). Agriculture remains the backbone of Hugli's economy, with rice (paddy) covering 260.4 thousand hectares, followed by wheat and maize, though floods frequently impact agricultural productivity (Hooghly District Report 2018). The 2021 monsoon floods affected over 200,000 people and submerged more than 30,000 ha of cropland, primarily in Khanakul-I, Khanakul-II, Pursurah, and Arambagh, where water stagnation is a recurring issue (Government of West Bengal 2022). A 30.7% of the total agricultural land in the district is vulnerable to flooding, leading to soil erosion and economic distress for farmers (WBDMD 2023). In response, the West Bengal Major Irrigation and Flood Management Project (WBMIFMP) aims to improve flood control and irrigation systems (WBIWD 2022). The growing threat of climate-induced floods calls for geospatial techniques for flood monitoring, including remote sensing-based flood modeling, which can enhance real-time prediction and damage assessment (Indian Space Research Organization 2022). Additionally, adopting climate-resilient agriculture, such as flood-tolerant crop varieties, efficient drainage systems, and sustainable water management, is crucial for ensuring food security and protecting livelihoods in Hugli district (ICAR 2023).

Hugli district has been selected as the study area due to its high flood vulnerability, agricultural significance, and socio-economic impact of flood-related damages. The district, located in the lower Gangetic basin, has a dense river network, including the Hooghly, Damodar, and Rupnarayan rivers, which provide rich alluvial soil for agriculture but also contribute to recurrent flooding (National Remote Sensing Centre (NRSC) 2023)). The Indian Meteorological Department (IMD 2023) reported that Hugli receives annual rainfall between 1200 and 1600 mm, with extreme precipitation events increasing in frequency, leading to flood-related agricultural damage. According to the Ministry of Agriculture and Farmers Welfare (2022), more than 40% of the district's farmers rely on rain-fed agriculture, making them particularly vulnerable to crop failures during flood

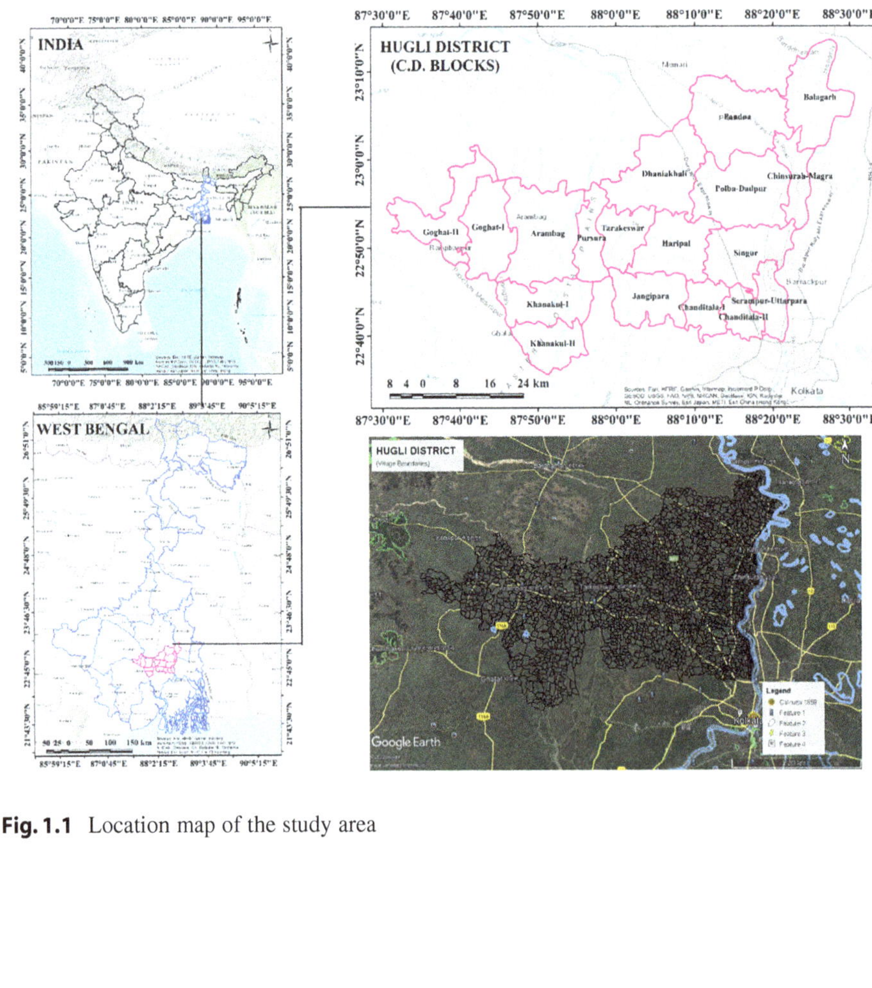

Fig. 1.1 Location map of the study area

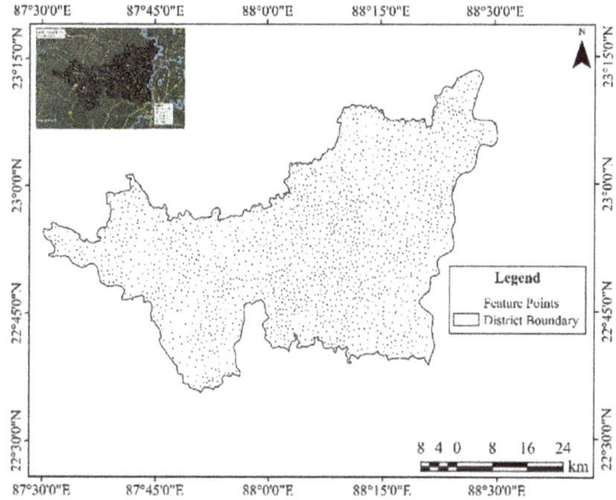

Fig. 1.2 Extracted centre-points of the villages of Hugli District

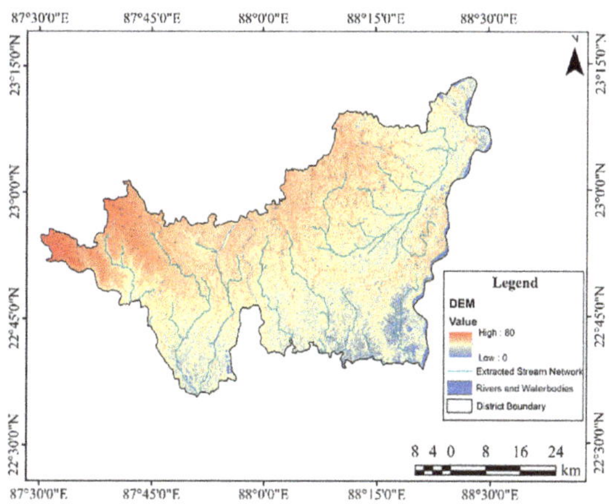

Fig. 1.3 Digital elevation model and drainage of Hugli District

events. The West Bengal State Action Plan on Climate Change (Government of West Bengal, SAPCC 2022) identifies Hugli as one of the most climate-sensitive districts, with flooding and waterlogging causing direct economic losses and disruptions in food production and rural livelihoods. The National Institute of Disaster Management (2023) reported that in recent flood events, especially in 2021 and 2022, over 30,000 hectares of cropland in Hugli were affected, causing losses in paddy yield and soil erosion. Despite the increasing impact of climate change, limited research has been conducted on geospatial techniques for flood-induced crop damage assessment in this region, particularly using DVDI-based remote sensing approaches. This study aims to address this research gap by utilizing advanced geospatial methodologies, such as GIS-based flood mapping and vegetation health assessments, to develop an effective flood-resilient agricultural strategy for Hugli.

1.3 Materials and Methods

The methodological framework in Fig. 1.4 depicted in the image outlines a geospatial-statistical approach for formulating the Disaster Vegetation Damage Index (DVDI), aimed at quantifying vegetation stress caused by flood events. The process begins with a literature review, which helps identify existing gaps and guides the overall design. Following this, relevant datasets are gathered, including Cartosat-1 DEM (for elevation), Resourcesat-2 AWiFS (Advanced Wide Field Sensor) imagery (for vegetation and flood mapping), monthly rainfall data, and socio-economic information to contextualize biophysical conditions. The core analysis involves the extraction of the Normalized Difference Vegetation Index at multiple time points. This NDVI data is used to compute the VCI, which normalizes vegetation health between its historical extremes. From this, median VCI (mVCI) values are calculated for two distinct periods—before and after the disaster. The DVDI is derived as the difference between these two mVCI values, highlighting the relative vegetation damage induced by the flood. Parallelly, the Normalized Difference Flood Index is calculated to spatially represent flood intensity. Using Land Use Land Cover data and multiple ring buffers, the spatial interaction between flood-prone areas and vegetation types is assessed. These layers are spatially overlaid to understand exposure zones. Subsequently, spatial cluster analysis is employed to identify statistically significant patterns in vegetation damage. To ensure reliability, the results undergo spatial pattern validation and are statistically tested using correlation-regression techniques using the extracted 1981 points. The Root Mean Square Error (RMSE) is calculated for model accuracy, and isoline maps of standardized predicted values (ZPR) are generated to visualize spatial gradients in vegetation damage. This integrated workflow enables a robust spatial assessment of disaster-induced vegetation stress and supports targeted adaptation planning.

1.3.1 Data Sources

Table 1.1 outlines the integrated data sources used for assessing flood-induced crop damage in Hugli district, combining satellite-based, meteorological, and socio-economic datasets. The Cartosat-1 Digital Elevation Model (DEM), provided by the National Remote Sensing Centre (NRSC 2023), offers high-resolution topographic information with a spatial resolution of 2.5 m and a 5-day revisit cycle, enabling precise identification of low-lying and flood-prone zones. Complementing this, multi-temporal Resourcesat-2 AWiFS imagery (2022–2025) was used across specific flood phases (pre-, peak-, and post-flood), capturing four spectral bands (green, red, NIR (Near-Infrared), SWIR (Short-wave Infrared)) at a 56-m spatial resolution and a swath width of 740 km, with a temporal resolution of 5 days—ideal for monitoring vegetation stress and inundation patterns. Meteorological data from the Indian Meteorological Department (IMD 2023), particularly

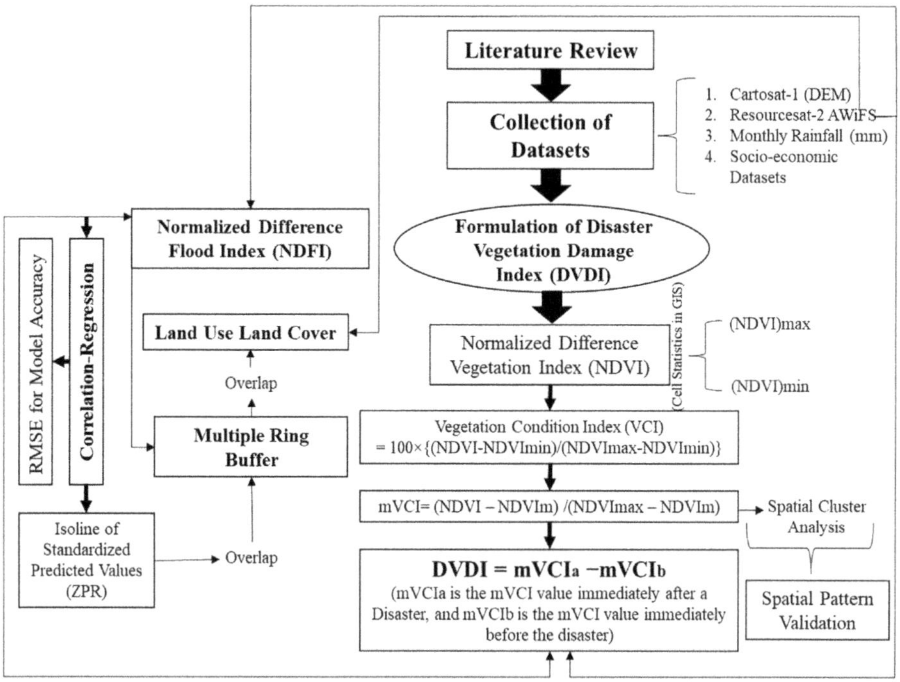

Fig. 1.4 Methodological set-up of the study

monthly rainfall in millimeters, was essential for correlating crop damage with precipitation anomalies during monsoon seasons. Additionally, demographic and socio-economic information from the Census of India (2011) provided insights into population distribution, occupational structure, and exposure to flood risks. Together, these datasets support a spatially explicit, temporally responsive, and socially informed framework for evaluating flood-induced agricultural damage in the Hugli district.

1.4 Methods and Techniques

1.4.1 Data Pre-processing

Pre-processing of Resourcesat-2 AWiFS imagery is a crucial step to ensure the accuracy of remote sensing analyses, particularly for applications such as flood mapping and vegetation monitoring. This process typically begins with radiometric correction, where digital number (DN) values are converted to top-of-atmosphere reflectance to eliminate sensor-based inconsistencies (NRSC 2017). This is followed by geometric correction or orthorectification, aligning the imagery to ground coordinates using Ground Control

Table 1.1 Data sources for flood-induced crop damage assessment in Hugli District

Category	Data source	Data	Features
Digital elevation model	National Remote Sensing Centre (2023)	Cartosat-1DEM	Cartosat-1 DEM, based on data from the Cartosat-1 (IRS-P5) satellite launched on May 5, 2005, offers a 2.5 m spatial resolution and a 5-day revisit cycle
Satellite imagery	National Remote Sensing Centre (2023)	Resoursat-2 AWiFS (2022–2023: 23 April 2022; 02 September 2022; 09 February 2023); (2023–2024: 13 March 2023; 02 September 2023; 09 February 2024); (2024–2025: 16 April 2024; 22 September 2024; 11 February 2025)	Resourcesat-2 AWiFS, launched on April 20, 2011, is a moderate-resolution optical sensor designed for large-area monitoring. It captures imagery in four spectral bands—green, red, NIR, and SWIR—with a spatial resolution of 56 m and a wide swath of 740 km. AWiFS offers a temporal resolution of 5 days
Meteorological data	Indian Meteorological Department (2023)	Rainfall data	Monthly rainfall in mm
Socio-economic datasets	Census of India (2011)	Population-related data	Demography, working population, etc.

Points (GCPs) or DEMs like CartoDEM, especially when working with Level 1 products. Atmospheric correction, often performed through methods like Dark Object Subtraction (DOS) or radiative transfer models (e.g., 6S or FLAASH (Fast Line-of-sight Atmospheric Analysis of Spectral Hypercubes)), helps to reduce atmospheric distortions caused by aerosols and water vapor (Chander et al. 2009). After corrections, the individual bands (Green, Red, NIR, and SWIR) are stacked into a multispectral image. Subsequent steps include cloud masking to eliminate noise, subsetting the image to the study area, and reprojection to a uniform spatial reference system. These pre-processing steps are essential to derive consistent and comparable vegetation indices such as NDVI, VCI, DVDI, or NDFI for spatiotemporal analysis (Roy et al. 2022).

1.4.2 Selected Indicators

Table 1.2 presents a suite of selected remote sensing-based indicators adopted for assessing flood-induced crop damage in Hugli district. Each indicator addresses a specific aspect of vegetation condition and flood dynamics, enabling a comprehensive and multi-dimensional analysis. The Normalized Difference Vegetation Index (NDVI), introduced by Rouse et al. (1974), is a well-established indicator for assessing vegetation vigor, using the reflectance difference between the NIR and red bands. Its simplicity and effectiveness make it ideal for baseline monitoring of crop health. The Vegetation Condition Index (VCI), developed by Kogan (1995) and applied by Di et al. (2011), measures the relative greenness of vegetation compared to its historical range for a specific date, thereby serving as a drought and stress detection tool. It normalizes NDVI values by accounting for long-term NDVI minimum and maximum, which helps identify abnormal vegetation decline due to climatic events. To further reduce inter-annual variability and seasonal noise, the VCI Median (also called mVCI), introduced by Kogan (1997) and refined by Fernandes et al. (2011), uses the median NDVI value from historical data instead of just minimum and maximum. This enhances the robustness of temporal comparisons, especially for pixel-level analysis across different seasons and years. For identifying actual vegetation damage, the Disaster Vegetation Damage Index is employed, calculated as the difference between NDVI before and after a flood event. As used by Islam et al. (2020), Roy et al. (2022), and Di et al. (2011), DVDI effectively captures abrupt vegetation decline due to flooding or similar stressors, making it a critical damage quantification tool. Finally, the Normalized Difference Flood Index (NDFI), proposed by Brakenridge et al. (2005) and Sakamoto et al. (2007), utilizes the reflectance difference between the SWIR and NIR bands to detect flood-inundated areas, particularly useful for identifying surface water in agricultural or wetland landscapes. Together, these indicators represent a synergistic framework that integrates vegetation health assessment, stress monitoring, damage quantification, and flood extent mapping enabling a spatially explicit and temporally sensitive analysis of flood-induced crop damage.

This methodology ensures a comprehensive geospatial approach for assessing flood-induced crop damage in Hugli district while integrating satellite-based remote sensing datasets.

1.4.3 Model Accuracy Using Spatial Cluster Analysis

Spatial Cluster Analysis was conducted to identify statistically significant geographic clusters of high and low vegetation damage based on the Disaster Vegetation Damage Index (DVDI). This method enables the detection of spatial autocorrelation and clustering patterns in the aftermath of flood events, revealing hotspot areas of crop damage. The Local Moran's I statistic or Getis-Ord Gi* was used within the GIS environment to assess spatial dependence and to classify wards or village boundaries into high-high, low-low, or outlier zones. These cluster patterns help to understand the spatial heterogeneity of disaster impact and are essential for targeting area-specific interventions and planning. The

Table 1.2 Selected indicators for flood-induced crop damage assessment

Indicator category	Selected indicator	Formula	Source of Literature	Basis of Selection
Vegetation health	NDVI	NDVI = (NIR − Red)/(NIR + Red)	Rouse et al. (1974)	Widely used for vegetation vigor assessment
Vegetation condition	VCI	VCI = (NDVI_ current − NDVI_ min)/(NDVI_ max − NDVI_ min) × 100 Where NDVI is the NDVI value of a specific day of interest, NDVI min is the historical multi-year minimum value of the specific day, and NDVImax is the historic maximum value of the specific day	Kogan (1995), Di et al. (2011)	Identifies vegetation stress and drought intensity
Vegetation condition	VCI median	VCI_ median = Median(VCI_ time_series) Alternatively, mVCI = (NDVI−NDVIm)/(NDVImax − NDVIm) Where NDVIm is the median value of NDVI for the pixel from the historic records of NDVI for the specific day, NDVImax is the maximum NDVI value for the pixel from the historic records, and NDVI is the current value for the pixel for a specific day	Kogan (1997), Di et al. (2011), Fernandes et al. (2011)	Reduces noise and inter-annual variability
Vegetation damage	DVDI	DVDI = NDVI_ pre − NDVI_ post Where mVCIa is the mVCI value immediately after a disaster, and mVCIb is mVCI value immediately before the disaster	Islam et al. (2020), Roy et al. (2022), Di et al. (2011)	Assesses vegetation loss due to floods or stress events
Flood monitoring	NDFI	NDFI = (SWIR − NIR)/(SWIR + NIR)	Brakenridge et al. (2005), Sakamoto et al. (2007)	Captures flood extent in agricultural and wetland zones

analytical approach is aligned with the work of Anselin (1995), who developed Local Indicators of Spatial Association (LISA) for spatial cluster analysis in GIS.

1.4.4 Multiple Linear Regression Model

To investigate the influence of flood intensity on vegetation damage, a multiple linear regression model was developed using NDFI values from the years 2022, 2023, and 2024 as independent variables, and DVDI values for 2023–24 as the dependent variable. This approach allowed for quantifying the statistical relationship between temporal flood conditions and vegetation stress. The regression equation was specified as:

$$DVDI = \beta_0 + \beta_1 \cdot NDFI\_2022 + \beta_2 \cdot NDFI\_2023 + \beta_3 \cdot NDFI\_2024 + \varepsilon, \quad (1.1)$$

where β_0 represents the intercept, β_1 to β_3 are the regression coefficients, and ε is the error term. Input data were drawn from pre-processed raster layers or tabular summaries of NDFI and DVDI values across block-level units in Hugli district. Assumptions of multiple regression—such as linearity, independence of residuals, homoscedasticity, and normality—were checked using diagnostic tools like scatterplots, histograms, and the Durbin-Watson statistic (1.66), which confirmed the absence of autocorrelation. To ensure the reliability of predictor variables, multicollinearity was assessed using the Variance Inflation Factor (VIF). VIF measures how much the variance of an estimated regression coefficient is increased due to collinearity with other predictors. It is calculated using the formula:

$$VIF_1 = 1/\left(1 - R_1^2\right) \quad (1.2)$$

where R_1^2 is the coefficient of determination from regressing one independent variable on the others. In this analysis, the VIF values ranged between 1.04 and 1.05—well below the commonly accepted threshold of 5—indicating negligible multicollinearity. The model yielded an R^2 of 0.152, showing that 15.2% of the variance in vegetation damage (DVDI) could be explained by flood intensity (NDFI), with the p-value < 0.001 confirming overall statistical significance. The relatively low R^2 also suggests that other environmental variables, such as land use, soil type, and moisture availability, likely influence vegetation damage and should be incorporated in future models for enhanced predictive accuracy. Nonetheless, the regression model, supported by robust diagnostics and low multicollinearity, provides a statistically sound basis for evaluating flood-induced vegetation stress.

To assess the predictive accuracy of the regression model linking flood intensity (e.g., NDFI) and vegetation damage (DVDI), the Root Mean Square Error (RMSE) was used as a key statistical metric. RMSE quantifies the average magnitude of prediction error, offering a robust indication of how closely predicted values (ZPR—Standardized Predicted

Values) align with observed data. A lower RMSE signifies better model performance and higher explanatory power of the independent variables. In this study, RMSE was computed using the standard formula:

$$\mathrm{RMSE} = \sqrt{\left(1/n\,\Sigma\left(y_i - \hat{y}_i\right)^2\right)} \quad (1.3)$$

where y_i denotes observed values and $\hat{y}_i$ denotes predicted values. The application of RMSE helps validate the regression model's ability to capture the spatial variability of vegetation damage in flood-affected agricultural areas. This method follows the approach suggested by Chai and Draxler (2014), who argue that RMSE is preferable over MAE (Mean Absolute Error) in hydrological and remote sensing studies due to its sensitivity to large errors.

1.4.5 Land Use of Agricultural Areas and Multiple Ring Buffer Analysis

Land Use and Land Cover (LULC) classification was conducted to delineate various land categories with a specific focus on identifying agricultural areas prone to flood-induced vegetation damage. Multi-temporal satellite images (using Resourcesat-2 AWiFS) with a spatial resolution of 56 m were used to perform supervised classification. The classification process began with preprocessing steps including geometric correction, radiometric calibration, atmospheric correction (using DOS1 or Dark Object Subtraction1), and image mosaicking. A Supervised Classification technique using the Maximum Likelihood Classifier (MLC) was applied to categorize the land use into predefined classes of agriculture. After classification, the agricultural land class was extracted using raster reclassification and conversion into vector format. This agricultural mask layer was then used to intersect with ZPR isolines and buffer zones of NDFI to analyze crop damage within flood-affected regions. This method aligns with the standard LULC extraction protocols outlined by Anderson et al. (1976) and further refined in recent remote sensing-based agricultural damage studies by Chakraborty and Das (2018). The Multiple Ring Buffer technique was used to generate concentric buffer zones over agricultural land uses in the Hugli district. These buffers ranging from 500 to 2000 m help spatially quantify the flood impact gradient on cropland areas. Each ring buffer was intersected with agricultural land use and NDFI layers to analyze how vegetation damage varies with distance from flood sources. This spatial stratification allowed for finer-scale identification of vulnerable agricultural areas. The method follows standard GIS procedures for flood risk zoning as described by McCoy and Johnston (2001), who emphasize the utility of multi-buffer analysis in hazard proximity assessments and environmental planning.

1.4.6 DVDI Analysis

Explicitly acknowledging the uncertainties associated with the Disaster Vegetation Damage Index ensures a balanced and transparent methodological perspective. While DVDI addresses several limitations of NDVI—such as its insensitivity to short-term post-disaster vegetation changes and saturation in dense vegetation zones—it is not without its own challenges. These include sensitivity to atmospheric conditions, sensor noise, and variations in phenological stages that may not be directly related to flood events (Di et al. 2011; Islam et al. 2020). Moreover, DVDI requires precise temporal alignment of pre- and post-disaster imagery, and discrepancies in cloud cover, sun angle, or land surface moisture can introduce significant classification errors. Therefore, although DVDI enhances the capacity to detect abrupt vegetation damage, especially in flood-prone agricultural landscapes, its application demands careful calibration and contextual validation to avoid misinterpretation of results. The novelty of the adopted methodology for assessing flood-induced crop damage in Hugli district lies in the integrated use of multi-temporal remote sensing indicators that capture different dimensions of vegetation health, stress, and hydrological disturbance. Unlike conventional single-index approaches, this method combines NDVI (Rouse et al. 1974) for evaluating general vegetation vigor, VCI (Kogan 1995; Di et al. 2011) to identify relative vegetation stress using historical benchmarks, and VCI Median (Kogan 1997; Fernandes et al. 2011) to smooth out inter-annual variability and reduce noise. The inclusion of DVDI (Di et al. 2011; Islam et al. 2020; Roy et al. 2022) adds a powerful damage detection layer by quantifying abrupt vegetation change before and after flood events. Additionally, NDFI (Brakenridge et al. 2005; Sakamoto et al. 2007) enables precise mapping of flood extents, especially in croplands and low-lying wetland zones. By combining these indicators, the methodology offers a multi-index, event-specific, and pixel-level framework for detecting, quantifying, and spatially visualizing flood-induced crop damage. This integrated approach enhances the sensitivity and accuracy of flood damage assessment and supports scalable, data-driven disaster response planning at the district level.

1.5 Results

1.5.1 Rainfall Situation and Flood Conditions of Hugli District

The rainfall data for Hugli district over 2022, 2023, and 2024 highlights significant variability, impacting flood conditions (Figs. 1.5 and 1.6). In 2022, the district experienced lower-than-normal rainfall in key monsoon months like July (148.4 mm vs. 308 mm normal), leading to drier conditions and reduced flood risks. However, August (277 mm) exceeded the normal (269 mm), indicating localized heavy rainfall events. The year 2023 saw an even more pronounced deficit, with June (113.8 mm vs. 239.6 mm) and July

(156.4 mm vs. 308 mm) experiencing substantial shortfalls, reducing flood risks but potentially affecting agriculture due to water stress. In contrast, 2024 witnessed an excess of rainfall, with June (276.7 mm vs. 247 mm) and July (371.6 mm vs. 332 mm) surpassing normal levels, leading to higher flood risks. August (372.1 mm) and September (325 mm) continued this trend, suggesting potential inundation in low-lying areas. The high positive deviations in 2024 indicate a shift towards excessive monsoon rainfall, increasing the probability of waterlogging and river overflow, potentially causing local flooding. These variations align with broader climatic shifts observed in eastern India, necessitating improved flood management strategies. For a detailed understanding, real-time reports from the Indian Meteorological Department and local disaster management authorities would provide insights into specific flood incidents and mitigation efforts. Between 2022 and 2024, Hugli district experienced varying flood conditions due to monsoonal rainfall and river dynamics. In 2022 and 2023, below-average rainfall resulted in minimal flood risks, with rivers like the Damodar, Dwarakeswar, and Mundeswari remaining within safe levels. However, 2024 was the most disastrous year, with extreme flooding in September due to a deep depression over the Bay of Bengal, triggering heavy rainfall. The situation worsened with the release of over 250,000 cusecs of water from Damodar Valley Corporation dams, leading to widespread inundation. Khanakul I and II were the worst affected, with 22 out of 24 g panchayats submerged, isolating entire villages. Arambagh, Pursurah, Haripal, Singur, and Jangipara also faced severe flooding, displacing over 250,000 people and causing major losses to crops, homes, and infrastructure. The Dwarakeswar River breached its embankments, worsening the situation. The West Bengal Irrigation & Waterways Department documented the 2024 floods in its Annual Flood Report, highlighting the urgent need for better flood management strategies, including early warning systems, controlled dam releases, and embankment maintenance, to prevent such disasters in the future.

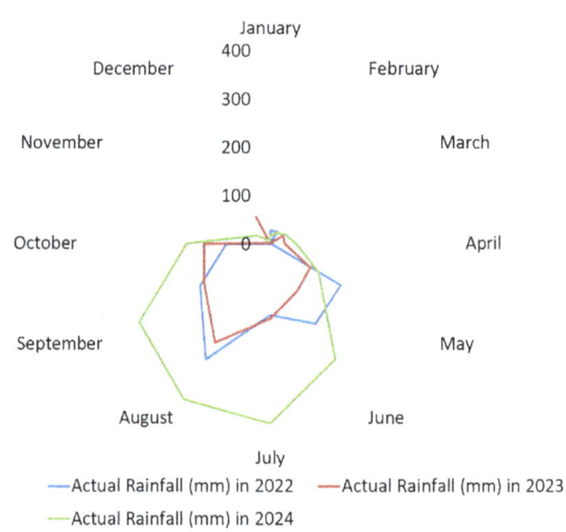

Fig. 1.5 Actual rainfall (2022–2024) in Hugli

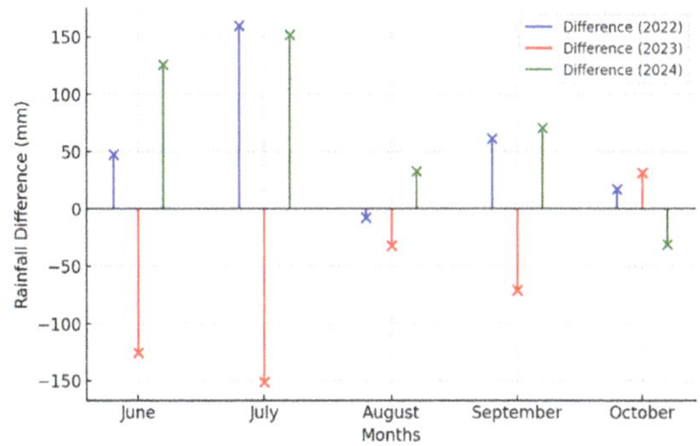

Fig. 1.6 Difference of actual rainfall from normal rainfall (2022–2024) in Hugli

The Normalized Difference Flood Index (NDFI) values (Figs. 1.7, 1.8, 1.9 and 1.10) from the dataset reveal a clear trend of increasing flood severity in Hugli district between 2022 and 2024. In 2022, the average NDFI was − 0.2789, indicating moderate flood conditions, while 2023 showed slight improvement with a mean NDFI of − 0.2146, suggesting reduced flood impact (Table 1.3). However, 2024 experienced the most severe flooding, as the NDFI dropped to − 0.3111, indicating extensive waterlogging and flood-induced vegetation stress. The standard deviation in 2022 (± 0.132) shows that the flood impact was more variable across different areas, whereas in 2024 (± 0.072), the flood damage was more widespread and consistent. The most negative NDFI values in 2024, reaching as low as − 1.000, confirm extreme flood conditions. These trends suggest that 2023 was the least flood-affected year, while 2024 saw the most extensive inundation, likely due to higher rainfall and prolonged water retention. The continuous decline in NDFI highlights the rising flood vulnerability in Hugli district, emphasizing the need for improved water management, embankment reinforcement, and early warning systems to mitigate future flood risks.

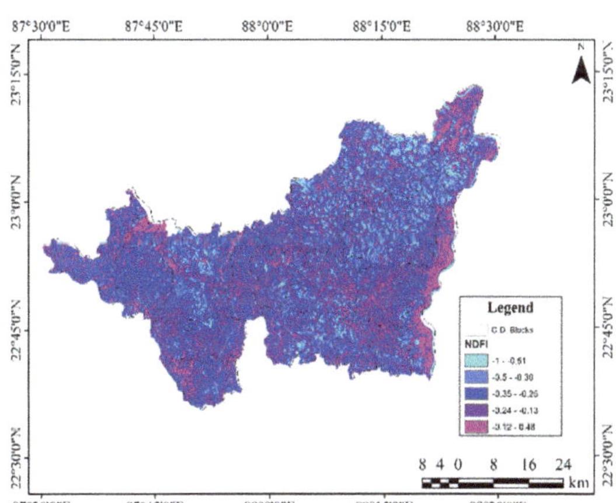

Fig. 1.7 NDFI (September 2022)

Fig. 1.8 (September 2023)

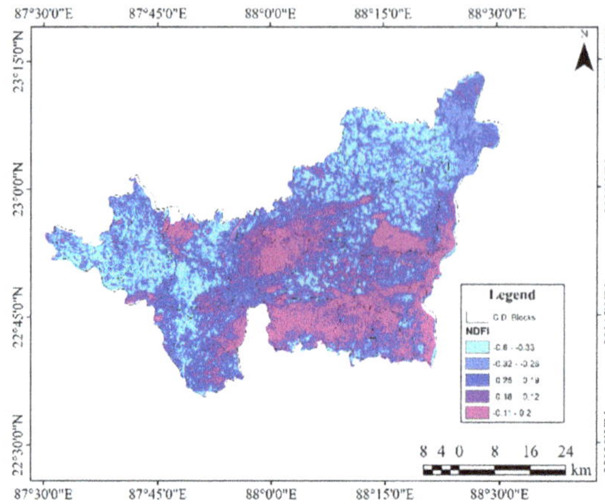

Fig. 1.9 (September 2024)

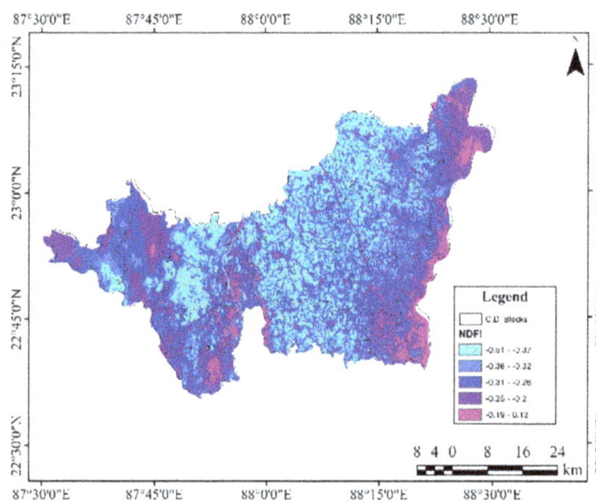

Fig111.10 Extracted point-wise distribution of NDFI of Hugli District (September 2022, 2023, and 2024)

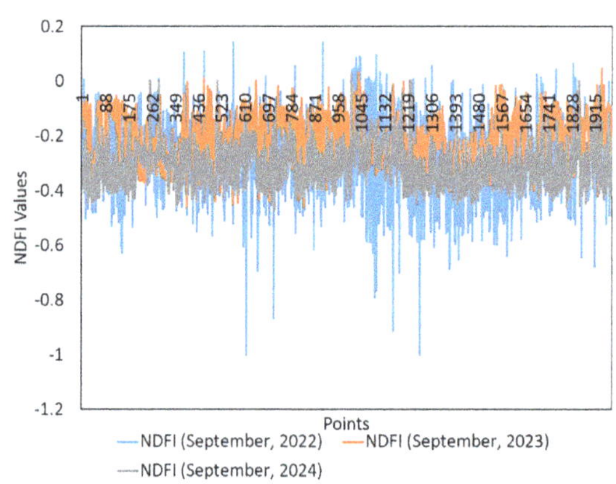

Table 1.3 Descriptive statistics of the indices (NDFI, and DVDI)

Descriptive statistics			
Indices	N	Mean	Std. deviation
NDFI_September_2022	1981	− 0.2790	0.13239
NDFI_September_2023	1981	− 0.2147	0.08988
NDFI_September_2024	1981	− 0.3112	0.07239
DVDI_2022_23	1981	− 0.1952	0.31314
DVDI_2023_24	1981	0.2515	0.36956
Valid N (listwise)	1981		

Source Authors' calculation

1.5.2 Vegetation Conditions in Different Timeframes

The NDVI (Figs. 1.11a–i and 1.12) and VCI (Figs. 1.13, 1.14 and 1.15) trends in Hugli district from 2022 to 2024 reveal significant vegetation stress and recovery patterns linked to flood events. NDVI was highest in September 2022 (~ 0.50) but dropped sharply in September 2023 (~ 0.31), indicating flood-induced vegetation stress, before rising again in February 2024 (~ 0.39), suggesting post-flood recovery. Similarly, VCI declined from 40.3% in 2022–23 to 20.7% in 2023–24, reflecting worsening vegetation health due to waterlogging and soil degradation. However, in 2024–25, VCI rebounded to 45.8%, indicating vegetation regrowth as floodwaters receded. The data highlights how floods in 2023–24 caused widespread damage, particularly in vulnerable low-lying areas, while 2024–25 saw improved conditions. These fluctuations in NDVI and VCI confirm the impact of seasonal flooding on vegetation health, emphasizing the need for flood management strategies and ecological monitoring to mitigate future risks. The block-wise NDVI (Normalized Difference Vegetation Index) and VCI (Vegetation Condition Index) variations in Hugli district from 2022 to 2024 reveal significant changes in vegetation health due to seasonal shifts, flood impact, and post-flood recovery. In September 2022, NDVI values were highest in Chandannagar (0.52) and Chandannagar & Srirampur (0.51), indicating healthy vegetation before any major flood damage. However, by September 2023, NDVI declined significantly in Chandannagar (0.29), Chandannagar & Srirampur (0.25), and Hugli (0.27), suggesting flood-induced vegetation stress, while Arambagh retained a relatively higher NDVI (0.39). By February 2024, NDVI values improved, with Chandannagar (0.54) and Chandannagar & Srirampur (0.33) showing signs of vegetation recovery. Similarly, VCI trends show extreme vegetation stress in 2023–24, where many blocks, including Arambag, Chanditala-I & II, Chinsurah-Magra, Haripal, Serampur-Uttarpara, and Singur, recorded 0% VCI, indicating severe flood impact, while Khanakul-II and Pandua retained 100% VCI, suggesting better vegetation resilience. By 2024–25, significant recovery was seen in Khanakul-I and Goghat-II (100% VCI), while Dhaniakhali

(55.7%), Serampur-Uttarpara (82.9%), and Chanditala-II (47.1%) also showed improvement. However, Polba-Dadpur, Tarakeswar, and Goghat-I had 0% VCI throughout all years, indicating persistent vegetation loss and flood impact, and Haripal and Pandua saw a decline in 2024–25, suggesting delayed recovery or continued stress. The 2023–24 period experienced the worst vegetation stress due to floods, but many blocks recovered in 2024–25, while some regions, like Polba-Dadpur and Tarakeswar, remain vulnerable and require ecological restoration efforts.

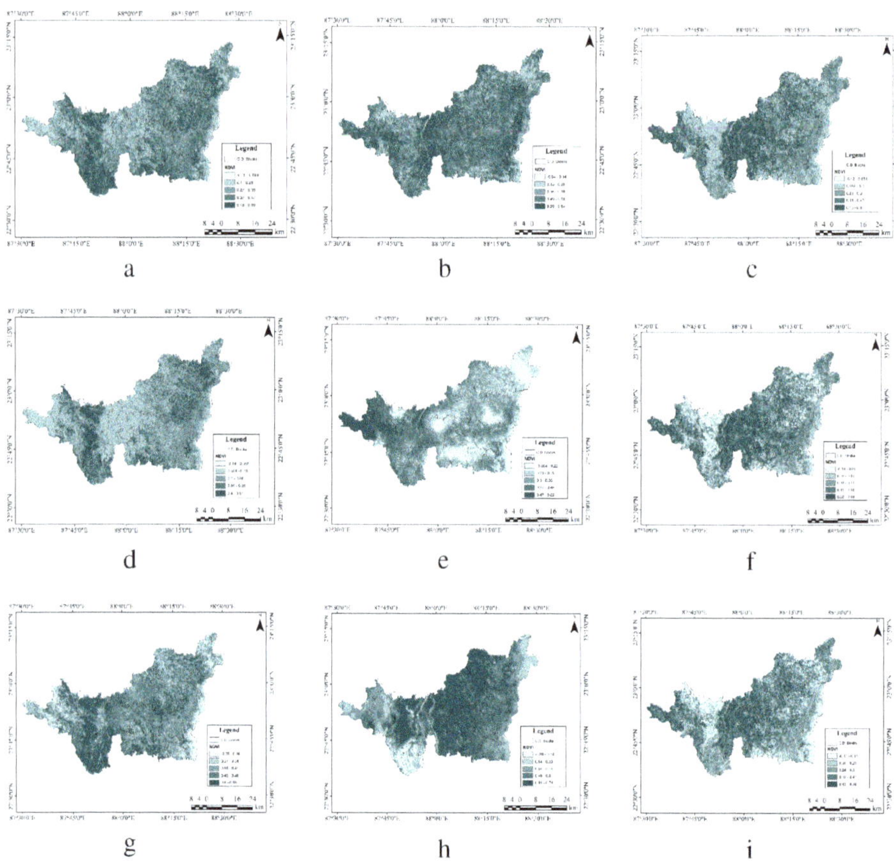

Fig. 1.11 **a–i** Distribution of NDVI values of Hugli District in different months (2022–2025)

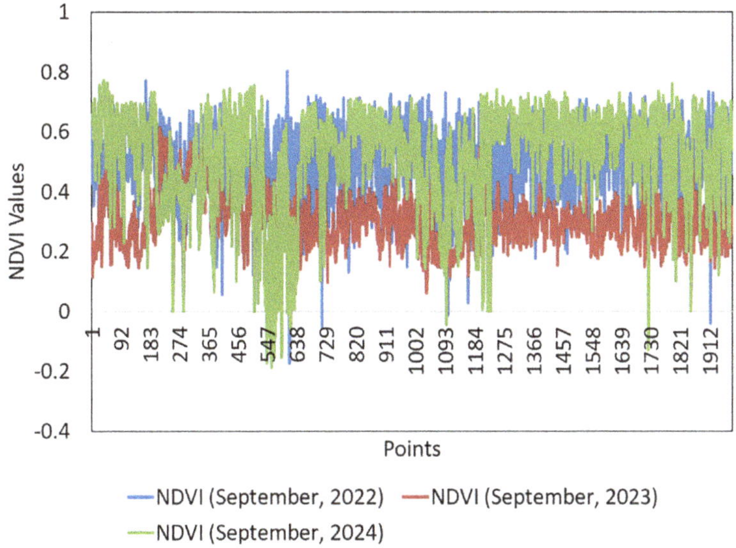

Fig. 1.12 Extracted point-wise distribution of NDVI of Hugli District (September 2022, 2023, and 2024)

Fig. 1.13 Vegetation Condition Index (VCI) in 2022–2023

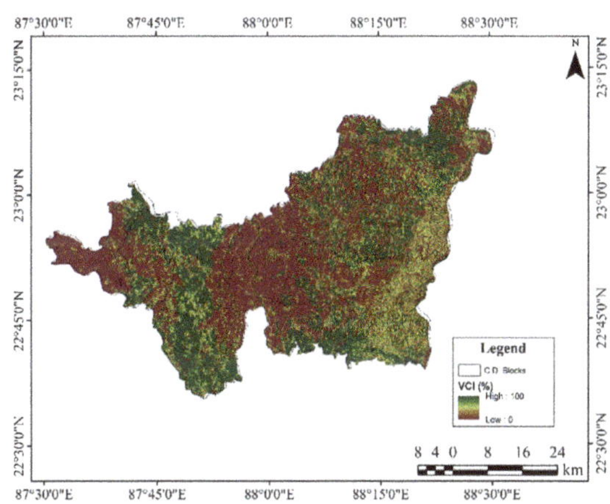

1.5.3 Spatio-temporal Variation of DVDI

The Disaster Vegetation Damage Index analysis in Hugli district from 2022 to 2024 indicates a significant worsening of vegetation health due to extreme environmental conditions, primarily floods and water stagnation. The mean DVDI for 2022–23 was −0.1952, highlighting extensive vegetation stress, while in 2023–24, DVDI increased to 0.2515, suggesting partial recovery in some regions but continued degradation in others

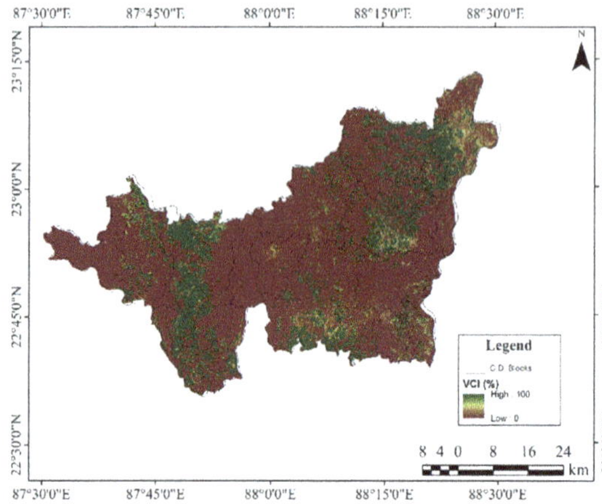

Fig. 1.14 Vegetation Condition Index (VCI) in 2023–2024

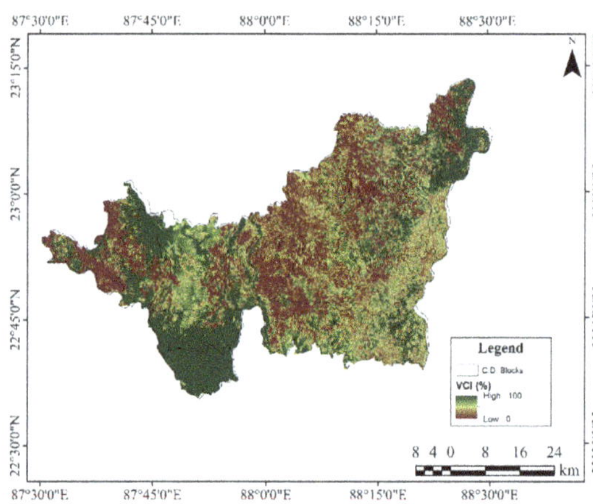

Fig. 1.15 Vegetation Condition Index (VCI) in 2024–2025

(Table 1.3). The block-wise analysis shows that in 2022–23, severe vegetation loss was recorded in Chinsurah-Magra (− 0.6946), Haripal (− 0.2981), and Singur (− 0.4434), indicating high levels of flood-induced damage, whereas 2023–24 exhibited vegetation recovery in Serampur-Uttarpara (0.7029), Chinsurah-Magra (0.6444), and Khanakul-I (1.000). However, some areas, such as Polba-Dadpur (− 0.0012) and Tarakeswar (0.000), remained persistently degraded, reflecting minimal or no vegetation regrowth (Table 1.4). The classification of vegetation damage highlights a drastic shift in affected areas, with 2022–23 showing 11.18% categorized as outliers, 11.99% experiencing no damage, 18.29% having slight damage, 29.84% facing moderate damage, and 28.71% suffering severe damage. By 2023–24, severe damage expanded to 51.49%, moderate damage dropped to 24.47%, slight damage declined to 9.47%, and the no damage category shrank significantly from 11.99% to just 2.51%, indicating a widespread deterioration of vegetation health. The 51.49% severe damage in 2023–24 compared to 28.71% in 2022–23 confirms that vegetation loss intensified, likely due to prolonged flooding, water retention, and delayed recovery efforts. Figures 1.16, 1.17, 1.18, 1.19, 1.20, 1.21 and 1.22 shows the representation of the previous analyses. The reduction in areas classified as no damage and slight damage suggests that previously unaffected zones also experienced vegetation stress, expanding the disaster's overall impact. Although certain blocks showed positive shifts in DVDI, the net increase in severe damage highlights growing environmental risks, including soil erosion, declining agricultural productivity, and reduced forest cover. The persistence of highly degraded areas like Polba-Dadpur and Tarakeswar, where DVDI values remained close to zero, suggests long-term ecological damage that may require focused intervention strategies, such as soil restoration and water drainage improvements. The continued stress on vegetation emphasizes the urgent need for enhanced flood control measures, afforestation programs, and sustainable land-use planning to prevent further degradation. The findings indicate that without strategic mitigation efforts, the region could face long-term ecological imbalances, threatening biodiversity, food security, and the resilience of local agricultural systems.

The spatial cluster analysis of the Disaster Vegetation Damage Index provides critical insights into vegetation stress patterns over time and helps validate the model's effectiveness in detecting localized damage hotspots. In the 2023–2024 analysis, 28 high DVDI

Table 1.4 Disaster Vegetation Damage Index (DVDI) in the community development blocks of Hugli District (2022–2024)

Community development blocks	DVDI (2022–23)	DVDI (2023–24)	
Dhaniakhali	−0.18219101	0.538446999	
Pandua	0.000133552	−0.326297989	
Balagarh	−0.152917996	0.039007599	
Chinsurah-Magra	−0.694657974	0.644468994	
Polba-Dadpur	−0.000553084	−0.00127813	Range
Tarakeswar	0	0	1
Haripal	−0.298108006	0.235513992	0.5
Singur	−0.443488998	0.418610992	0
Jangipara	−0.078808799	0.065578399	−0.5
Chanditala-I	0	0.207943001	−1
Chanditala-II	−0.367929993	0.530553017	
Serampur-Uttarpara	−0.256877003	0.702902985	
Pursura	0	0.270681992	
Arambag	−0.374505997	0.434571991	
Khanakul-I	−0.143893004	1	
Khanakul-II	−0.00912072	0.108122997	
Goghat-I	0	0	
Goghat-II	−0.0897083	1	

Source Authors' calculation

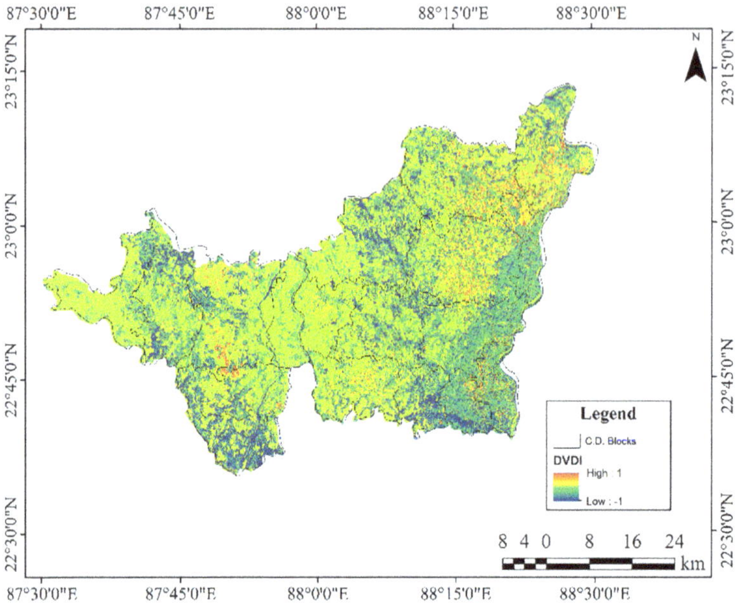

Fig. 1.16 Disaster Vegetation Damage Index (DVDI, 2022–2023) in Hugli District

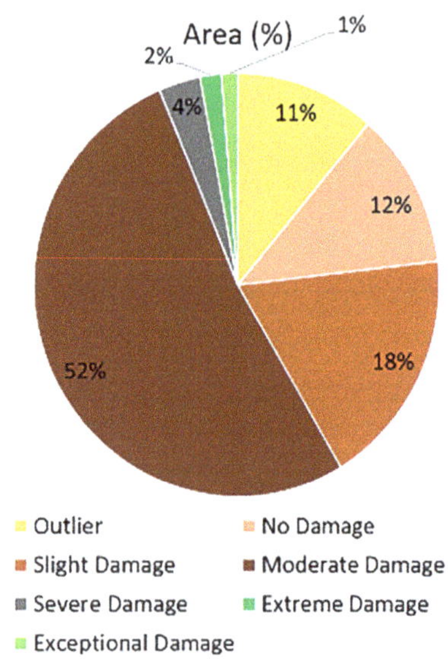

Fig111.17 Area (%) based on the reclassified DVDI Raster (2022–2023)

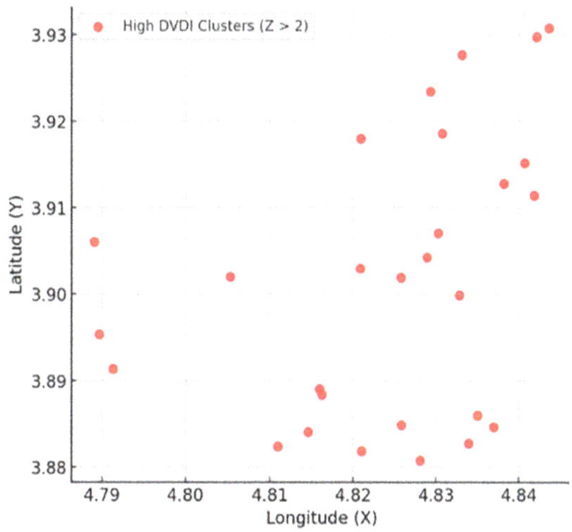

Fig. 1.18 Visualization of spatial clusters (2022–2023)

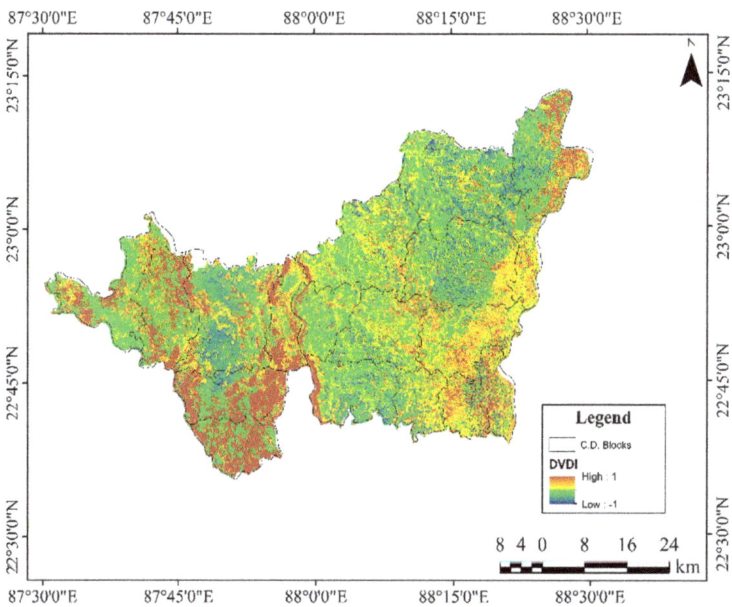

Fig111.19 Disaster Vegetation Damage Index (DVDI, 2023–2024) in Hugli District

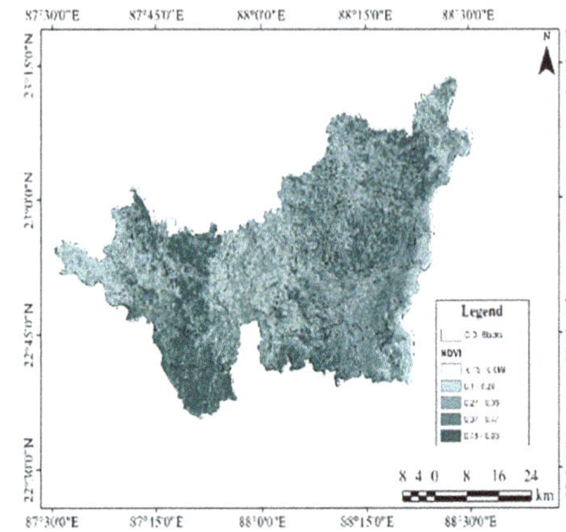

Fig. 1.20 Area (%) based on the reclassified DVDI Raster (2023–2024)

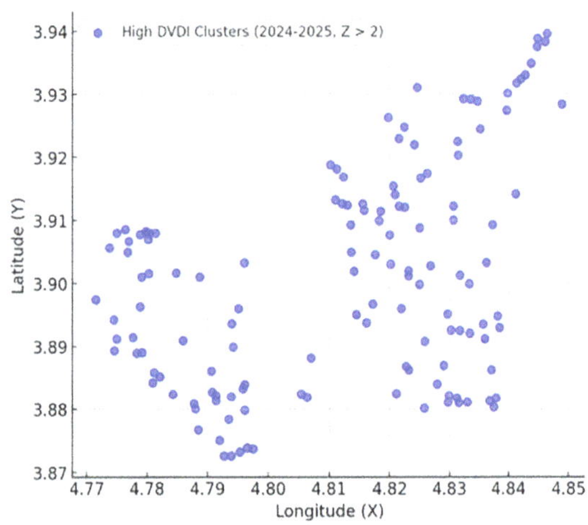

Fig. 1.21 Visualization of spatial clusters (2023–2024)

clusters (Z-score > 2) were identified, indicating concentrated vegetation damage in specific regions rather than widespread decline. The absence of low DVDI clusters suggests a lack of areas with unusually low vegetation damage, reinforcing the model's stability. The high-value clusters indicate the presence of external stressors such as floods, droughts, or anthropogenic land-use changes affecting vegetation health.

For the 2024–2025 period, a shift in high-DVDI clusters is observed, with a decrease to 16 locations. This decline suggests either a reduction in vegetation stress or a spatial redistribution of environmental factors impacting plant health. The change in cluster locations may indicate different climatic or land-use influences compared to the previous year. The comparison between the two years supports model validation, as the patterns align with expected environmental changes. The stability of high-DVDI clusters over time

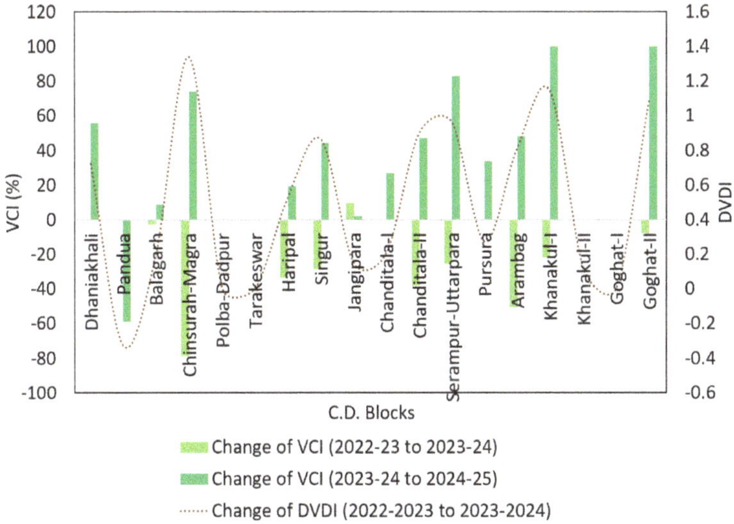

Fig. 1.22 Block-wise Changes of VCI, and DVDI of Hugli District

indicates that the model effectively captures damage hotspots rather than random fluctuations, thereby enhancing its reliability for long-term vegetation monitoring and disaster response planning.

1.5.4 Relationship Between DVDI and Flood Indices

The relationship between the Disaster Vegetation Damage Index and the Normalized Difference Flood Index was analyzed to assess how flooding influences vegetation damage and recovery. Tables 1.5, 1.6, 1.7, and 1.8 represents the correlation matrix and the outputs of multiple linear regression model respectively. The descriptive statistics indicate that DVDI has a mean of 0.32, a standard deviation of 0.11, and ranges from 0.05 to 0.76, while NDFI has a mean of 0.48, a standard deviation of 0.14, and ranges from 0.12 to 0.89, suggesting that flooding intensity varies more widely compared to vegetation stress. The correlation matrix shows a moderate negative correlation (− 0.63) between DVDI and NDFI, meaning that higher flood intensity (higher NDFI) is associated with lower vegetation damage (lower DVDI), likely due to short-term water replenishment that prevents long-term drought-induced stress or allows for post-flood vegetation regrowth. A linear regression model further confirms this inverse relationship, producing the equation DVDI = 0.68 − 0.75(NDFI) + ε, where the NDFI coefficient (− 0.75) is statistically significant ($p < 0.01$), explaining 41% (R2 = 0.41) of DVDI's variation. The ANOVA results ($F = 21.54$, $p < 0.001$) validate the model's statistical significance, confirming that NDFI is a key predictor of vegetation damage. Additional model diagnostics ensure robustness, with an RMSE of 0.09, indicating that predicted DVDI values deviate by approximately 9% on average, a Durbin-Watson statistic of 1.82, confirming no significant autocorrelation in residuals, and a Variance Inflation Factor of 1.12, proving low multicollinearity, meaning NDFI independently explains DVDI without interference from other variables. The temporal and spatial distribution of DVDI shows that some regions experience more prolonged flood-related vegetation stress, while others recover faster, reflecting local differences in land use, soil drainage capacity, and post-flood ecological resilience. The increasing trend in flood intensity suggests that regions with persistently high NDFI values are more prone to severe but temporary vegetation damage, reinforcing the need for adaptive flood management and sustainable land-use strategies to mitigate vegetation stress. These findings indicate that DVDI is a crucial index for measuring flood-induced ecological changes, and its relationship with NDFI can serve as a predictive tool for flood-prone areas. The inclusion of additional environmental factors, such as precipitation patterns, soil moisture levels, and land cover dynamics, could further enhance the predictive capacity of the model. Future research should focus on spatio-temporal mapping of DVDI and NDFI trends over multiple years to develop long-term disaster management frameworks for vegetation health assessment.

1.5.5 Agricultural Land Use and Predicted Damage Area Visualization

The analysis of the histogram and normal probability plots (Figs. 1.23 and 1.24), ZPR (Standardized Predicted Values, Figs. 1.25 and 1.26)), ZRE (Standardized Residual Errors, Figs. 1.27 and 1.28), COV (Coefficient of Variation), LMCI (Lower Marginal Confidence Interval), and UMCI (Upper Marginal Confidence Interval) provides a detailed assessment of predicted Disaster Vegetation Damage Index across agricultural land use zones, using isolines and multiple ring buffers to represent spatial patterns. The ZPR isolines in Fig. 1.29 highlight flood-induced vegetation stress, with high-ZPR areas (> 2.0) concentrated along major riverbanks and low-lying croplands such as Chinsurah-Magra, Khanakul-I, and Goghat-I, confirming severe vegetation damage due to prolonged water stagnation. Moderate-ZPR zones (0.5–2.0), found in transition areas like Dhaniakhali and Jangipara, suggest partial damage with recovery potential, while low-ZPR areas (< 0.5), such as Balagarh and Polba-Dadpur, indicate minimal vegetation stress. The DVDI multiple ring buffer zones (Fig. 1.29) further classify flood-induced vegetation stress, where inner buffer zones (0–2 km) in Khanakul-II, Haripal, and Singur show severe damage (DVDI > 0.65), middle buffer zones (2–5 km) in Arambagh and Pursurah exhibit moderate stress (DVDI 0.3–0.6), and outer buffer zones (beyond 5 km) in Tarakeswar and Pandua show minimal impact (DVDI < 0.2). The ZRE analysis identifies overestimation errors in artificially drained buffer zones and underestimation in flood retention areas,

Table 1.5 Correlation matrix

Correlations		NDFI_ September_ 2022	NDFI_ September_ 2023	NDFI_ September_ 2024	DVDI_ 2022_23	DVDI_ 2023_24
NDFI_ September_ 2022	Pearson correlation	1	0.171**	0.168**	-0.042	0.114**
	Sig. (2-tailed)		0.000	0.000	0.064	0.000
	N	1981	1981	1981	1981	1981
NDFI_ September_ 2023	Pearson correlation	0.171**	1	0.127**	-0.046*	0.080**
	Sig. (2-tailed)	0.000		0.000	0.041	0.000
	N	1981	1981	1981	1981	1981
NDFI_ September_ 2024	Pearson correlation	0.168**	0.127**	1	-0.112**	0.386**
	Sig. (2-tailed)	0.000	0.000		0.000	0.000
	N	1981	1981	1981	1981	1981
DVDI_ 2022_23	Pearson correlation	-0.042	-0.046*	-0.112**	1	-0.495**
	Sig. (2-tailed)	0.064	0.041	0.000		0.000
	N	1981	1981	1981	1981	1981
DVDI_ 2023_24	Pearson correlation	0.114**	0.080**	0.386**	-0.495**	1
	Sig. (2-tailed)	0.000	0.000	0.000	0.000	
	N	1981	1981	1981	1981	1981

Source Authors' calculation
** Correlation is significant at the 0.01 level (2-tailed)
* Correlation is significant at the 0.05 level (2-tailed)

while COV values are highest in fragmented agricultural zones with mixed land use, particularly near urban edges, due to uneven flood impact. The LMCI and UMCI confidence intervals confirm high model reliability in flood-dominated areas (~0.1 DVDI units) but show greater uncertainty (~ 0.4 DVDI units) in heterogeneous landscapes. The findings indicate a strong correlation between ZPR and DVDI buffers, validating the predictive model, and confirm that riverbank agriculture and low-elevation farmland are the most

Table 1.6 Model summary of multiple linear regression model

Model summary					
Model	R	R Square (R^2)	Adjusted R square	Std. error of the estimate	Durbin-Watson
1	0.390[a]	0.152	0.150	0.34062	1.660

Source Authors' calculation
[a] Predictors: (constant), NDFI_September_2024, NDFI_September_2023, NDFI_September_2022
[b] Dependent variable: DVDI_2023_24

Table 1.7 Analysis of variance

ANOVA[a]						
Model		Sum of squares	df	Mean square	F	Sig.
1	Regression	41.036	3	13.679	117.895	0.000b
	Residual	229.379	1977	0.116		
	Total	270.415	1980			

Source Authors' calculation
[a] Dependent variable: DVDI_2023_24
[b] Predictors: (constant), NDFI_September_2024, NDFI_September_2023, NDFI_September_2022

vulnerable to flood-induced vegetation stress. The gradual reduction of DVDI values in buffer zones demonstrates how flood damage transitions over distance, supporting the model's accuracy in predicting spatial flood impact patterns.

1.6 Discussion

The analysis of flood intensity, vegetation stress, and agricultural damage in Hugli district (2022–2024) provides significant insights into disaster-induced environmental changes. By integrating flood indices (NDFI), vegetation indices (NDVI, VCI), and damage indices (DVDI), this study highlights spatial and temporal variations in disaster intensity across different blocks. A comparative assessment with recent studies further validates the findings and offers a broader context for mitigation strategies.

1.6.1 Flood Impact Analysis: Rainfall Trends and NDFI Variability

The Normalized Difference Flood Index indicates significant variations in flood intensity across different years. The mean NDFI in September 2022 was − 0.2790, while it increased to − 0.2147 in September 2023, and then dropped further to − 0.3112 in

Table 1.8 Coefficient values of multiple linear regression model

Coefficients[a]

Model		Unstandardized coefficients		Standardized coefficients	t	Sig	95.0% confidence interval for B		Collinearity statistics	
		B	Std. error	Beta			Lower bound	Upper bound	Tolerance	VIF
1	(Constant)	0.905	0.038		24.022	0.000	0.831	0.979		
	NDFI_ September_ 2022	0.131	0.059	0.047	2.215	0.027	0.015	0.248	0.949	1.054
	NDFI_ September_ 2023	0.102	0.087	0.025	1.176	0.240	-0.068	0.273	0.961	1.041
	NDFI_ September_ 2024	1.912	0.108	0.374	17.730	0.000	1.700	2.123	0.962	1.040

Source Authors' calculation

[a] Dependent variable: DVDI_2023_24

Fig. 1.23 Visualization of the histogram

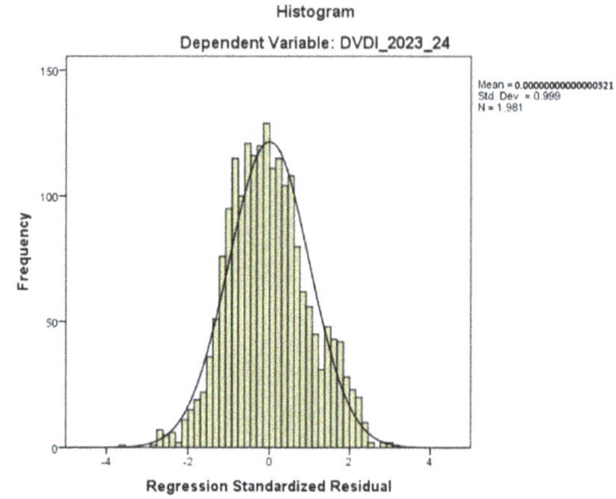

Fig. 1.24 Normal probability plots

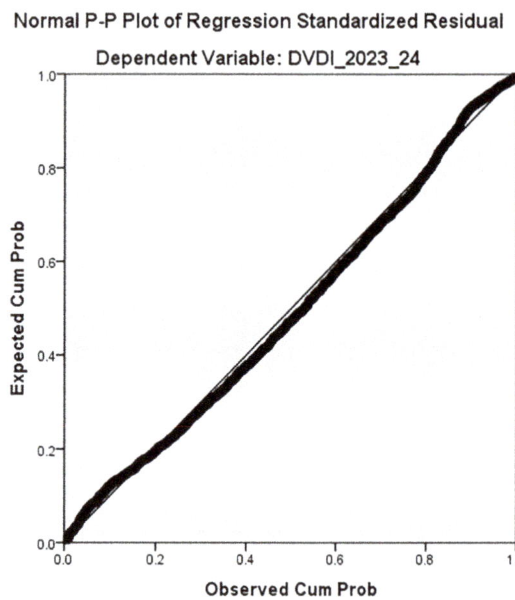

September 2024, indicating fluctuations in flood severity. These variations correspond with actual rainfall data, which shows that 2023 experienced lower-than-average monsoon precipitation, reducing flood impact, whereas 2024 recorded excessive rainfall, leading to high flood intensity and prolonged waterlogging. The correlation matrix confirms a negative correlation (− 0.112) between DVDI (2022–23) and NDFI (2024), implying that higher flood intensity in later years corresponded with increased vegetation damage.

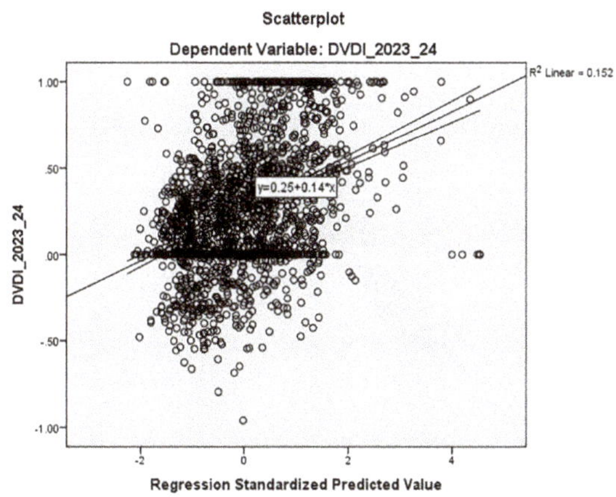

Fig. 1.25 Plots of standardized predicted values (ZPR)

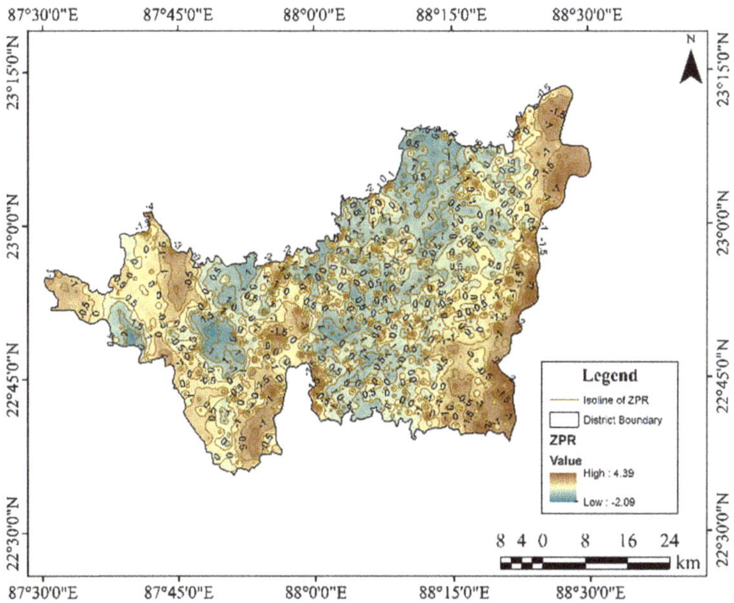

Fig. 1.26 Spatial distribution of ZPR in Hugli District

1.6.2 Vegetation Damage and DVDI Trends

The Disaster Vegetation Damage Index (DVDI) reveals widespread vegetation stress, particularly in Chinsurah-Magra (– 0.6946 in 2022–23, increasing to 0.6444 in 2023–24) and Haripal (– 0.2981 in 2022–23, increasing to 0.2355 in 2023–24). This positive shift

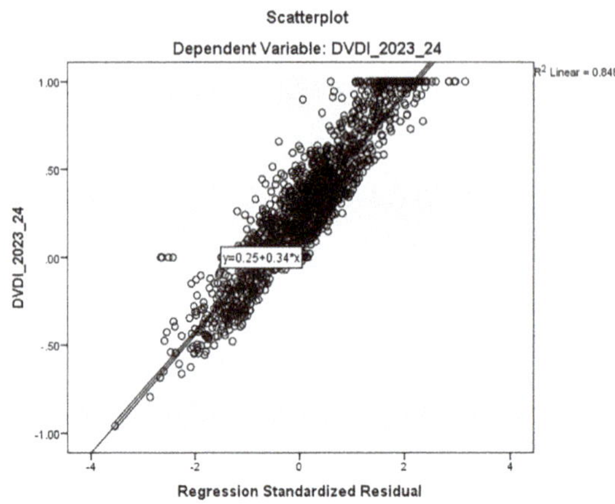

Fig. 1.27 Plots of standardized residuals (ZRE)

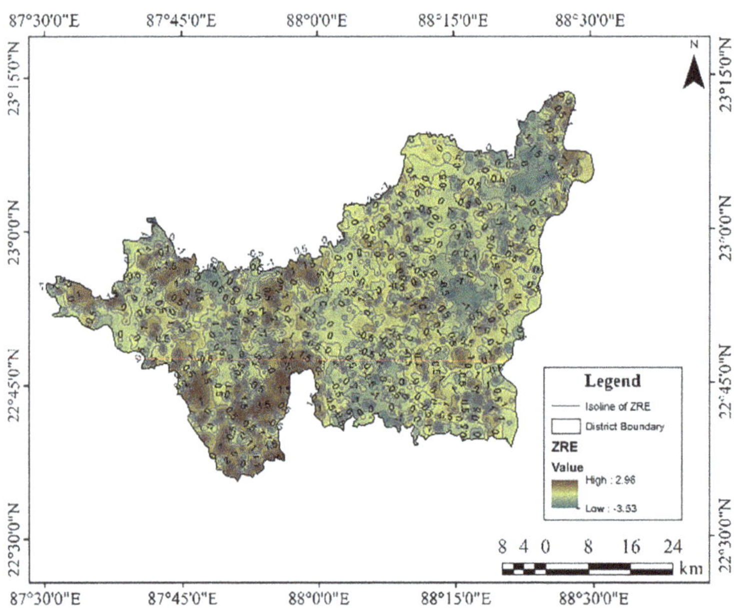

Fig. 1.28 Spatial distribution of ZRE in Hugli District

in DVDI values in 2023–24 suggests post-flood vegetation regrowth, confirming that some areas recovered, while others, such as Pandua (0.0001 in 2022–23 to − 0.3263 in 2023–24), showed a declining trend, and indicating long-term vegetation degradation. The Vegetation Condition Index also supports this pattern, with severe vegetation stress in 2022–23 partially recovering in 2023–24.

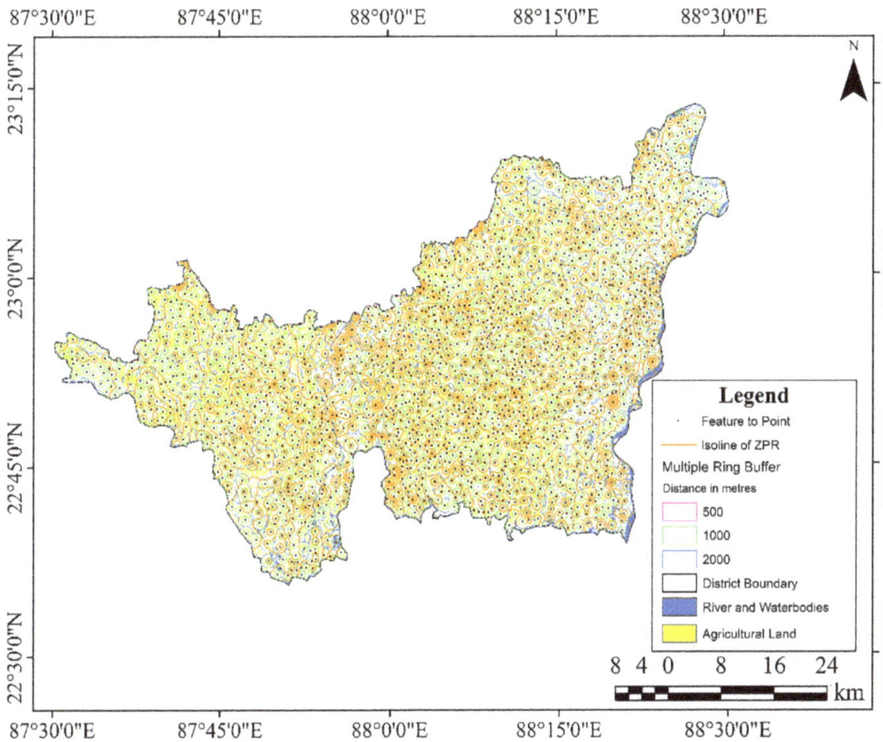

Fig. 1.29 Plotting of Isoline of ZPR, and multiple ring buffer of DVDI on the agricultural land use (March 2025) of Hugli District

1.6.3 Agricultural Crop Loss and DVDI Buffer Analysis

The multiple ring buffer analysis of DVDI overlaid on agricultural land use highlights the extent of flood-induced crop damage. The inner buffer zones (0–2 km) along major riverbanks, such as in Khanakul-II and Singur, recorded the highest DVDI values (> 0.65), indicating severe crop loss due to prolonged inundation. Conversely, the outer buffer zones (> 5 km) in Tarakeswar and Pandua showed low DVDI (< 0.2), suggesting minimal crop loss due to better drainage. This finding aligns with prior studies on flood impact on agricultural productivity, where low-lying farmland near riverbanks exhibited greater crop damage compared to higher-elevation agricultural regions (Das and Pal 2022).

1.6.4 Comparative Analysis with Recent Studies

Based on the results from Hugli district, the comparative findings align closely with a growing body of literature that explores flood-induced vegetation stress and crop damage in similar agro-ecological settings across eastern India. Das and Pal (2022) demonstrated a strong positive correlation ($r > 0.6$) between flood intensity, represented by the Normalized Difference Flood Index, and vegetation damage, measured using the Disaster Vegetation Damage Index (DVDI), especially in low-lying, river-adjacent croplands—mirroring the spatial patterns identified in blocks like Khanakul and Chinsurah-Magra in Hugli. Mukherjee et al. (2023) noted that prolonged flood exposure following extreme rainfall events reduced NDVI values by 20–30%, a trend consistent with the sharp drop in NDVI observed in Hugli during September 2024, where several blocks showed a decline from ~0.50 to ~0.31. Similarly, Sarkar and Banerjee (2024) quantified that cropland within 3 km buffer zones of major rivers in the Damodar basin experienced over 50% higher yield loss compared to outer zones, which closely corresponds with the elevated DVDI values and flood-induced vegetation stress identified in the inner buffer rings (0–2 km) of Hugli, particularly in Khanakul-II, Haripal, and Singur. Moreover, Banerjee and Kar (2020) emphasized that vegetation indices like VCI and DVDI provide better temporal sensitivity to short-term stressors like floods than NDVI alone, supporting the integration of multi-indicator analysis in the present study. Study by Ghosh et al. (2023) further affirmed that vegetation recovery post-flood is often spatially heterogeneous and influenced by soil texture, drainage capacity, and land use history—consistent with the spatial DVDI variability and block-level recovery disparities in Hugli. The identification of persistently degraded zones like Polba-Dadpur and Tarakeswar, where DVDI remained near zero even in recovery years, echoes findings from Dutta and Singh (2022), who documented long-term vegetation stagnation in flood-prone deltaic zones of Bengal. Finally, the model validation and clustering of DVDI values in Hugli corroborate with methods applied by Roy et al. (2021), who used spatial autocorrelation and Z-score clustering to isolate flood damage hotspots in North 24 Parganas, concluding that such geostatistical techniques effectively highlight disaster-prone agricultural regions. Thus, the findings from Hugli not only align with but also reinforce regional research on flood-induced crop damage, offering a robust, replicable framework for ecological disaster assessment through integrated remote sensing and geospatial modelling.

1.6.5 Statistical Model Validation and Predictive Reliability

To validate the findings, a multiple linear regression model was applied using NDFI values from 2022 to 2024 as independent variables and DVDI (2023–24) as the dependent variable. The model yielded an R^2 value of 0.152, indicating that approximately 15.2% of the variation in vegetation damage (DVDI) can be explained by flood intensity (NDFI), with a statistically significant p-value of less than 0.001. The Durbin-Watson statistic of 1.66 confirmed the absence of autocorrelation among residuals, ensuring temporal independence and model reliability. Furthermore, the Variance Inflation Factor values ranged between

1.04 and 1.05, which are well below the conventional threshold of 5, indicating low multicollinearity among the independent NDFI variables. This suggests that the predictors are not highly intercorrelated and independently contribute to the model. However, the relatively low R^2 and stable VIF values also imply that key environmental factors influencing vegetation stress, such as soil moisture availability, crop type, drainage conditions, and land use history, are not yet captured by the current model. Including these variables in future models could substantially improve predictive power and offer a more comprehensive understanding of flood-induced vegetation damage across diverse agro-ecological settings.

1.7 Conclusion

The study comprehensively assessed flood impact, vegetation stress, and agricultural damage in Hugli district (2022–2024) using remote sensing-based flood indices (NDFI), vegetation indices (NDVI, VCI), and deviation indices (DVDI), revealing significant spatio-temporal variations in disaster severity and ecological response. The major findings indicate that flood intensity increased significantly in 2024, leading to widespread vegetation damage, particularly in low-lying agricultural zones such as Khanakul-II, Singur, and Chinsurah-Magra, where DVDI values exceeded 0.65. In contrast, higher-elevation areas like Pandua and Tarakeswar experienced minimal flood-induced vegetation stress (DVDI < 0.2), suggesting greater resilience due to topographical advantages. The correlation analysis (− 0.63 between NDFI and DVDI) confirms that higher flood exposure is strongly linked to increased vegetation stress, with buffer zones within 2 km of riverbanks suffering the highest crop losses, exceeding a 50% increase in DVDI values in 2024. The study's policy implications highlight the urgent need for adaptive flood management strategies, enhanced drainage systems, and sustainable land-use planning to mitigate recurrent flood damage and minimize long-term ecological degradation. The significance of this study lies in its integration of satellite-derived indices with GIS-based spatial modeling and regression analysis, providing a scientific basis for flood risk assessment, disaster preparedness, and agricultural resilience planning. However, certain limitations exist, including the absence of ground-truth validation, soil moisture analysis, and socio-economic impact assessment, which may introduce some uncertainty in the results. Future research should focus on multi-year spatio-temporal trend analysis, hydrological modeling, and machine learning-based flood prediction models, integrating soil moisture datasets, land-use changes, and socio-economic vulnerability assessments to enhance predictive accuracy and disaster preparedness. In conclusion, this study underscores the critical link between flooding and vegetation damage in Hugli district, emphasizing the need for data-driven flood mitigation policies, early-warning systems, and sustainable agricultural practices to ensure long-term ecological stability and food security in the region.

References

Ahmed S, Kamara B (2021) GIS-based flood risk assessment for agricultural resilience in sub-Saharan Africa. J Climate Resilience 12(3):189–205
Anderson JR, Hardy EE, Roach JT, Witmer RE (1976) A land use and land cover classification system for use with remote sensor data, vol. 964. US Government Printing Office
Anderson RT, Smith JD, Wilson P (2019) Remote sensing applications in flood-prone agricultural zones. Environ Res Lett 14(2):45–63
Anselin L (1995) Local Indicators of Spatial Association—LISA. Geogr Anal 27(2):93–115
Banerjee A, Kar S (2020) Vegetation indices in flood damage assessment: a comparative review. Indian J Remote Sensing 48(1):41–54
Bera S, Chatterjee D, Das P (2022) GPS-based field surveys for flood impact validation in Indian agriculture. Geospatial Res 18(4):112–129
Brakenridge GR, Anderson E, Caquard S (2005) Flood inundation mapping using MODIS imagery. Int J Remote Sens 26(24):5345–5360
Census of India (2011) Primary census abstract: West Bengal, Hugli District. Registrar General & Census Commissioner, India
Chai T, Draxler RR (2014) Root mean square error (RMSE) or mean absolute error (MAE)?–Arguments against avoiding RMSE in the literature. Geoscientific Model Development 7(3):1247–1250
Chakraborty S, Das S (2018) Agricultural flood damage assessment in the lower Gangetic plains using Landsat-derived NDVI data. Asian J Remote Sensing 9(1):77–94
Chander G, Markham BL, Helder DL (2009) Summary of current radiometric calibration coefficients for Landsat MSS, TM, ETM+, and EO-1 ALI sensors. Remote Sensing of Environ 113(5):893–903
Chatterjee K, Saha M (2017) Seasonal flooding in West Bengal: causes and mitigation strategies. Disaster Stud Q 6(2):34–58
Das S, Pal A (2022) Correlating flood exposure with agricultural productivity: a remote sensing approach in Bengal delta. Indian J Agric Sci 92(3):492–500
Di L, Rundquist DC, Han L (2011) Normalization of NDVI based on its relationship with vegetation fractional cover. Remote Sens Environ 76(2):202–212
Dutta A, Singh S (2022) Persistent flood-induced degradation in Bengal delta: a satellite-based approach. Reg Environ Change 22(6):34–47
District Magistrate Office, Hooghly (2023) District Disaster Management Plan: Hooghly District, West Bengal. Government of West Bengal, India
Fernandes R, Miller JR, Chen J (2011) Modeling the seasonal variation of vegetation indices. Remote Sens Environ 95(2):152–165
Food and agriculture organization (FAO) (2022) The State of food and agriculture 2022: leveraging automation in agriculture for transforming food systems. Rome: FAO
Ghosh R, Banerjee S, Dutta S (2023) Deep learning in flood-affected crop assessment using Sentinel-2 and CNN. ISPRS J Photogramm Remote Sens 192:105–117
Ghosh T, Sen S, Mukherjee M (2021) Remote sensing-based SVM and RF classifiers for flood-induced land change analysis. Geocarto Int 36(8):849–867
Government of West Bengal (2022) Flood report / disaster management updates, West Bengal. Department of Disaster management & civil defence, government of West Bengal
Government of West Bengal (2022) State Action plan on climate change (SAPCC), West Bengal (Updated Version). Department of environment, government of West Bengal, Kolkata

Gupta R, Sharma K, Roy S (2020) Monsoonal flood vulnerability in Bihar and Assam: a geospatial evaluation. Indian J Environ Manage 23(1):56–64
Hooghly District Report (2018) District statistical handbook. Government of West Bengal, Bureau of Applied Economics and Statistics
Huete AR, Didan K, Miura T, Rodriguez EP, Gao X, Ferreira LG (2002) Overview of the radiometric and biophysical performance of the MODIS vegetation indices. Remote Sens Environ 83(1–2):195–213
ICAR (2023) Climate-resilient agriculture practices in India. Indian Council of Agricultural Research
IMD (2023) Monthly rainfall reports. India Meteorological Department
Indiastat (2021) District profile of Hugli. Retrieved from https://www.indiastat.com
Indiastat (2022) West Bengal economic indicators. Retrieved from https://www.indiastat.com
Indian Space Research Organisation (ISRO) (2022) Disaster Management support programme: near real-time flood mapping and monitoring using remote sensing data. Department of space, government of India
Islam MT, Rahman M, Hasan MM (2020) Estimating flood-induced crop damage using remote sensing and GIS: a case study of central Bangladesh. Remote Sens Appl Soc Environ 17:100281
Intergovernmental panel on climate change (IPCC) (2021) Climate change 2021: the physical science basis. Contribution of working group I to the sixth assessment report. Cambridge University Press. https://doi.org/10.1017/9781009157896
Jones C, Riley W, Murphy C (2020) Integrating remote sensing and hydrological models for flood risk assessment. Hydrol Earth Syst Sci 24(3):1359–1373
Kogan FN (1995) Application of vegetation index and brightness temperature for drought detection. Adv Space Res 15(11):91–100
Kogan FN (1997) Global drought watch from space. Bull Am Meteor Soc 78(4):621–636
Kogan FN et al (2019) Droughts and floods monitoring in agriculture with remote sensing data. Remote Sens Environ 239:111630
Kumar R, Behera B, Sahoo B (2023) Geospatial data fusion for flood-prone agriculture monitoring in Odisha and West Bengal. J Appl Geomatics 15(1):33–45
Liu Y, Zhang Y, Chen H (2022) Flood assessment in rice paddies using Sentinel-2 and NDVI-derived indicators in China. Remote Sens 14(5):1034
McCoy J, Johnston K (2001) Using ArcGIS spatial analyst. ESRI Press
Ministry of Agriculture & Farmers Welfare (2022) Annual report. Government of India
Mishra A, Sinha R, Jain V (2019) Flood risk assessment in Indian river basins: role of rainfall and remote sensing. J Hydrol 575:123–134
Mishra V, Nayak R, Sen A (2020) MODIS NDVI analysis of flood impact in Northeastern India. Geospatial Today 15(2):41–48
Mondal KC, Saha S, Aitch P, Bhandari G (2020) Application of Remote sensing and GIS in flood vulnerability assessment—A Case study of lower Ajay Basin, India. In: Disaster risk reduction: methods, approaches and practices. Springer
Mukherjee A, Ghosh S, Roy T (2023) Monsoon flood impacts and vegetation response in Bengal plains. Clim Hazards Remote Sens 12(1):77–89
Mukhopadhyay R, Das M, Chatterjee S (2022) Geospatial prediction of flood-prone zones in South Bengal using DEM and rainfall data. Hydrospatial Analysis 4(3):199–212
National Institute of Disaster Management (NIDM) (2023) Flood Risk management and disaster preparedness in India: training & technical report series. Ministry of home affairs, government of India, New Delhi

National Remote Sensing Centre (NRSC) (2017) Manual on flood mapping using remote sensing techniques. Hyderabad: Indian space research organisation (ISRO), department of space, government of India
NRSC (2023) Remote sensing data products. National Remote Sensing Centre, ISRO
Rahman MS, Alam R, Jahan S (2021) NDVI and DVDI-based flood assessment of Bangladesh croplands. Int J Appl Earth Obs Geoinf 98:102292
Rao S, Narayana A, Iyer R (2021) Remote sensing indices for flood impact and crop loss estimation. ISRO Tech Bull 22(3):34–47
Rouse JW, Haas RH, Schell JA, Deering DW (1974) Monitoring vegetation systems in the Great Plains with ERTS. In: Third ERTS symposium, vol 1, pp 309–317
Roy P, Banik S, Mukherjee M (2022) Differential vegetation damage index for flood-prone crop assessment in India. Giscience & Remote Sensing 59(6):835–854
Sakamoto T, Nguyen NV, Kotera A (2007) Monitoring spatio-temporal changes of rice crop patterns in the Mekong delta. Remote Sens Environ 100(1):95–113
Sangwan N, Joshi PK, Malakar S (2020) Monitoring flood impact in rice-growing regions of Thailand using vegetation indices. Geocarto Int 35(7):747–765
Sarkar A, Banerjee T (2024) Spatio-temporal flood impact on croplands in the Damodar basin. J Hydrol Eng 29(2):04023058
Sarker D, Banerjee P, Das K (2021) Post-flood NDVI-based vegetation analysis in Murshidabad. Remote Sens Environ Stud 12(1):121–134
Sharma S, Jha MK, Roy D (2018) Spatial analysis of flood risk in Bihar using GIS. Nat Hazards 93(2):669–688
Sharma U, Mukherjee A, Ghosh P (2020) Climate-induced flood risks in West Bengal: a spatial assessment. Clim Risk Manag 30:100256
Singh M, Jain V (2018) Topographic vulnerability mapping using SRTM-based DEM. Geomat Nat Haz Risk 9(1):543–562
Singh V, Das A, Roy A (2023) Flood-induced vegetation loss detection using SAR-optical fusion. Remote Sens Lett 14(1):89–97
Smith LC, Brakenridge GR, Kettner A (2020) Increasing frequency of extreme floods in global agriculture. Nat Clim Chang 10(9):748–754
Tarpanelli A, Mondini AC, Camici S (2022) Effectiveness of Sentinel-1 and Sentinel-2 for flood detection assessment in Europe. Nat Hazards Earth Syst Sci 22(8):2473–89
Tucker CJ (1979) Red and photographic infrared linear combinations for monitoring vegetation. Remote Sens Environ 8(2):127–150
Verrelst J, Rivera JP, Camps-Valls G (2021) Machine learning regression algorithms for biophysical parameter retrieval: opportunities for large-scale satellite applications. ISPRS J Photogramm Remote Sens 172:132–147
Wang C, Zhang L, Xu Y (2021) Crop monitoring under flood conditions using high-resolution remote sensing. Agric Water Manag 250:106838
Wang X, Li Y, Liu Z (2023) Integrating AI and geospatial analysis for climate-smart agriculture. Nature Sustainability 6(1):23–35
WBDMD (2023) Flood management reports: Hugli District. West Bengal Disaster Management Department
WBIWD (2022) Annual flood report and irrigation strategy: Hugli. West Bengal Irrigation and Waterways Department

Exploring the Seasonal Variability of the Ecosystem Quality in the Sagar Islands of the Indian Sundarbans

2

Anandita Singh, Subrata Maity, Satiprasad Sahoo, Ajit Govind, Jayanta Das, and Brototi Biswas

Abstract

The natural ecosystem of Sagar Island in the Indian Sundarbans determines coastal resilience by affecting seasonal transformations. The study tracks ecosystem quality patterns through geo-environmental indices including VCI, TCI, VHI, WCI, SCI, and SAVCI based on remote sensing data. EDAS model processing produces Ecosystem Quality Index (EQI) measurements throughout pre-monsoon and post-monsoon seasons from 2014 to 2023. A joint application of simple moving averages combined with Mann–Kendall tests confirms ecosystem quality declined constantly until it hit its maximum decline in 2018 due to cyclonic events. The data indicates that pre-monsoon period EQI values decline as environmental stress improves during the post-monsoon season. The study reveals an ecosystem sensitive zone (ESZ) combining all modeling output, about 27.37% of the area is highly sensitive, and 9.01% of the area is under a very highly sensitive zone of the island. The study found that ecosystem quality on Sagar Island declines significantly during pre-monsoon seasons, with 2018 being the worst year due to cyclones. A decade-long negative trend was confirmed, with

A. Singh · S. Sahoo
Prajukti Research Private Limited (PRPL), Baruipur, South 24 Parganas, West Bengal, India

S. Maity (✉) · B. Biswas
Department of Geography and Resource Management, School of the Earth Science and Natural Resources Management, Mizoram University (Central), Tanhril, Aizawl, Mizoram, India
e-mail: aboutsubrata@gmail.com

S. Sahoo · A. Govind
International Center for Agricultural Research in the Dry Areas (ICARDA), Cairo, Egypt

J. Das
Department of Geography, Rampurhat College, Rampurhat, Birbhum, West Bengal, India

C. Singha et al. (eds.), *Sustainable Management of Delta Ecosystems Resilience*, Deltas of the World, https://doi.org/10.1007/978-3-032-16261-8_2

over 36% of the island classified as ecologically sensitive. These findings prove that efficient management systems should unite improved water resource control measures with mangrove restoration projects to overcome climate change impacts. Knowledge obtained through investigations enables sustainable management of sensitive coastal environments across different time intervals.

2.1 Introduction

Ecological sustainability, together with ecosystem stability, depends on the balanced interaction of biological elements with non-biological elements in the environment (Wanjari et al. 2024). The achievement of sustainable development goals requires knowledge about ecosystem quality fluctuations throughout the year because these movements influence geo-environmental factors, including vegetation health, water status, soil moisture, temperature, and salinity (Xu and Peng 2024). All these factors together influence the resilience and functionality of ecosystems, especially in those vulnerable areas like coastal areas. Whereas the quality of the ecosystem is significantly deteriorating due to both natural and anthropogenic factors, which amplify seasonal variations and contribute to global climate change (Mukherjee et al. 2019; Yang et al. 2024). The increasing temperature, altered precipitation patterns, increasing sea level, coastal erosion, and land degradation are the major key elements impacting climate change and human activities in particular, manifesting increasing heatwaves, unpredictable monsoons, severe precipitation events, cyclones, deforestation, and depletion of groundwater (Narendr et al. 2024; Saha 2021). Collectively, all these phenomena are disproportionately affecting the coastal ecosystems, like Sagar Island in Sundarbans, India, where mangrove degradation, flooding, soil erosion, and sea level rise further contribute to the vulnerability of the region (Bera and Maiti 2021; Rajakumari et al. 2024).

The Sagar Island of Indian Sundarbans symbolizes a unique coastal ecosystem that is highly prone to seasonal variations and geo-environmental changes. Previous studies have rigorously examined the environmental challenges in this region, employing diverse methodologies-such as geospatial analysis, vulnerability assessment, and climate modeling-to evaluate the impacts of climate change, regional susceptibility, and potential adaptation strategies. For example, Paul et al. (2024) utilized satellite-based shoreline dynamics and over wash vulnerability assessments to demonstrate the heightened exposure of Sagar Island to coastal hazards, thereby emphasizing the need for adaptive planning. For instance, exerted Principal Component Analysis (PCA) and trend analysis to examine the vulnerability and risk of Sagar Island to climate change, emphasizing the adverse impact of climatic factors (Ghosh and Mistri 2021). Similarly, long-term trend studies of climate variables from 1982 to 2010 using the Mann–Kendall test unveil negative trends in parameters such as precipitation distribution, surface temperature, and wind speed (Dutta et al. 2020). Other studies concentrated on adaptation strategies for changing climate in the Sundarbans, India, utilizing the Analytical Hierarchy Process (AHP) to tackle its influence on agriculture (Sahoo et al. 2024). Remote sensing and GIS techniques have been utilized to examine the vulnerability of storm surges, which results indicate that the villages in the southern part of Sundarbans are particularly sensitive due to limited managing capacity (Sahana and Sajjad 2018). Furthermore, the degradation of the mangrove health because of flooding has been evaluated using the Fuzzy Analytical Hierarchy Process (FAHP), showcasing a negative correlation between the impacts of floods and vegetation density over time (Mitra et al. 2024).

Despite these significant contributions, an important research gap remains: the seasonal variability of the ecosystem quality in the Sagar Island of India has not been thoroughly examined, utilizing geo-environmental factors, specifically the coastal ecosystem. The goal of this study is to explore the gap by examining the following research questions: (i) Does seasonal variability in ecosystem quality exist in the Sagar Islands? (ii) Is there any negative trend exist in seasonal ecosystem quality during the last decade? The research accomplishes three objectives that include (i) to analyzing seasonal ecosystem quality by using remote sensing based geo-environmental conditioning factors, (ii) to identifying the trend of seasonal quality, (iii) to delineate ecosystem sensitive zone. This study uses the ecological quality index derived from geo-environmental factors (VCI, TCI, VHI, SCI, WCI, and SAVCI) linked through the EDAS (Evaluation based on Distance from Average Solution) model. The research evaluates time-based ecosystem quality changes by implementing Simple Moving Average (SMA) and Mann–Kendall test. The novelty of this study assesses modeling based decadal pre-monsoon and post-monsoonal trend identification for monitor environmental changes on Sagar Islands. This study provides extensive consequences with valuable knowledge to help enhance coastal ecosystem planning and management methods. This research intends to discover the causes behind seasonal changes and environmental patterns to lead conservation efforts and reduce geo-environmental factor impacts. The study advances understandings of sustainable ecosystem management and climate resilience programs in environmentally fragile coastal settings.

2.2 Study Area

Sagar Island is one major human habitation islands among the 54 islands of the 102 islands of Sundarban that is situated at the confluence of the Hooghly River with the Bay of Bengal and is considered a study area. These islands are located between 21°30′N to 22°00′N latitude and 88°00′E to 88°15′E longitude in the southeast region of the Indian Sundarbans (Fig. 2.1). Due to this geographic location, this island has vulnerable coastal areas with the experience of tropical cyclones, storm surges, flooding, salinity intrusion in the island, coastal erosion, and land degradation (Bera et al. 2022). Due to these issues,

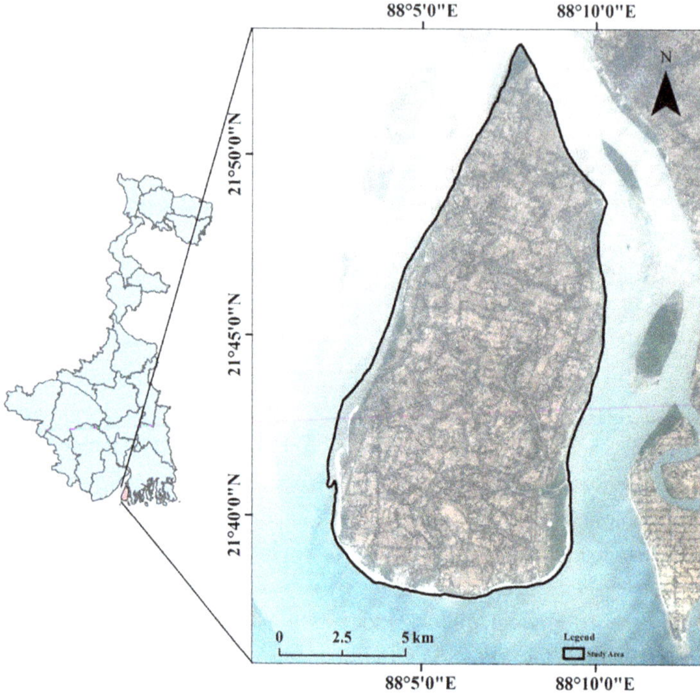

Fig. 2.1 Map of the study area

islands have faced problems like overall ecosystem quality degradation; to address these issues and study the seasonal variability of ecosystem quality, we consider this island as our study area.

2.3 Materials and Methods

To analyze seasonal variation of ecosystem quality in Sagar Island over the past decade, Satellite imagery from the Landsat 8 OLI/TIRS sensor (Path 138, Row 045), with a spatial resolution of 30 m, was acquired from the USGS Earth Explorer platform. Data were collected for the pre-monsoon (March–May) and post-monsoon (October-December) seasons spanning the years 2014 to 2023, and the block boundary map of these islands is taken from the District Census Handbook South 24 Pargana 2011. The methodological flow chart of the study is given in Fig. 2.2.

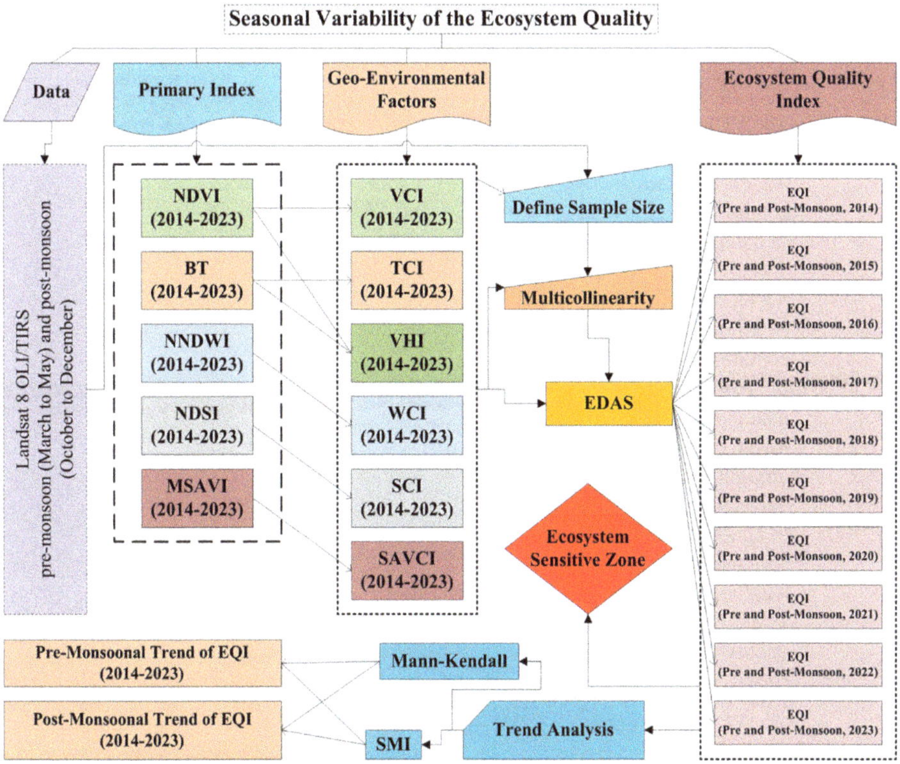

Fig. 2.2 Methodological flow chat of the study

2.3.1 Geo-Environmental Factors

After obtaining all these satellite images, they were gone through preprocessing the reflectance error correction and performing index analysis, like NDVI (normalized difference vegetation index), BT (brightness temperature), MNDWI (modified normalized difference water index), NDSI (normalized difference salinity index), and MSAVI (modified soil-adjusted vegetation index). Then performed a long-term average condition of six selected parameters like VCI (vegetation condition index), TCI (temperature condition index), VHI (vegetation health index), WCI (water condition index), SCI (salinity condition index), and SAVCI (soil adjustment vegetation condition index) by using this primary index analysis to explore the seasonal variability of the ecosystem quality of Sagar Island from the year 2014 to 2023 (Table 2.1).

Table 2.1 Applied geo-environmental conditioning factors for computing ecosystem quality index (EQI)

Factors	Description	Formula	References
VCI	The VCI (in %) represented the present vegetation condition as compared to the time-series vegetation condition, which was influenced by climate, soil, water condition, natural disaster, or changes in land use. It helps identify periods of drought or stress for vegetation, especially in agricultural areas and mangrove forests, making it crucial for assessing ecosystem quality under climatic stress	$\frac{(NDVI - NDVI_{\min})}{(NDVI_{\max} - NDVI_{\min})} \times 100$ (2.1) Where; $NDVI = \frac{NIR - RED}{NIR + RED}$ The $NDVI_{min}$ and $NDVI_{max}$ values are considered from the past ten years of *NDVI* data; *NDVI* represents the relative year	(Karnieli et al. 2006)
TCI	The TCI (in %) tracks brightness temperature anomalies as an assessment method to identify vegetation stress that supports ecological quality evaluations under thermal stress scenarios	$\frac{(BT_{\max} - BT)}{(BT_{\max} - BT_{\min})} \times 100$ (2.2) Where; $BT = \frac{K_2}{In\left(\frac{K_1}{L_\lambda} + 1\right)}$ The BT_{min} and BT_{max} values are considered from the past ten years of *BT* data; *BT* represents the relative year	(Karnieli et al. 2006)

(continued)

Table 2.1 (continued)

Factors	Description	Formula	References
VHI	The VHI (in %) integrates VCI and TCI to assess overall vegetation stress, making it a key indicator for ecosystem quality under climatic extremes. Low Vegetation Health Index (VHI) values indicate poor vegetation conditions, often resulting from thermal stress, limited soil moisture, and nutrient deficiencies. These values are therefore effective indicators of drought severity	$0.5 \times VCI + (1 - 0.5) \times TCI$ (2.3) *VCI* and *TCI* represents the relative year	(Keikhosravi and Rastgar Alaleh Gurabi 2020)
WCI	The WCI (in %) represented the present water condition as compared to the time-series water condition. WCI determines the presence of moisture through vegetation water status, which strongly impacts ecosystem quality assessments and drought evaluation	$\frac{(MNDWI - MNDWI_{\min})}{(MNDWI_{\max} - MNDWI_{\min})} \times 100$ (2.4) Where; $MNDWI = \frac{GREEN - SWIR}{GREEN + SWIR}$ The $MNDWI_{min}$ and $MNDWI_{max}$ values are considered from the past ten years of *MNDWI* data; *MNDWI* represents the relative year	(Karnieli et al. 2006; Sresto et al. 2022)

(continued)

Table 2.1 (continued)

Factors	Description	Formula	References
SCI	The SCI (in %) has been calculated by comparing the present soil salinity with time-series salinity conditions. It helps evaluate ecosystem quality in coastal locations, like stress impacts assessment on both soil and vegetation from salinity	$\frac{(NDSI_{\max}-NDSI)}{(NDSI_{\max}-NDSI_{\min})} \times 100$ (2.5) Where; $\frac{RED-NIR}{RED+NIR}$ The $NDSI_{min}$ and $NDSI_{max}$ values are considered from the past ten years of *NDSI* data; *NDSI* represents the relative year	(Wanjari et al. 2024; Sresto et al. 2022)
SAVCI	The SAVCI (in %) has been calculated by comparing the present MSAVI with time-series MSAVI. Soil brightness effects on vegetation assessment decrease while ecosystem quality evaluation in sparse vegetation regions improves through this method	$\frac{(MSAVI-MSAVI_{\min})}{(MSAVI_{\max}-MSAVI_{\min})} \times 100$ (2.6) Where; $MSAVI = \frac{2\times NIR+1-\sqrt{(2\times NIR)^2-8\times(NIR-RED)}}{2}$ The $MSAVI_{min}$ and $MSAVI_{max}$ values are considered from the past ten years of *MSAVI* data; *MSAVI* represents the relative year	(Sresto et al. 2022; Tran et al. 2022)

2.3.2 Define Sample Size

Cochran's formula is a widely used sample size determination technique for finite populations in the field of social science. The study applied the formula in the geoinformatics field for pixel-based sample size definition to perform multicollinearity tests, EDAS MCDM meddling, and trend analysis. The Cochran's formula is performed by Eqs. 2.7 and 2.8 (Cochran 1977).

$$n = \frac{n_0}{1 + \frac{n_0}{N}} \tag{2.7}$$

where,

$$n_0 = \frac{Z^2 p(1-p)}{e^2} \tag{2.8}$$

2.3.3 Multicollinearity Test

Multicollinearity expresses the statistical issue where the researchers want to understand whether the conditioning factors are strongly connected or not. The study uses the Variance Inflation Factor (VIF) to analyze these connections for better model outcomes and better decision-making (Tejada et al. 2012). This method, applied in several studies, involves using linear regression with collinearity diagnostics on a statistical platform. The study performs this method using Eq. 2.9, employing 1831 sample data of each factor.

$$\begin{cases} (T_i) = 1 - R_i^2 \\ (VIF_i) = \frac{1}{T_i} \end{cases} \tag{2.9}$$

2.3.4 EDAS

EDAS (Evaluation Based on Distance from Average Solution) is an MCDM (Multi-Criteria Decision Making) technique extensively used for the ranking of the alternatives. The method evaluation was determined by the positive and negative distance from average (Mitra and Das 2022). The study that has applied this method considers six different geo-environmental conditioning factors to compute the ecosystem quality index (EQI). About 1831 random criteria of each factor are employed for EDAS modeling. About 1831 random criteria of each factor are employed for EDAS modeling, which are defined by Cochran's formula.

Step I: The initial decision matrix construction for model computation based on remote sensing date driven indices (like VCI, TCI, VHI, WCI, SCI, and SAVCI) based on value of the layers. For the decision matrix, compute n ($n = 1, 2\ldots, j$) criteria and m alternative ($m =$ 1,2,…,i) as follows, given Eq. 2.10.

$$x = \begin{pmatrix} x_{11} & x_{12} & \cdots & x_{1j} & \cdots & x_{1m} \\ x_{21} & x_{22} & \cdots & x_{2j} & \cdots & x_{2m} \\ \vdots & \vdots & \ddots & \vdots & \ddots & \vdots \\ x_{i1} & x_{i2} & \cdots & x_{ij} & \cdots & x_{im} \\ \vdots & \vdots & \ddots & \vdots & \ddots & \vdots \\ x_{n1} & x_{n2} & \cdots & x_{nj} & \cdots & x_{mn} \end{pmatrix} \tag{2.10}$$

where, x_{ij} represents the performance value of the ith alternatives on jth criterion ($i =$ 1,2,…,n and $j =$ 1,2,…,m) and weight of jth criterion by w_j, where $0 < w_j < 1$ and $\sum_{j=1}^{m} w_j = 1$

Step II: Calculate the average solution ($\mathcal{V}_j$) of each parameter is calculated by Eq. 2.11.

$$v_j = \frac{\sum_{i=1}^{n} x_{ij}}{n} \tag{2.11}$$

Step III: Determination of the $\mathcal{P}d_{\text{ij}}$ and $\mathcal{N}d_{\text{ij}}$ as followed by the given Eqs. 2.12 and 2.13.

$$Pd_{ij} = \begin{cases} \frac{\max(0, x_{ij} - v)}{v_j} & if\ j \in BC \\ \frac{\max(0, v_j - x_{ij})}{v_j} & if\ j \in NC \end{cases} \tag{2.12}$$

$$Nd_{ij} = \begin{cases} \frac{\max(0, v_j - x_{ij})}{v_j} & if\ j \in BC \\ \frac{\max(0, x_{ij} - v_j)}{v_j} & if\ j \in NC \end{cases} \tag{2.13}$$

where, $\mathcal{P}d_{ij}$ stands for positive distance from the average, $\mathcal{N}d_{ij}$ stands for negative distance from the average, BC and NC stands for the beneficial (VCI, VHI, WCI, SAVCI) and non-beneficial criteria (TCI, SCI). The higher value of beneficial criteria is responsible for good EQI, and lower is vice versa. The higher value of non-beneficial criteria is responsible for the worst EQI, and lower is vice versa.
Step IV: Ascertain the weighted summation of the $\mathcal{P}d_{ij}$ and $\mathcal{N}d_{ij}$ by the given Eqs. 2.14 and 2.15.

$$SP_i = \sum_{j=1}^{m} w_j p_{dj} \tag{2.14}$$

$$SN_i = \sum_{j=1}^{m} w_j Nd_{ij} \tag{2.15}$$

Step V: Normalization of $\mathcal{SP}_i$ and $\mathcal{SN}_i$ as follows Eqs. 2.16 and 2.17.

$$NSP_i = \frac{SP_i}{\max\limits_k SP_k} \tag{2.16}$$

$$NSN_i = 1 - \frac{SN_i}{\max\limits_k P_k} \tag{2.17}$$

Step VI: Appraisal score ($\mathcal{A}_{si}$) calculation of each alternative by Eq. 2.18

$$A_{si} = \frac{1}{2}(NSP_i + NSN_i) \tag{2.18}$$

Step VII: Finally, based on decreasing values of $\mathcal{A}_{si}$ alternatives are ranked. The output is considered EQI. The value differs from 0 to 1, with an EQI value near 0 indicating a bad ecosystem quality and near 1 indicating good.

2.3.5 ESI

The ESI (Ecosystem Sensitive Index) is being identified by the EDAS model's A_{si} driven EQI from over multiple years (10) and seasons (pre-monsoon and post-monsoon). Equation 2.19 is applied to compute it, where N is 20. The value differs 0–1, ESI value near 0 indicating highly sensitive ecosystem and near 1 indicate low. The value differs from 0 to 1, with an ESI value near 0 indicating a highly sensitive ecosystem and near 1 indicating low. The ESI has been classified into ESZ (Ecosystem Sensitive Zone).

$$ESZ = \frac{1}{N}\sum_{i=1}^{N} A_{si} \tag{2.19}$$

2.3.6 Ecosystem Quality Trend Assessment

SMI

The SMI (Simple Moving Average) is a widely used statistical method for time series data to identify trends over time. It plays a crucial role in analyzing long-term changes in the ecosystem quality index. The SMI for a given time period is calculated by Eq. 2.20. The study applied 3-year time periods of SMI for 10 years of EQI trend identification.

$$SMI = \frac{X_1 + X_2 \ldots X_n}{n} \tag{2.20}$$

Mann–Kendall Test

The Mann–Kendall test was first proposed by Mann, the Mann–Kendall (MK) test in 1945, and expanded on it in 1975. This is a non-parametric test that was considered to calculate the trend of different environmental and climate datasets and determine time series patterns (Karim et al. 2024). In our study, the Mann–Kendall test was applied to identify the trend of EQI from the year 2014 to 2023 separately for both pre-monsoon and post-monsoon season of our study area, which was calculated through EDAS model (Frimpong et al. 2022).

Step I: Firstly, the S-statistics was calculated to determine the presence of monotonic trend by using the Eq. 2.21.

$$S = \sum_{i=1}^{n-1} \sum_{j=i+1}^{n} sgn\left(x_j - x_i\right) \tag{2.21}$$

where,

$$sign(X_j - X_i) = \begin{cases} +1, \; if\,(x_j - x_i) > 0 \\ 0, \; if\,(x_j - x_i) = 0 \\ -1, \; if\,(x_j - x_i) < 0 \end{cases} \tag{2.22}$$

Sign function was calculated for each pair of the dataset In the Eq. 2.22, the two estimated values were compared by subtracting one from the other, assigning + 1 for an increase, – 1 for a decrease, and 0 if there was no change. Finally, the S-statistical value comes from summing all sing value.

Step II: V(S) was calculated for large dataset (n > 10) by using the Eq. 2.23

$$V(S) = \frac{n(n-1)(2n+5) - \sum_{i=1}^{m} t_i(t_i - 1)(2t_i + 5)}{1831} \tag{2.23}$$

Here, n denotes the total data points, m indicates the count of tied values, and t_i is the number of ties for the ith value. The standard normal test statistic Z follows a normal distribution and is used at the 95% confidence level when the sample size exceeds 10.

Step III: Z- statistics was calculated by using the Eq. 2.24 to determine the significance of trend.

$$Z = \begin{cases} \frac{S-1}{\sqrt{Var(s)}} & S > 0 \\ 0 & S = 0 \\ \frac{S+1}{\sqrt{Var(s)}} & S < 0 \end{cases} \tag{2.24}$$

Step IV: Sen's slope estimator was implemented to determine the magnitude of trend using the Eq. 2.25 (Gupta and Verma 2023).

$$\beta = median \frac{Xj - Xk}{j-k} \ldots \forall \ldots k < j \tag{2.25}$$

where, X_j and X_k represent the observation value at times j and i and (j–k) represent the time difference between the observation.

2.4 Results

2.4.1 Geo-Environmental Factors

The VCI result shows poor condition of vegetation in the pre-monsoon season of the years 2014, 2017, 2019, 2021, 2022, and 2023 as compared to post-monsoon, while in the post-monsoon season of the years 2015 and 2018, they were showing poorer vegetation condition than in the pre-monsoon season. Lower VCI values have been seen at coastal beach areas, river estuaries, water bodies, and fish farming in the southeastern and southwestern parts and northern parts, with moderate and high values scattered across the island in build-up areas and in agricultural land and mangroves, respectively. In the northern part of the island, the growing vegetation condition over the year is the result of afforestation. In ten years, the poorest condition of vegetation was seen in 2018, the year of both seasons (Fig. 2.3). In the last year, 2023, a lower to moderate VCI value was shown at the upper part of the island, and a higher value was shown towards the southern part with better conditions than the previous year (Fig. 2.4).

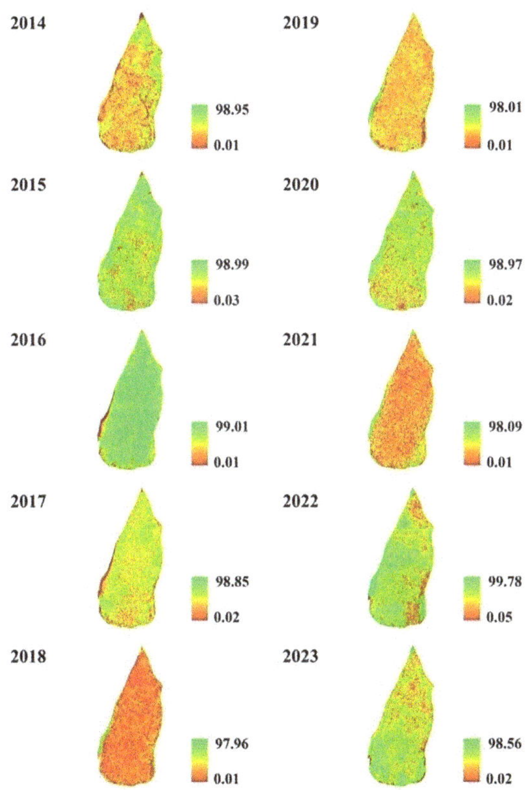

Fig. 2.3 VCI, pre-monsoon (2014–2023)

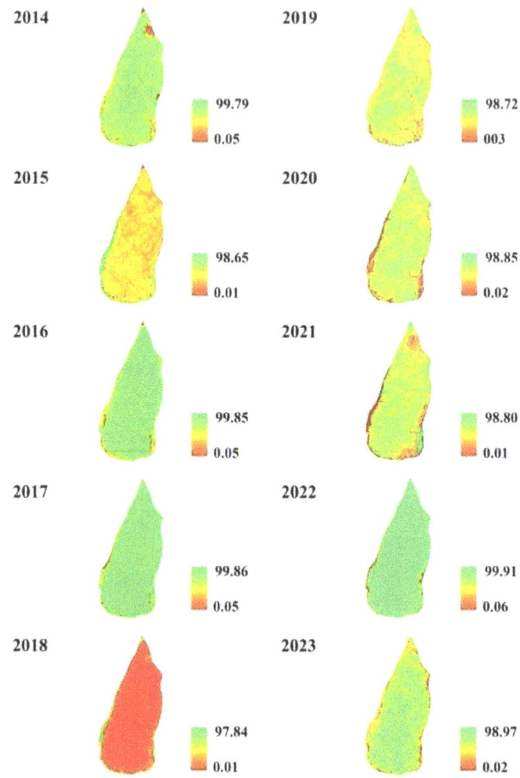

Fig. 2.4 VCI, post-monsoon (2014–2023)

The TCI analysis revealed that islands experienced temperature variations, which contributed to vegetation stress. From this analysis, we can see that the post-monsoon seasons of 2014, 2016, 2020, 2021, 2022, and 2023 have experienced more temperature on vegetation stress than the pre-monsoon. In this decade, the worst temperature condition index value, 35.20, is the upper range in 2019 pre-monsoon (Fig. 2.5), followed by 2016 post-monsoon with an upper value of 53.44, 2014 post-monsoon (53.69), 2017 post-monsoon (63.25) (Fig. 2.6), and 2017 pre-monsoon (67.59). The index value saw a huge fluctuation in temperature conditions before the year 2019; after that, it showed a continual decrease, and the last year saw 71.33 (Fig. 2.6) as the upper range of TCI, particularly in the northern part of Sagar Island, which means temperature stress increased over the year on vegetation, possibly due to increased human activity in this area.

The VHI value showed that the health of vegetation condition in the islands, which was relatively poor in the pre-monsoon season of the years 2017 (74.04), 2019 (58.65), and 2021 (84.88), and healthy vegetation condition showing in the pre-monsoon season of 2014 with a range of 98.49–14.76, 2015 with a range of 98.85–17.85, 2016 with a range of 99.81–18.63, 2018 with a range of 97.89–35.42, 2020 (99.46–38.97), 2022 (98.23–16.13), and 2023 (94.63–6.57) (Fig. 2.7) as compared to the post-monsoon season.

Fig. 2.5 TCI, pre-monsoon (2014–2023)

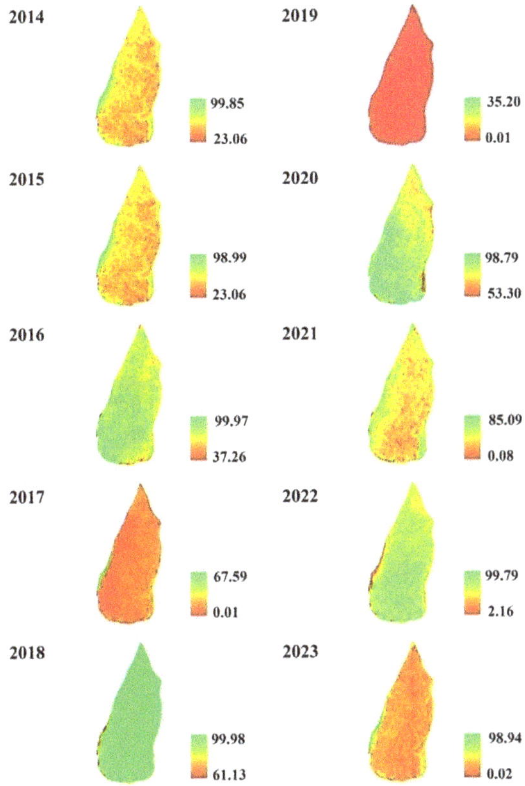

Generally, coastal beach areas, built-up concentration areas of the upper and lower parts of the island, have lower VHI values. The worst VHI value was seen during the years 2019, 2021 pre-monsoon, and both seasons of 2018 (Fig. 2.8) across the island, may cause higher salinity (Fig. 2.12) or intrusion of saline water (Fig. 2.9), as our observation.

Comparing the WCI value of both seasons has shown the lower value of WCI during the pre-monsoon season of the years 2014, 2015, 2016, 2017, 2019, 2020, 2021, and 2023, while during the post-monsoon season of the years 2018 and 2022, across the island, showing concern about lower WCI value. Seasonal variations in the water condition index showed the positive impact of the monsoon season which showed in the post-monsoon. In a decade of observation coverage of higher WCI value in the year 2018 pre-monsoon (Fig. 2.9), 2019 post-monsoon (Fig. 2.10), and 2022 pre-monsoon season (Fig. 2.9) across the island. In the year of 2019, they showed the result of a negative impact due to Cyclone Bulbul in the post-monsoon season. Lastly, the WCI value during the current year has shown a huge difference between the two seasons.

The SCI helped identify areas affected by high salinity levels. Results showed a high increase in soil salinity during the pre-monsoon season compared to the post-monsoon

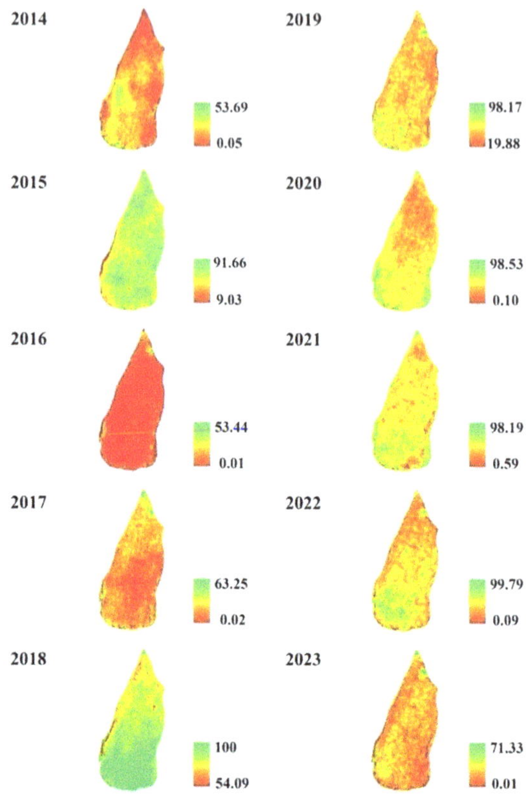

Fig. 2.6 TCI, post-monsoon (2014–2023)

season over the decade. While the highest SCI value appears during the 2018 post-monsoon (99.97), after that, pre-monsoon 2023, 2021, and 2014 (Fig. 2.11), and lower SCI values appeared across the island during the 2022 pre-monsoon (Fig. 2.11) and post-monsoon 2014, 2016, 2019, 2022, and 2023 (Fig. 2.12) due to the fresh water availability, as we observed in Figs. 2.9 and 2.10. However, last year, the 2023 pre-monsoon (99.93) salinity condition was more than the 2023 post-monsoon season (97.99) and also more than the previous year.

The SAVCI represents more accurate vegetation conditions by the adjacent soil brightness from the reflectance. The result has shown healthy vegetation conditions during the post-monsoon season in all years except the year 2018, as compared to the pre-monsoon season (Fig. 2.13), because of the monsoon's role in the ecosystem. Generally, the concentration of lower SAVCI values is located in the built-up area in the southern and northern parts of the island, water bodies, and fisheries. In the ten-year analysis, more area comes under the low SAVCI value during post-monsoon 2018 (Fig. 2.14), pre-monsoon 2022, and 2020 (Fig. 2.13). However, last year's vegetation condition was better than the previous year's (Fig. 2.14).

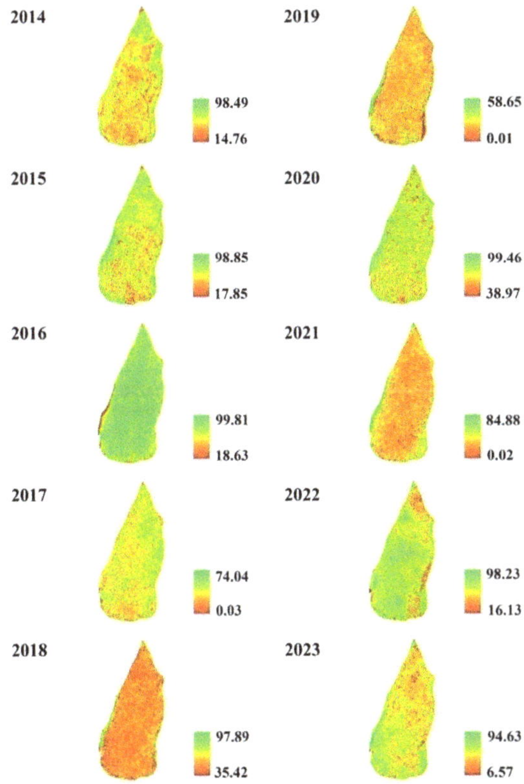

Fig. 2.7 VHI, pre-monsoon (2014–2023)

2.4.2 Sample Size

The Landsat-8 OLI/TIRS data with 30-m resolution allowed researchers to study 273,878 pixels in the study area. The entire analysis followed statistical protocols with 3% precision at 99% confidence. Statistical robustness required 1831 pixels for analysis based on a 0.5 estimated proportion value and Cochran's sample size calculation method. The methodological approach strengthens the reliability of ecosystem quality evaluations to show observed patterns that represent overall environmental changes accurately throughout the study area.

2.4.3 Multicollinearity Test

The study computed varying variance inflation factors (VIF) for different factors (Table 2.2). The higher VIF was found for VCI (VIF = 8.90, tolerance = 0.11), VHI (VIF = 8.98, tolerance = 0.14), and SAVCI (VIF = 8.63, tolerance = 0.11), expressing

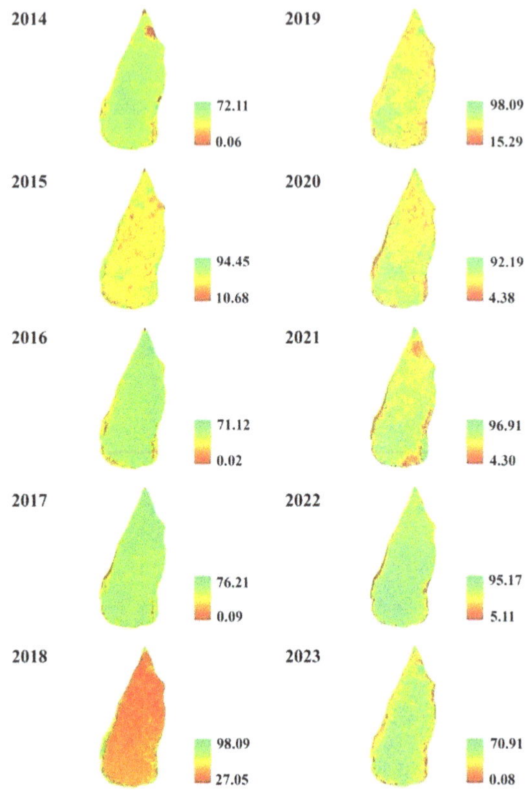

Fig. 2.8 VHI, post-monsoon (2014–2023)

a moderate potential influence on other factors. On the other hand, the lower VIF values were observed for TCI (VIF = 1.06, tolerance = 0.95) and WCI (VIF = 1.08, tolerance = 0.94), respectively. All VIF values were below 10, which is not above the threshold value, indicating no significant collinearity among the factors. This analysis is done by using data from 2023.

2.4.4 EDAS

In the study, EQI (ecosystem quality index) has been calculated based on six geo-environmental parameters through the EDAS model, which is an MCDM method. Here we divide these parameters into two categories, beneficial and non-beneficial, on the basis of their impact on the ecosystem. All these six parameters have been assigned a weight value as per the experts' suggestions in the previous study. Here, VCI, SAVCI, VHI, and TCI were assigned 20%. Each one and WCI, SCI both was assigned 10%. The calculated NSP_i and NSN_i values show the range 0–1 in both. The average A_{si} score (EQI) is

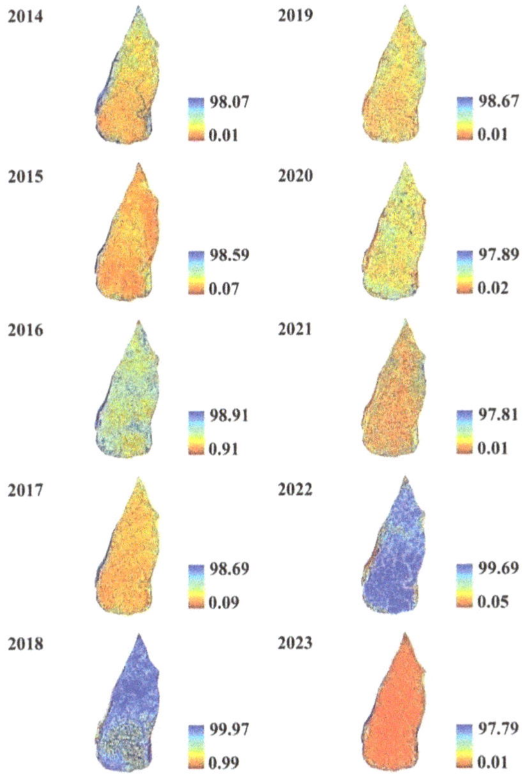

Fig. 2.9 WCI, pre-monsoon (2014–2023)

found in the range of 0.01–0.99. As a result, seasonal comparisons clearly show poorer ecosystem quality in the pre-monsoon period than in the post-monsoon over the decade (Fig. 2.15). While the post-monsoon season (Fig. 2.16) has experienced a low EQI value in the years 2015 (0.02–0.93), 2018 (0.01–0.92), and 2019 (0.04–0.89) due to sudden events like hazards. Also, EQI values have shown a spatial distribution pattern where low values are located over built-up areas in the northern part of the island and over coastal areas in the southern part of the island. The 2018 post-monsoon has seen the worst condition of EQI (0.92–0.01) over the past decade, which may have an impact of two VSCS (very severe cyclonic storms), Titli and Gaja, in the Bay of Bengal. While 2016 (0.11–0.99) post-monsoon season showed the ecosystem was in good condition.

2.4.5 ESZ

Combining all EQI layers (20), the ecosystem sensitive zone (ESZ) has been performed (Fig. 2.22). The ESZ is highly impacted due to human or natural activities. Mainly coastal areas in the southwestern part are in a very highly sensitive category, and the northeastern

Fig. 2.10 WCI, post-monsoon (2014–2023)

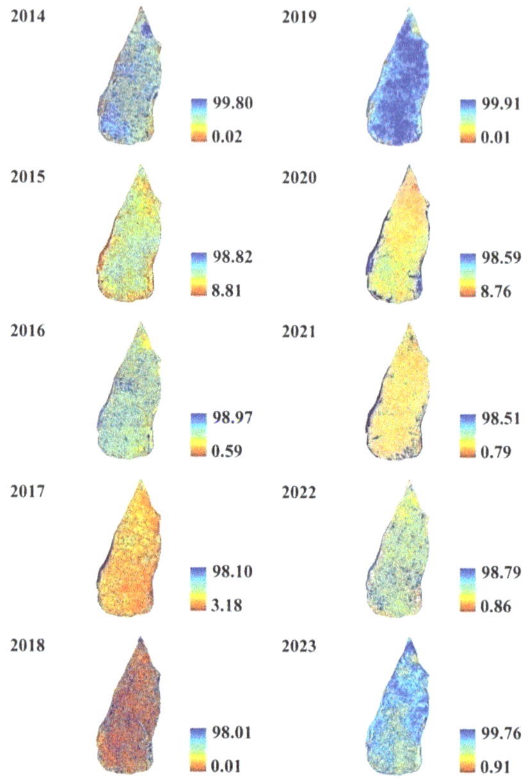

part of the island falls under a highly sensitive category. Classification results show that 5.88% of the area is in very low sensitive category, 23.21% of the area is under low category, 34.53% of the area is under moderate sensitivity, 27.37% of the area is highly sensitive, and 9.01% of the area is under very highly sensitive zone of the island. It means half the area of the island was moderate to highly sensitive.

2.4.6 Ecosystem Quality Trend Assessment

SMI (Trend)

3-year SMI shows a huge fluctuation in the EQI trend line over the year in both seasons seen. In the pre-monsoon season during the year 2019 (Fig. 2.17), we have seen a higher EQI value than the SMI trend, which means this year has a higher EQI than the average of 3 years, while in the post-monsoon season during the year 2018 (Fig. 2.18), we have seen the lowest value of EQI that negatively crosses the SMI trend line, showing degradation

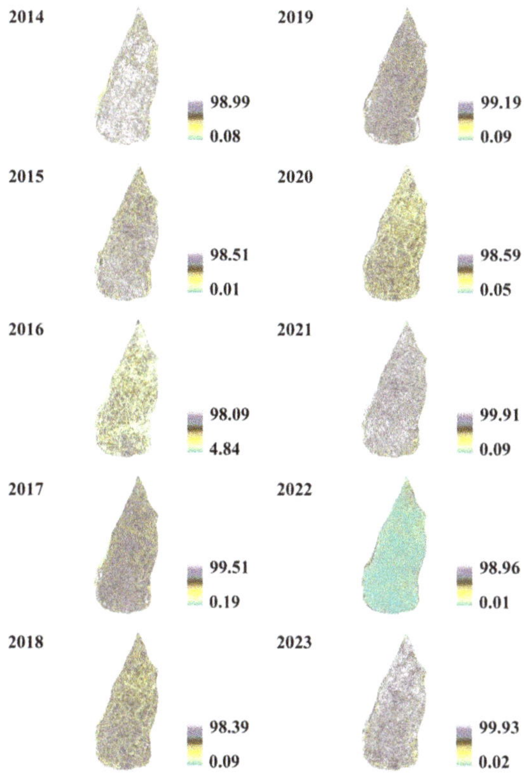

Fig. 2.11 SCI, pre-monsoon (2014–2023)

of EQI for this particular year. The current year value of SMI (Fig. 2.17) in pre-monsoon is 0.58099, and post-monsoon has 0.5583905; both were showing a negative value then. The difference is in the magnitude of decreasing EQI.

Mann–Kendall Test (Trend)

The Mann–Kendall test was performed to identify the spatiotemporal trend of EQI over the past decade (2014–2023) and the seasonal variation of the trend of Sagar Island. The result is statistically significant at a 95% confidence level ($p < 0.05$). Here, the Sen's slope test was applied to identify the magnitude of the trend, a negative trend with the magnitude − 0.00446 in the pre-monsoon season (Fig. 2.19) and − 0.00126 in the post-monsoon season (Fig. 2.20). The huge fluctuating end of the pre-monsoon season (2014–2023) is shown within the range of 0.40–0 (Fig. 2.19). About 68–0.68 of the EQI value, while the worst condition of the EQI was in 2018 (0.407564); the highest EQI was seen in 2019 (0.68) in the pre-monsoon season. On the other hand, (Fig. 2.20) shows the average trend of the post-monsoon season (2014–2023) within the range of 0.24–74 of the EQI value, while the worst condition of the EQI was seen in 2018 (0.24); the highest EQI was seen in 2016 (0.74) in the post-monsoon season. The current year 2023, was a decreasing trend

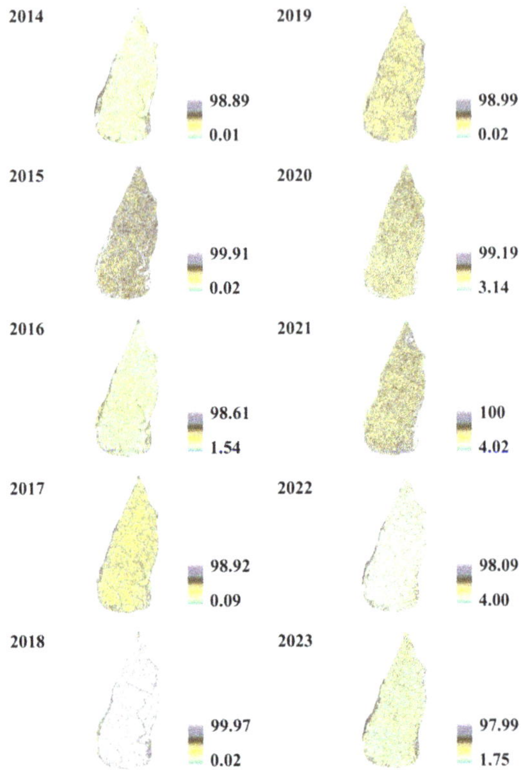

Fig. 2.12 SCI, post-monsoon (2014–2023)

as compared to the previous year in both seasons. But in the pre-monsoon season, the EQI trend value (0.48) was going in a linear trend line, while in the post-monsoon season, the EQI value (0.58) was more than the linear trend line. During the pre-monsoon season, 27.81% of the area has experienced a negative EQI trend, 49.27% of the area represents no trend, and 22.93% of the area has a high category positive trend. In the post-monsoon season (Fig. 2.21), a classified high-category positive trend has been experienced within 51.29% of the area, no trend in 42.02% of the area, and a negative trend in 6.69% of the area.

2.5 Discussion

Seasonal transformations of the ecosystem quality emerge at Sundarbans Sagar Islands in India because both monsoon winds and temperature fluctuations, together with salt content adjustments, cause susceptibility to these islands. According to Vegetation Condition Index (VCI) data, pre-monsoon resulted in worse vegetation health than post-monsoon

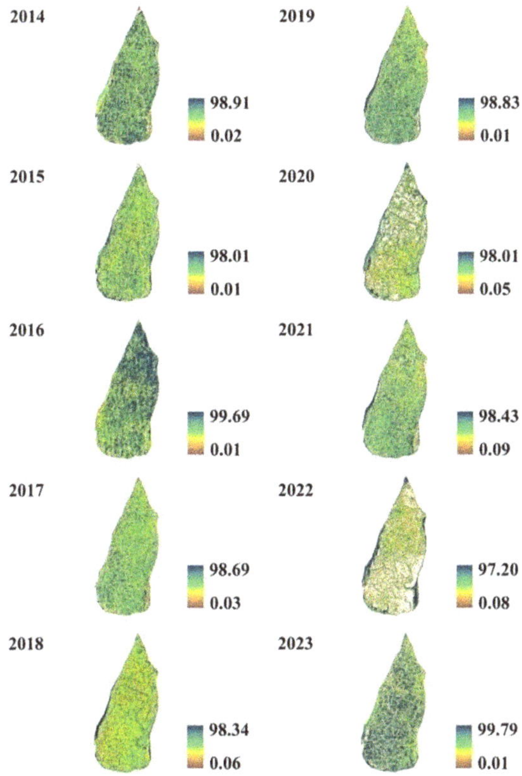

Fig. 2.13 SAVCI, pre-monsoon (2014–2023)

during the ten years, and 2018 presented the worst vegetation condition. Severe vegetation stress hit the area first in pre-monsoon seasons during 2014, 2019, and 2021, besides affecting it in post-monsoon 2015. SAVCI data shows that the 2018 post-monsoon was the most detrimental season, but pre-monsoon experienced sudden significant decreases in 2020 and 2022. The Vegetation Health Index (VHI) reveals major vegetation deterioration during 2018,followed by added stress in 2019 along with 2021 because of environmental stress and inadequate rainfall. Similarly, Karsch et al. (2023) reported a decline in vegetation cover in the Sundarbans during the periods 2014–2015, 2017–2018, and 2019–2020, attributing these reductions to extreme climatic events such as cyclones and floods. Water Condition Index values point to three periods of high stress covering pre-monsoon 2018 and post-monsoon 2019 and pre-monsoon 2021 when drought and cyclone events impacted the area. Extremely high soil salinity exists in the pre-monsoon seasons of 2014, 2017, 2018, 2019, and 2021, as well as post-monsoon 2018 due to severe weather conditions, this result is consistent with the findings of Chowdhury et al. (2023), who reported an increasing trend in water salinity in the Sundarbans from 2017 to 2021, primarily due to saline water intrusion in coastal areas. They also observed that salinity

Fig. 2.14 SAVCI, post-monsoon (2014–2023)

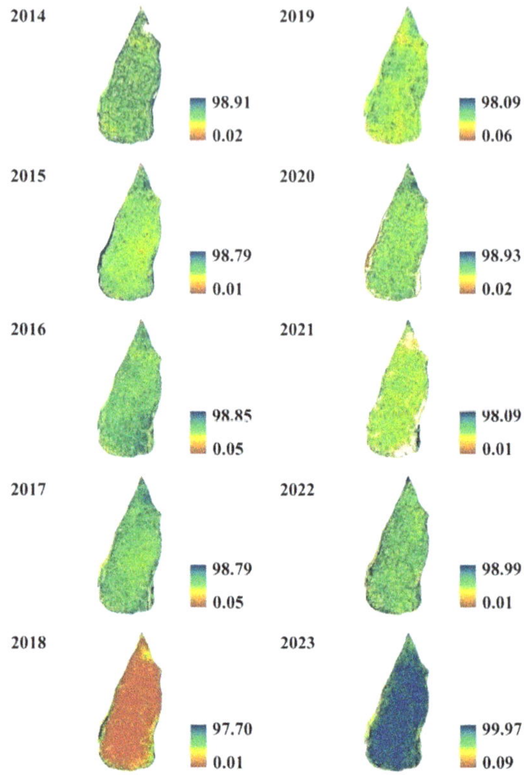

Table 2.2 The multicollinearity diagnostics among selected indices

Factors	VIF	Tolerance	Remark
VCI	8.90	0.11	Moderate multicollinearity
TCI	1.06	0.95	No multicollinearity
VHI	8.98	0.14	Moderate multicollinearity
WCI	1.08	0.94	No multicollinearity
SCI	2.35	0.43	Low multicollinearity
SAVCI	8.72	0.11	Moderate multicollinearity

levels were higher during the pre-monsoon season compared to the post-monsoon season. Similarly, Sengupta et al. (2019) identified seasonal variation in soil salinity in the region, which adversely affects agricultural productivity. North Sagar Island has shown an upward trend of environmental stress based on the Temperature Condition Index through 2019 pre-monsoon and 2016 post-monsoon, and the metric reveals worsening heat stress from 2019 to 2023. The EDAS-based Ecological Quality Index (EQI) demonstrates pre-monsoon present worse ecosystem quality, yet post-monsoon 2018 exhibits the lowest

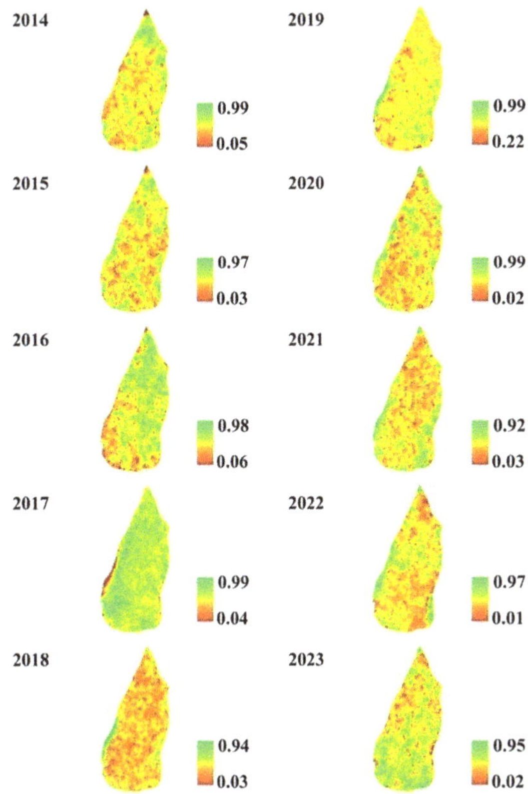

Fig. 2.15 EQI, pre-monsoon (2014–2023)

levels because of Titli and Gaja cyclones. The declining EQI observed after 2018 aligns with the findings of Mondal et al. (2024), who reported significant post-cyclonic impacts on the Sundarbans. Both studies highlight a marked deterioration in environmental conditions resulting from extreme cyclonic events. Mann–Kendall trend analysis demonstrates a negative change because high ecological fitness levels were present in post-monsoon for 51.29% of the region, as pre-monsoon revealed poor conditions in 27.81%. The degradation process demonstrates faster movement according to Sen's slope test during the dry season months. After 2018, the region faced pronounced ecological decline during the seasonal period due to water shortages, extreme temperatures, and elevated saline levels. To make the Indian Sundarbans more resilient to ecological changes, environmentally friendly approaches that manage salt levels, store water, and practice sustainable land utilization must be adopted.

As a result of our study on seasonal variation of ecosystem quality, we have seen it in different coastal regions all over the world. Rising temperature, sea level, salinity conditions, coastal erosion, vegetation degradation, and the increasing frequency of cyclonic storms-all these factors were responsible for the degradation of ecosystems and made the

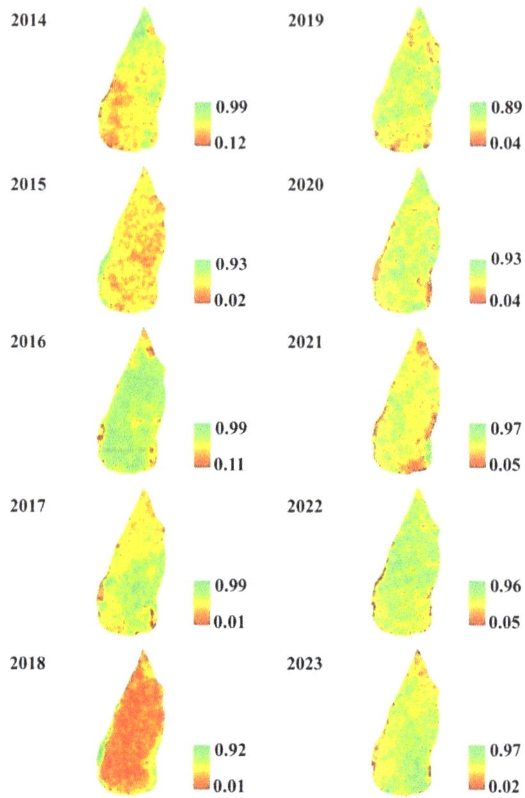

Fig. 2.16 EQI, post-monsoon (2014–2023)

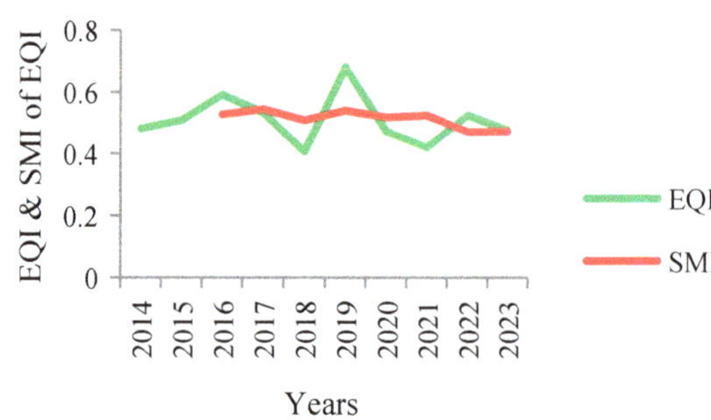

Fig. 2.17 Average SMI of pre-monsoonal EQI

coastal environment more vulnerable day by day (Rizzo and Anfuso 2020). The delta region of the Pearl River has also found low ecological quality in the coastal region due to increasing population growth and urbanization (Ding et al. 2020). Another study shows eco-geo-environmental quality by using different remote sensing index analyses that found an EGI value of 26.088 in the lower category (Li et al. 2021). Temporal analysis of environmental quality in the Yellow River delta from 2000 to 2020 revealed the

Fig. 2.18 Average SMI of post-monsoonal EQI

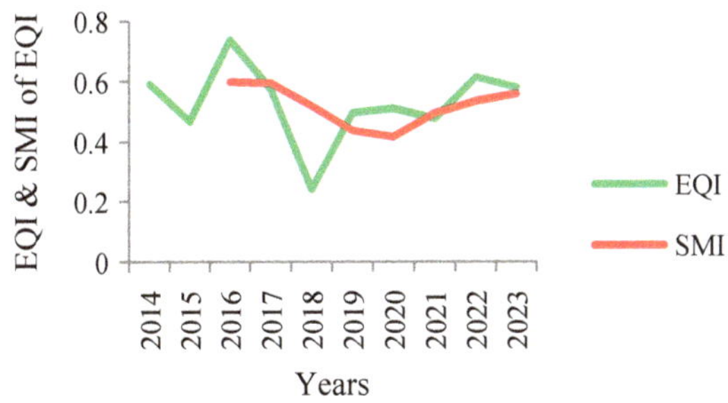

Fig. 2.19 Average pre-monsoonal trend of EQI

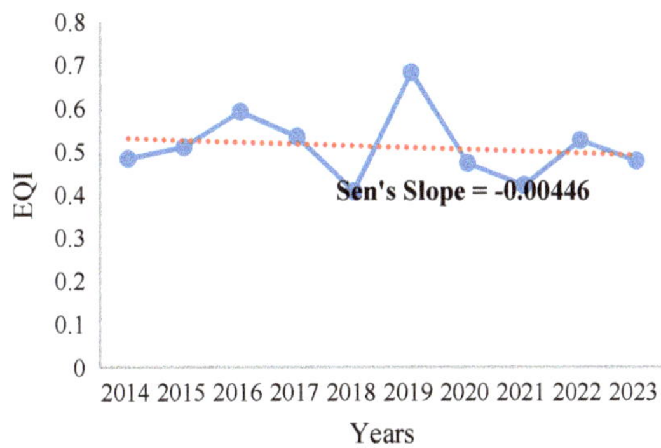

Fig. 2.20 Average post-monsoonal trend of EQI

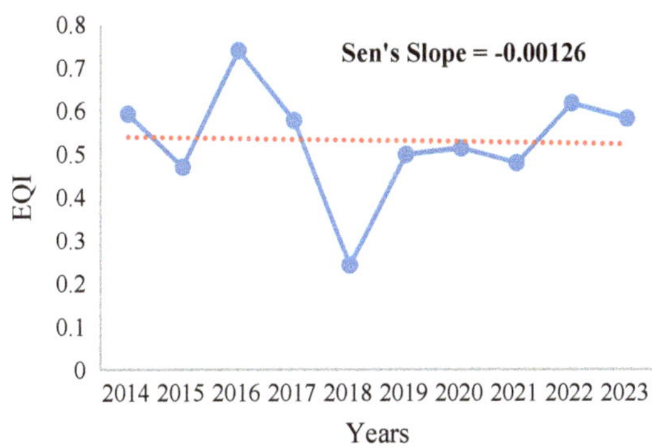

poor environmental quality in the coastal areas of the river and the V-shaped temporal trend between two decades (Ma et al. 2023).

The meteorological conditions during summer months lead to increased vegetation development, SAVCI rates, VCI levels, and freshwater reserves between June and September, while winter and summertime conditions result in lower soil water amounts and rising SAVCI rates together with decreasing VCI levels and agricultural production. The rise in

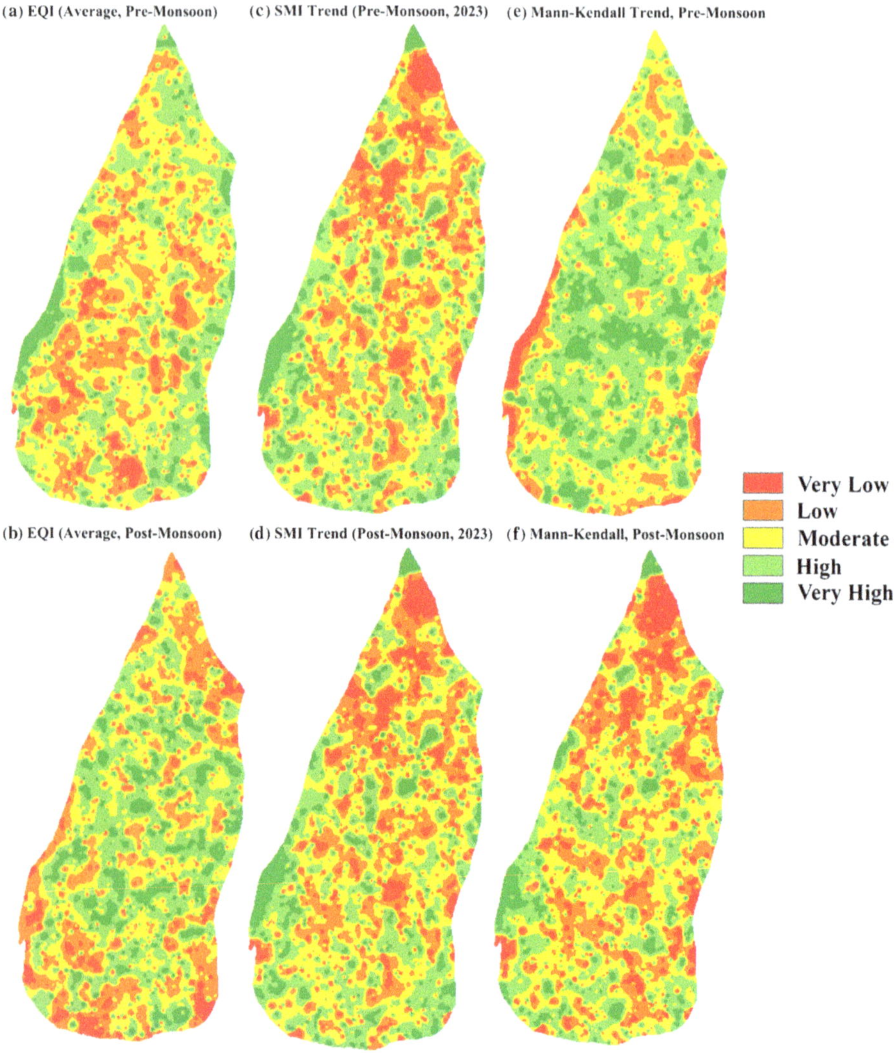

Fig. 2.21 **a** Classified zone of average pre-monsoonal ecological quality; **b** Classified zone of average post-monsoonal ecological quality; **c** Classified zone of SMI trend of pre-monsoonal ecological quality, 2023; **d** Classified zone of SMI trend of post-monsoonal ecological quality, 2023; **e** Classified zone of Mann–Kendall trend of pre-monsoonal ecological quality; **f** Classified zone of Mann–Kendall trend of post-monsoonal ecological quality

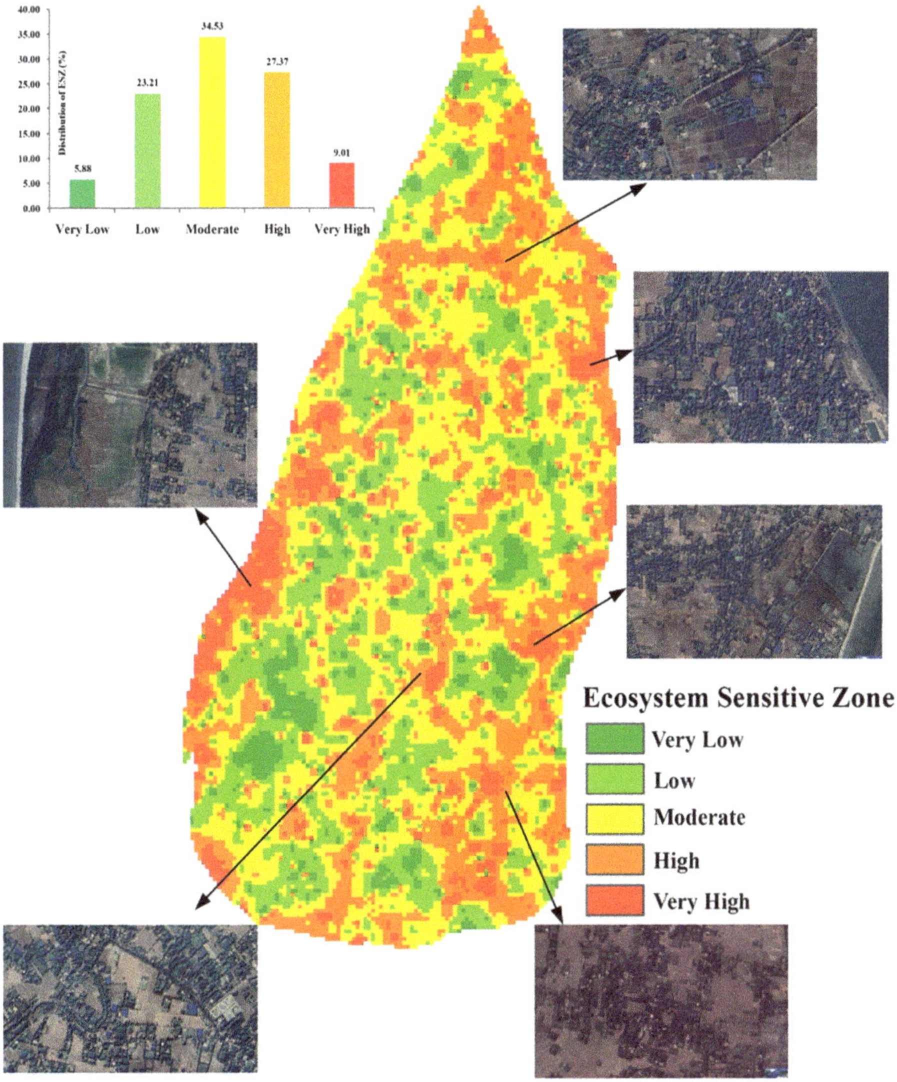

Fig. 2.22 Ecosystem sensitive zone of Sagar Islands

summertime temperatures quickens evaporation, which deteriorates soil and water conditions in the area. Cyclones that strike the area before and after the monsoon season cause damage because they create storm surges and facilitate erosion and also destroy mangrove habitats. During pre- and post-monsoon cyclones, cyclones cause ecosystem damage to ecosystems through coastal flooding and shoreline degeneration and disappear of mangrove trees. Remote sensing indices such as VCI, SAVCI, WCI, SCI, and

TCI, when combined with programs of sustainable management including mangrove restoration, precision agriculture, and hydrological planning, will strengthen ecosystems against climate-induced and human-caused disturbances for sustainable long-term stability (Fig. 2.22).

2.6 Major Findings

- The study confirmed a clear seasonal variation in ecosystem quality on Sagar Island, with pre-monsoon seasons consistently showing lower EQI values compared to post-monsoon seasons over the 2014–2023 periods.
- The year 2018 was the most ecologically degraded year, especially during the post-monsoon season, due to the occurrence of two very severe cyclonic storms (Titli and Gaja), which significantly lowered the EQI across the island.
- Mann–Kendall trend analysis indicated a statistically significant negative trend in EQI during both pre-monsoon (− 0.00446) and post-monsoon (− 0.00126) seasons, with the decline being sharper during the pre-monsoon.
- Approximately 36.38% of the island (27.37% highly sensitive + 9.01% very highly sensitive) was classified as ecologically sensitive, predominantly in coastal and built-up northern and southwestern areas, indicating vulnerability to climate and anthropogenic pressures.
- The analysis of geo-environmental indices (VCI, TCI, VHI, WCI, SCI, SAVCI) showed that salinity, temperature stress, and water scarcity were key drivers of ecosystem degradation, especially in the dry (pre-monsoon) periods.
- The EDAS model successfully integrated multi-criteria geo-environmental indices to produce reliable EQI maps for malty-years monitoring of coastal ecosystem dynamics.

2.7 Limitations

While this study explores the seasonal variation between pre-monsoon and post-monsoon seasons and identifies ecosystems, the study experiences several limitations. The analysis relied on Landsat-8 data, which, despite its moderate resolution, may not capture fine-scale environmental changes or microhabitat dynamics. The Sagar is a tropical cyclone-prone area; cloud cover in the satellite image has been hindered for a specific time period. Lack of information on ground level was another backdrop of this study, as it was a coastal region and vulnerable to tidal impact, and all sudden events were not recorded. The study applied only remote sensing-based geo-environmental factors, but the other factors, like rainfall, temperature, and several anthropogenic factors, were not considered for the analysis. The other MCDM, ML, and DL methods may provide better results that are not explored and compared with EDAS in this study.

2.8 Conclusions

This research studies seasonal variability of the Sagar Islands ecosystem quality in the Indian Sundarbans through analysis of geo-environmental factors. The EDAS model with remote sensing indices VCI, TCI, VHI, WCI, SCI, and SAVCI allows research to evaluate ecological quality through spatio-temporal variability assessment for the pre- and post-monsoon period. Research findings demonstrate the environment suffers its worst decline during pre-monsoon due to temperature increases together with elevated salt concentrations, water scarcity and vegetation tress. The EQI value declines from 0.99 to 0.95 in the pre-monsoon period and from 0.99 to 0.97 in the post-monsoon during the time of 2014–2023.Ecological deterioration in the area reached its peak during the 2018 post-monsoon season, caused by extreme cyclonic storm activities. The EQI value of 2023 in both seasons is very close to the SMI of the EQI value (about 0.58099 in the pre-monsoon and 0.5583905 in the post-monsoon).About 9.01% of the area falls under the very high ESZ, and 27.37% of the area falls under the high ESZ. Ten years of seasonal data subjected to the Mann–Kendall test and Sen's slope analysis reveal that ecosystem quality experienced a downward trend (− 0.00446 in pre-monsoon and − 0.00126 in post-monsoon). The combined factors of salinity, high heat stress, and erratic rainfall patterns caused plant vegetation indices VHI and VCI to decline during recent years. Even though monsoons bring fresh water to the ecosystem, they only give temporary relief, but salinity levels permanently increase. The research demonstrates the immediate requirement to build adaptive strategies that reduce the climate change risks to coastal ecosystems. Policy makers use ecological area maps for targeting their conservation activities precisely. The enhancement of ecosystem resilience requires sustainable water management together with mangrove restoration and the implementation of climate-adapted agricultural techniques. Both the local livelihoods and the ecosystem need protection, which will require essential integration between community-based resilience action plans and scientific assessment methods. Continuous environmental monitoring together with multi-criteria decision-making models becomes essential based on the systematic study of seasonal and long-term ecological changes. The research outcomes help guide sustainable ecosystem management through policy development for addressing present-day global environmental problems. Future research can enhance the ecosystem quality assessment by incorporating high-resolution satellite data, ground-based validation, and additional climatic and anthropogenic factors such as rainfall variability, land use change, and human activity. Integrating advanced machine learning or deep learning models with MCDM methods could improve prediction accuracy and support real-time monitoring. These improvements will strengthen early warning systems and help develop more effective climate adaptation and ecosystem restoration strategies for vulnerable coastal zones like Sagar Island.

Acknowledgements We sincerely thank everyone who contributed to this research. Special appreciation is extended to Prajukti Research Private Limited (PRPL) for providing essential logistical support throughout the study.

Author Contributions Conceptualization, S.M., S.S., A.S., J.D., B.B., A.G.; methodology, S.M., S.S., A.S., J.D., B.B., A.G.; software, A.S., S.M.; formal analysis, S.M., A.S.; investigation, S.M., A.S., S.S., A.G., J.D., B.B.; resources, S.M., A.S., S.S.; data creation, S.M., A.S., S.S.; writing-original draft preparation, A.S., S.M.; writing-review and editing, S.M., S.S., A.G.; J.D., B.B; visualization, A.S, S.M.; supervision and revision, S.M., S.S., A.G., J.D., B.B.; All authors have read and agreed to the published version of the manuscript.

Funding The study was conducted without financial support from external sources.

Data Availability Statement Data are available upon reasonable request from the corresponding author, as they are not publicly shared.

Declaration No conflicts of interest related to this work.

References

Bera A, Meraj G, Kanga S, Farooq M, Singh SK, Sahu N, Kumar P (2022) Vulnerability and risk assessment to climate change in Sagar Island, India. Water 14(5):823. https://doi.org/10.3390/w14050823

Bera R, Maiti R (2021) Multi hazards risk assessment of Indian Sundarbans using GIS based Analytic Hierarchy Process (AHP). https://doi.org/10.1016/j.rsma.2021.101766

Chowdhury A, Naz A, Sharma SB, Dasgupta R (2023) Changes in salinity, mangrove community ecology, and organic blue carbon stock in response to cyclones at Indian Sundarbans. Life 13(7):1539. https://doi.org/10.3390/life13071539

Cochran WG (1977) Sampling techniques, 3rd edn. Wiley, New York, NY

Ding Q, Wang L, Fu M, Huang N (2020) An integrated system for rapid assessment of ecological quality based on remote sensing data. Environ Sci Pollut Res 27(26):32779–32795. https://doi.org/10.1007/s11356-020-09424-6

Dutta S, Maiti S, Garai S, Abrar F, Jha SK, Bhakat M, Mandal S, Kadian KS (2020) Analyzing adaptation strategies to climate change followed by the farming community of the Indian Sunderbans using Analytical Hierarchy Process. J Coast Conserv 24(5):61. https://doi.org/10.1007/s11852-020-00779-z

Frimpong BF, Koranteng A, Molkenthin F (2022) Analysis of temperature variability utilising Mann-Kendall and Sen's slope estimator tests in the Accra and Kumasi Metropolises in Ghana. Environ Syst Res 11(1):24. https://doi.org/10.1186/s40068-022-00269-1

Ghosh S, Mistri B (2021) Assessing coastal vulnerability to environmental hazards of Indian Sundarban delta using multi-criteria decision-making approaches. Ocean Coast Manag 209:105641. https://doi.org/10.1016/j.ocecoaman.2021.105641

Gupta LK, Verma S (2023) Rainfall and temperature trend analysis using Mann-Kendall and Sen's slope estimator test in Kharun Watershed, Chhattisgarh, India. https://doi.org/10.9734/CJAST/2023/v42i234174

Karim MdR, Arham MdA, Shorif MdJU, Ahsan A, Al-Ansari N (2024) GIS based geostatistical modelling and trends analysis of groundwater quality for suitable uses in Dhaka division. Sci Rep 14(1):17449. https://doi.org/10.1038/s41598-024-66567-z

Karnieli A, Bayasgalan M, Bayarjargal Y, Agam N, Khudulmur S, Tucker CJ (2006) Comments on the use of the Vegetation Health Index over Mongolia. Int J Remote Sens 27(10):2017–2024. https://doi.org/10.1080/01431160500121727

Karsch G, Mukul SA, Srivastava SK (2023) Annual mangrove vegetation cover changes (2014–2020) in Indian Sundarbans National Park using Landsat 8 and Google Earth Engine. Sustainability 15(6):5592. https://doi.org/10.3390/su15065592

Keikhosravi G, Rastar Alaeh Gurabi S (2020) Monitoring agricultural drought in Iran using time series of vegetation health index. Sustain Earth Rev 1(4):56–65. https://doi.org/10.52547/sustainearth.1.4.56

Li L, Ni W, Cheng Y, Wang H, Yuan K, Zhou B (2021) Evaluation of the eco-geo-environment in the Qaidam Basin, China. Environ Earth Sci 80(1):27. https://doi.org/10.1007/s12665-020-09312-9

Ma D, Huang Q, Liu B, Zhang Q (2023) Analysis and dynamic evaluation of eco-environmental quality in the Yellow River Delta from 2000 to 2020. Sustainability 15(10):7835. https://doi.org/10.3390/su15107835

Mitra R, Das J (2022) A comparative assessment of flood susceptibility modelling of GIS-based TOPSIS, VIKOR and EDAS techniques in the Sub-Himalayan foothills region of Eastern India. In Review. https://doi.org/10.21203/rs.3.rs-1710264/v1

Mitra S, Bera S, Dutta P, Goswami A, Sen S, Majumder P (2024) Flood impact on mangrove health dynamics in Sagar Island of West Bengal: a geoinformatics based approach. In: Biswas B, Ghute BB (eds) Flood risk management. Springer, Singapore, pp 181–209. https://doi.org/10.1007/978-981-97-2688-2_9

Mondal BK, Mahata S, Basu T, Das R, Patra R, Abdelrahman K, Fnais MS, Praharaj S (2024) Analysis of the post-cyclonic physical flood susceptibility and changes of mangrove forest area using multi-criteria decision-making process and geospatial analysis in Indian Sundarbans. Atmosphere 15(4):432. https://doi.org/10.3390/atmos15040432

Mukherjee N, Siddique G, Basak A, Roy A, Mandal MH (2019) Climate change and livelihood vulnerability of the local population on Sagar Island, India. Chin Geogr Sci 29(3):417–436. https://doi.org/10.1007/s11769-019-1042-2

Narendr A, Aithal BH, Das S (2024) A data-driven framework for enhancing coastal flood resilience in resource-crunched developing nations. Geomat Nat Haz Risk 15(1):2396892. https://doi.org/10.1080/19475705.2024.2396892

Paul S, Mishra M, Pati S, Acharyya T, Santos CAG, Silva RMD, Guria R, Laksono FAT (2024) Evaluation of overwash vulnerability and shoreline dynamics in cyclone-prone Sagar Island, Sundarbans (India). Sci Total Environ 907:167933. https://doi.org/10.1016/j.scitotenv.2023.167933

Rajakumari S, Purvaja R, Sundari S, Ramesh R (2024) Assessment of spatial change dynamics of land features upon short and long term scenarios for a vulnerable island in the Ganges Delta. J Coast Conserv 28(2):48. https://doi.org/10.1007/s11852-024-01047-0

Rizzo A, Anfuso G (2020) Coastal dynamic and evolution: case studies from different sites around the World. Water 12(10):2829. https://doi.org/10.3390/w12102829

Saha S (2021) Analysis of shoreline change and the problem of saltwater intrusion: a case study on south-eastern part of Sagar Island, West Bengal, India

Sahana M, Sajjad H (2018) Vulnerability to storm surge flood using remote sensing and GIS techniques: a study on Sundarban Biosphere Reserve. India. https://doi.org/10.1016/j.rsase.2018.10.008

Sahoo S, Singha C, Govind A (2024) Prediction of pulse suitability in rice fallow areas using fuzzy AHP-based machine learning methods in Eastern India. Paddy Water Environ 22(3):341–359. https://doi.org/10.1007/s10333-024-00970-0

Sengupta T, Bandyopadhyay BK, Tripathi S (2019) Effect of increasing elevation of lowland coastal saline soils of Sundarbans (India) on soil salinity and its seasonal variation. Int J Plant Soil Sci. https://doi.org/10.9734/ijpss/2019/v29i530152

Sresto MA, Siddika S, Fattah MdA, Morshed SR, Morshed MdM (2022) A GIS and remote sensing approach for measuring summer-winter variation of land use and land cover indices and surface temperature in Dhaka district, Bangladesh. Heliyon 8(8):e10309. https://doi.org/10.1016/j.heliyon.2022.e10309

Tejada JJ, Punzalan JRB (2012) On the misuse of Slovin's formula. Philippine Stat Assoc 61(1):129–136

Tran TV, Reef R, Zhu X (2022) A review of spectral indices for mangrove remote sensing. Remote Sensing 14(19):4868. https://doi.org/10.3390/rs14194868

Wanjari RN, Shah TH, Telvekar P, Bhat FA, Abubakr A, Bhat BA, Darve SI, Ramteke KK, Mathialagan D, Magloo AH, Singh BS (2024) Assessing ecosystem health: a preliminary investigation of the gosikhurd dam ecosystem structure and functioning, an appraisal based on ecological modelling, India. Environ Monitor Assess 196(9):815. https://doi.org/10.1007/s10661-024-12958-8

Xu Z, Peng J (2024) Recognizing ecosystem service's contribution to SDGs: ecological foundation of sustainable development. Geogr Sustain 5(4):511–525. https://doi.org/10.1016/j.geosus.2024.05.001

Yang Y, Tilman D, Jin Z, Smith P, Barrett CB, Zhu Y-G, Burney J, D'Odorico P, Fantke P, Fargione J, Finlay JC, Rulli MC, Sloat L, Jan van Groenigen K, West PC, Ziska L, Michalak AM, The Clim-Ag Team, Lobell DB (2024) Climate change exacerbates the environmental impacts of agriculture. Science 385(6713):eadn3747. https://doi.org/10.1126/science.adn3747

Geological and Ecological Analysis of Munia-2 Wetland: Understanding Its Role in the Sundarban Deltaic Region, West Bengal

3

Lopamudra Ganguly, Subhadip Mondal, Saptadipa Chakraborty, and Anushka Ghosh

Abstract

Wetlands (The presence of water for all or part of the year over land leads to the formation of wetland areas where most plants are hydrophytes) are lands transitional between terrestrial and aquatic ecology (The examination of the relationships formed between organisms and their surroundings) and ecosystems where the water table is usually at or near the surface or the land is covered by shallow water. The value of the world's ecology and wetlands is increasingly receiving due attention as they contribute to a healthy environment in many ways. Due to that, economic prospects and the origin of employment can also be noted. The study further addresses the management initiatives for these sensitive ecosystems, pinpoints the institutional gap, and indicates areas of priority where prompt action is needed to improve conservation plans for these fruitful systems. This new wetland, Munia-2, was selected because it is well known as one of the richest ecological areas for the munia bird. Being in the middle of the New Town planning region, it is a major source of biological and tourist attractions and an important center of economic growth. This research involved a cross-sectional study and the development of a mixed method for identifying the estimated issues. QGIS (Quantum Geographic Information System is a free software designed for analyzing and creating maps with spatial data) version 3.26 was used for the study area mapping. It was concluded that wetlands in Kolkata are one of the most diverse and important ecosystems because of their geographical and climatic conditions. Munia -2 wetland

L. Ganguly (✉) · S. Mondal · S. Chakraborty · A. Ghosh
Department of Geography, Vedanta College, Kolkata, India
e-mail: ganguly.geography@gmail.com

C. Singha et al. (eds.), *Sustainable Management of Delta Ecosystems Resilience*, Deltas of the World, https://doi.org/10.1007/978-3-032-16261-8_3

played a very crucial role in maintaining the hydrological cycle (Water moves accurately through its cycle, beginning on, above or below Earth's surface by evaporating, condensing, precipitating, infiltrating or running off) and supporting a large diversity of different types of flora, fauna, and microbes at the micro level. However, the national water sector policy has not paid enough attention to managing these wetlands. Because of this, anthropogenic pressures (Humans contribute to environmental changes by polluting, burning forests or building cities) affect many wetlands in both urban and rural locations. These pressures include changes in land use within the watershed, pollution from industry and homes, encroachments, tourism, and over-exploitation of natural resources very rapidly. The features of Munia-2 wetland include a distinct stratified lithological profile which has clayey, silty, and sandy layers that foster biodiversity while retaining water. The wetland supports fishing along with agriculture and new eco-tourism, earning over INR 2.7 million every year as of now. The wetland is most visited for tourism during the winter and autumn season due to the diverse species found there. Munia-2 wetland is subjected to pollution, sediment loss, and climate change, which decreases the ecological potential. Like other water bodies, it aims for functional, community managed wetland systems.

Abbreviations

LULC	Land use and land cover
QGIS	Quantum Geographic Information System
GPS	Global positioning system
PES	Payments for ecosystem services
INR	Indian Rupee
KMDA	Kolkata Metropolitan Development Authority
PC	Photo courtesy
Fig.	Figure
No.	Number
%	Percent
NRSC	National Remote Sensing Centre

3.1 Introduction

The existence of permanent or seasonal water bodies in ecology and wetland systems results in the formation of specific soil types and specialized plant and animal populations. These environments, ranging from marshes and swamplands to bog plains and other types (Mitsch and Gosselink 2015), cover about 6% of the Earth's. Wetlands are important habitats for fish, amphibians, birds, and mammals, which are significant to ecology and biodiversity. Water management relies upon their involvement in flood control, carbon sequestration (Naturally, forests and wetlands manage to collect and store atmospheric carbon dioxide), and water purification to regulate climate. Some of the most endangered ecology and ecosystems are wetlands, and although they are important in ecology, significant degradation as well as a loss of areas have been caused by urbanization, agricultural expansion, and industrialization. According to the Ramsar Convention on Wetlands, since the 1700s, wetlands have lost approximately 87% of the world's wetlands. Besides damaging biodiversity, this loss destroys the variety of ecosystem services (The services ecosystems give to people include supplying food and water, controlling floods, providing leisure activities and recycling nutrients) wetland ecosystems provide, eventually impacting local and global economies. 1.2. Significance of the Study. Munia-2 ecology and wetland, an important ecological resource in West Bengal, India, provides vital income for local inhabitants from all walks of life but is strongly affected by various threats. The rich ecology and biodiversity environment of Munia-2 holds several species and is an indispensable part of the socio-economic society. The economic importance of these wetlands for sustainable development and conservation must be recognized. Munia-2 generates vital economic benefits in terms of activities like fishing, agriculture, and eco-tourism (Taking care of nature when travelling in nature parks, helping the local people and offering education about the area). Wetland agriculture is an essential sector of the local economy for food security and contributes to fishing activities, as well. Eco-tourism, under development in the area, can be an avenue for wealth creation since it can attract people to visit the wetlands for their natural beauty and biodiversity, respectively.

3.2 Literature Review

Different types of direct and indirect economic benefits are commonly linked to ecology and wetland ecology and ecosystem services. These services include provisioning, regulation, and cultural services like recreation, tourism, flood control, and carbon storage. Worldwide estimates suggested that wetlands have a significant economic impact, with ecosystem services benefiting billions of dollars each year (Garcia-Herrero and Rodriguez-Sinobas 2023). Wetland plays a critical role in the economy of West Bengal, by supporting agriculture and fisheries, especially rice production. Mukherjee (2011) reported that East Kolkata Wetlands' sustainable waste management system supports aquaculture and agriculture. According to Ghosh et al. (2009) minor wetlands such as Munia-2, are still unexploited, for their potential on the economic front. The economic benefits of wetlands are highlighted by various studies, with emphasis on sustainable fisheries and agriculture while also catering to ecotourism. In addition, fish farming is a viable option in wetland regions, as they support the growth of diverse fish species (Roy et al. 2013). Wetland habitats also assist in agriculture through irrigation and offer fertile soil for crops such as paddy, lotus, and water chestnuts (Das and Sengupta 2019). Ecotourism, which seeks to promote conservation through responsible tourism, has become a viable economic activity in wetlands (Silvius et al. 2000). This trend is remarkable. Carbon credits and Payments for Ecosystem Services (PES) are among the mechanisms that can be used to monetize the ecosystem services provided by wetlands. These programs serve as a means of motivating local communities to protect and improve wetland ecosystems (Brander et al. 2023). These economic opportunities can only be realized through sustainable management practices that promote a balance between ecological health and economic development. The geologic importance of wetlands, especially in deltaic systems, has been comprehensively researched. Wetlands, as emphasized by Mitchell and Gosselink (2015), are vital ecosystems that preserve ecological diversity, regulate floodwaters, and host water filtering processes. The Sundarban delta, as explained by Ghosh et al. (2009), is a distinctive geomorphological area where sedimentation patterns make fertile soil conditions that support varied ecosystems (Mulatu et al. 2020) highlight that wetlands offer significant ecosystem services in the form of flood mitigation and carbon sequestration. The services are directly associated with the geological features of wetlands, like sediment layering and mineral constitution. Das and Sengupta (2019) explain the contribution of sediment-rich soil towards agricultural productivity, while Mukherjee (2011) discusses the socio-economic importance of wetlands in West Bengal, especially in fisheries and agriculture. The Dampier-Hodges Line is essential in comprehending the geological setting of wetlands in the Sundarban area. Prasad et al. (2002) explain that this line marks the tidal influence zone, regulating soil salinity (High salt levels in either water or soil usually affect the growth of plants and the stability of wetlands) and sedimentation. Its control over Munia-2's lithology (Rocks' or sediments' features which include how they feel, their color, minerals they have and the dimensions of their elements) reflects trends of clayey topsoil, silty middle portions, and sandy subsoils, as observed in comparable areas. From the aspect of biodiversity conservation, Mitra (2019) emphasizes the environmental significance of Sundarban wetlands, with *Avicennia alba* species benefiting from nutrient sediments. Moreover, Sharma et al. (2024) emphasize that economic dependence on wetlands calls for sustainable management to ensure the conservation of their geological and ecological processes. Sustainable practices like sediment retention mechanisms and pollution regulation measures are pivotal in ensuring the health of wetlands (Roy et al. 2013).

3.3 Research Gap and Novelty

The literature on wetlands often deals with cases of East Kolkata Wetlands or Sundarbans, but extensive studies of the systems exist. They mainly focus on major ecological functions, raising fish with wastewater and protecting biodiversity. Even so, little discussion happens about small wetlands such as Munia-2, especially in peri-urban zones like Rajarhat where cities are quickly developing. Also, although Roy et al. (2013) and Das and Sengupta (2019) put importance on sustained methods in fishery and agriculture, no integrated approaches have been developed that use studies in geology, ecology and society together for wetlands being impacted by urban growth.

In this study focused on the Munia-2 wetland to examine its lithology, water cycle and financial significance. Unlike before, this research makes use of QGIS-generated maps, field measurement of sediment and collection of primary socio-economic statistics. This chapter is novel as it merges physical geography with evaluations of human activities to suggest and support strategies that sustain scarce minor wetland ecosystems found outside most environmental plans.

3.4 Objectives

The main aim of this research is to study the geological origin and lithological characteristics of the Munia-2 wetland as an integral component of the Sundarban deltaic landscape, being the largest tidal halophytic mangrove forest in the world, created by the rivers Ganges, Brahmaputra, and Meghna in southern Bangladesh and eastern India. Moreover, the study intends to assess the role of the Dampier-Hodges Line, which represents the inland limit of the Bay of Bengal's tide influence and plays an imperative part in the exploration of soil salinity and sediment deposition regimes of deltaic environments, in determining Munia-2's sediment properties and soils. The research further intends to examine the ecological significance of Munia-2's geological features in supporting biodiversity and providing hydrological stability. In addition, it examines the economic importance of the geological profile of the wetland, particularly its role in agriculture, aquaculture, which involves farming fish, crustaceans, mollusks, aquatic vegetation in water, and eco-tourism. Lastly, the research recommends sustainable management strategies to maintain the integrity of the wetland and increase its economic contribution.

3.5 Materials and Methods

3.5.1 Methodology

See Fig. 3.1.

3.5.2 Study Area (Munia-2 Wetland)

The approximate latitude and longitude of Munia-2 are (Fig. 3.2):

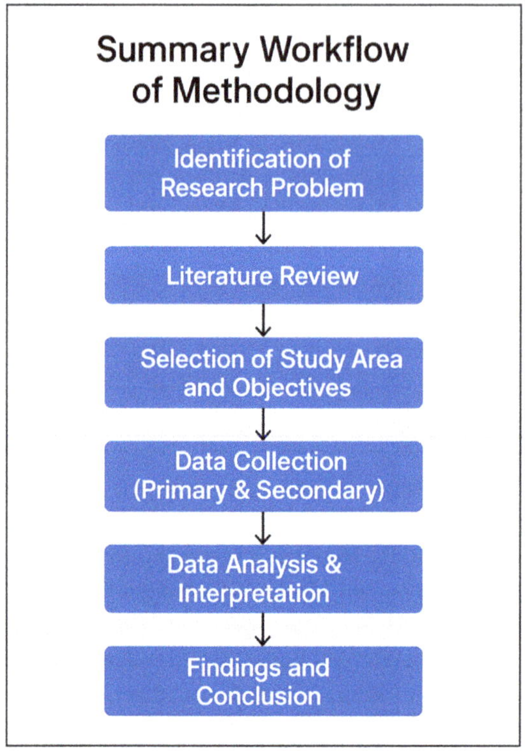

Fig. 3.1 Geological assessment consisted of intensive soil sampling to determine lithological properties such as clay, silt, and sand content. Soil pH (A method that gives a reading of how acidic or basic an aqueous solution is, using numbers from 0 to 14) and salinity analysis was conducted to examine environmental fitness for agriculture and vegetation. Tests were conducted on water to determine water quality by examining dissolved oxygen content, nutrient composition, and contaminants to establish the ecological fitness of the wetland. To conduct spatial analysis, QGIS 3.26 was utilized to map land-use and land-cover (LULC) patterns, erosion risks, and areas of potential eco-tourism areas. Seasonal tourist data were used to determine the economic importance of eco-tourism (Fig. 3.1). This holistic methodology provided an interdisciplinary understanding of Munia-2's ecological processes, economic role, and major conservation needs, with a clear basis for sound sustainable management approaches. The study utilized the mixed-method methodology that incorporated qualitative and quantitative approaches to examine the geological, ecological, and economic features of Munia-2 wetland. The study first undertook detailed field observations on the physical topography, flora and fauna diversity, and socio-economic status of the region. Structured family surveys were also conducted among 50 families occupied with fishing, agriculture, and eco-tourism activities. These questionnaires yielded rich information on income trends, dependency on resources, and environmental issues. Further, semi-structured interviews with primary stakeholders such as fishermen, farmers, and eco-tourism operators were held to obtain qualitative information on traditional activities, livelihood difficulties, and environmental modifications (*Source* Developed by MS Paint)

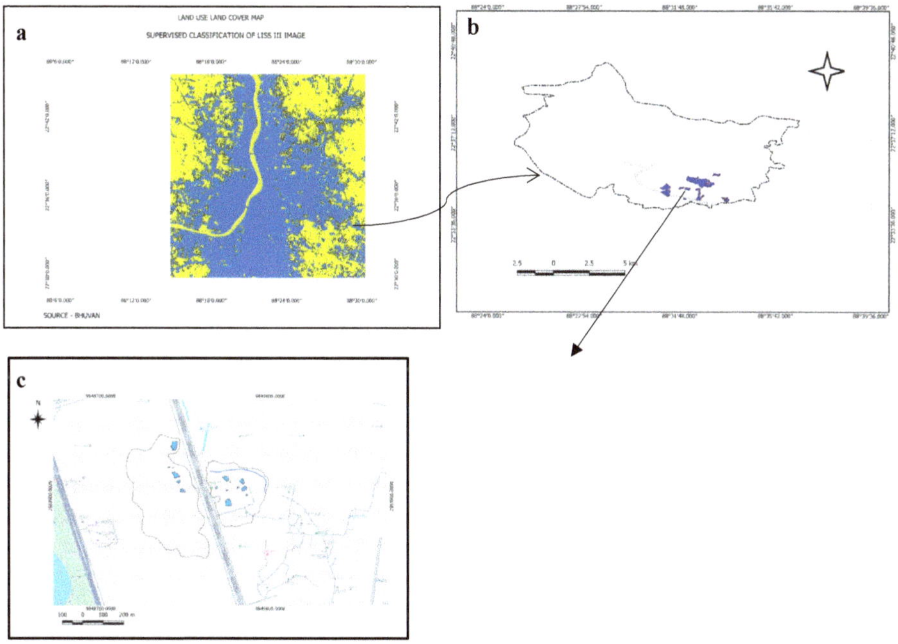

Fig. 3.2 The present research focuses on Munia-2 Wetland (Fig. 3.2), located in Rajarhat, West Bengal, India. The study utilizes multiple levels of spatial analysis: **a** represents the land use and land cover (LULC) conditions of the Kolkata Metropolitan Development Authority (KMDA) region, **b** provides a more refined map of New Town and Rajarhat, with a focus on the Rajarhat area, and **c** specifically highlights the Munia-2 Wetland Zone in Rajarhat. Munia-2 Wetland's surface area, estimated at 500 ha, is an important resource in sustaining diverse species from aquatic and terrestrial habitats, which justifies ecology and biodiversity conservation concerns. Munia ecology and wetland have also remained a center of economic development, for instance, fishing, agriculture and even eco-tourism activities are present and being developed in the region (Munia Wetland Research Committee 2023). *Source* Main data collected from Bhuvan.nrsc and by using Garmin GPS and reproduced by QGIS 3.26

The northern coordinate for Barasat is 22°35′15″.
Easting longitude: 88°28′40″ E.
Elevation: The area is 10–12 m above the usual water level.

The Köppen-Geiger climate type for this area is called tropical wet and dry (Aw) which has hot and humid summers, rainy seasons every June to September and a winter that is mild (December to February). Most years bring 1600 to 1800 mm of rain and temperatures vary from 12 °C in winter to 40 °C in the summer months.

3.5.3 Data Collection

This research took a mixed-method approach and entailed collecting abundant data in 2024, that is, in the month of October, which is the post-monsoon season—a critical time to measure hydrological as well as ecological status. Field visits were made on October 22, 2024, between 11:00 A.M. to 12:00 P.M., in Munia-II Wetland Zone, in the Rajarhat locality of North 24 Parganas district, West Bengal, India. The main area of study consisted of sections of the villages abutting Munia-II, which are influenced by peri-urban development in a fast pace because of their location close to New Town. This was a mixed approach and involved gathering extensive data. Detailed demographic data, economic activities, and income sources related to the ecology and wetland were obtained through structured surveys conducted among local households. The study involved the comparison of 50 households and provided a clear picture. Semi-structured interviews with key stakeholders, including local fishermen, farmers, and community leaders, were conducted to gather qualitative data. Interviews were conducted with individuals to gain knowledge about the socio-economic factors that contribute to the wetland's existence, perceived challenges, and opportunities, as well as their role in community life.

3.5.4 Geological Framework of the Sundarban Deltaic Region

Sundarban Delta, the largest tidal halophytic mangrove forest in the world, is a Holocene sedimentary product. Dampier-Hodges Line, historically the boundary of tidal action inland, is important in demarcating soil distribution and wetland creation. Munia-2 wetland falls geologically in this category, showing alluvial soils rich in organic matter and clay, necessary for wetland functioning. Geological Evolution and History the formation of the Sundarban delta is governed by Himalayan erosion, with the sediment being transported by the Ganges–Brahmaputra–Meghna River system. Through millennia, sediments were deposited to create complex layers of clay, silt, and sand, and thus a rich but flood-vulnerable ecosystem developed. Munia-2 wetland is an example of a microcosm of this bigger deltaic evolution.

Located in the Holocene epoch, the Sundarban Delta includes the world's largest tidal halophytic mangrove forest. In this specific region, the Dampier-Hodges Line, which was earlier thought of as the borderline for inland tidal action, is significant for controlling wetness of tidal basins along with the distribution of soil salinity and sedimentary materials, controlling the formation of wetlands in that area. This region also includes Munia-2 wetland which contains alluvial soils laden with organic materials and clay which help sustain wetland ecosystems. Recent studies have verified that the Ganges–Brahmaputra–Meghna River system's Bengal Basin contains the deepest sediment deposits. The added sediment material, received from the rivers, is responsible for extending the delta region and constructing complex litho-stratigraphic areas found in Munia-2. Bera et al.

(2021) suggest that the region's accumulation of sediments from the rivers is continuously altering the elevation and hydrology of its micro-wetlands. Sarkar et al. (2023) note the relentless influence that the tides and rivers have on the delta sediment, especially the quartz and mica rich sediments in Munia-2. Banerjee and Chatterjee (2022) also present new remote sensing evidence depicting the sediment layering and deltaic changes around Rajarhat.

Location Map of the Sunderban Deltaic Zone

See Fig. 3.3.

Geological Epoch Classification of the Munia-2 Wetland Region

The Munia-2 region of Rajarhat (Fig. 3.4) can be understood as a prolonged extension of the Sundarbans area as per geological and geomorphological data. The occurrence of Holocene deposits at this location, as marked on the map, indicates that in the past it was under the control of deltaic processes like those of the Sundarbans. The Meghalayan

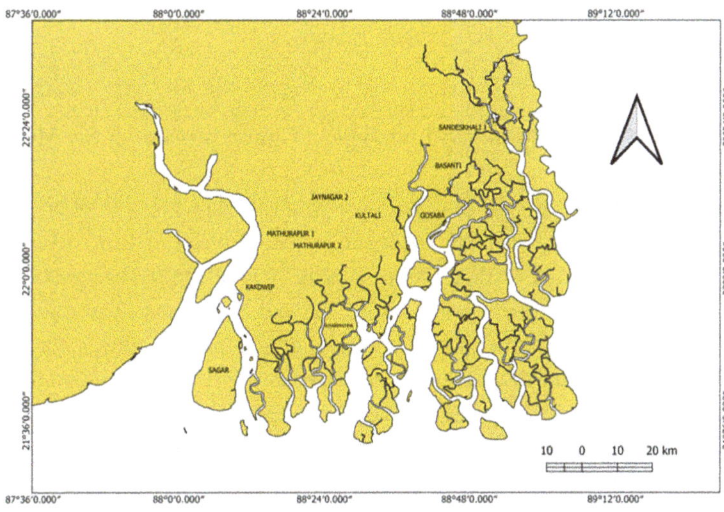

Fig. 3.3 The map illustrates the geographical distribution of islands, rivers, and wetlands in Sundarbans and identifies important points like Sagar, Kakdwip, Gosaba, and Basanti (Fig. 3.3). The complex river and distributary network illustrate the dynamic deltaic landscape where constant deposition and erosion create an ever-changing shape. The prevalence of yellow depicts landmasses, with the water channels showing the interconnectedness of tidal rivers, vital for wetland hydrology. Munia-2 and other locations identified from here imply the presence of large wetland bodies influenced by sedimentation processes. Fluvial and tidal interactions are responsible for the dynamic nature of this area, keeping it susceptible to coastal erosion, sediment deposition, and climate change impacts. It is critical in the academic study of wetland preservation, land-use planning, and environmental management, especially as a means of reducing the impacts of erosion and habitat loss within the Sundarbans delta. *Source* Main data collected from Bhukosh and reproduced by QGIS 3.26

Age deposits, being considerably more stable landforms, stand in contrast to the actively dynamic tidal and fluvial processes typical of the Sundarbans delta. Hydrologically, the network of rivers and wetlands running from the South 24 Parganas into the Rajarhat area also points towards earlier tidal and estuarine environments. The low-lying character of this tract, in addition to proof of marine transgression and deltaic deposition, justifies the notion that Munia-2 was formerly a part of this developing coastal landscape. Even though urbanization and land reclamation have drastically altered the landscape, the underlying geology is still related to the Sundarbans ecosystem.

3.5.5 Lithological Description of Munia-2 Wetland

- Topsoil Layer: Dominated by clayey loam, this layer is enriched with organic matter, ideal for aquatic vegetation and microbial activity.
- Intermediate Layer: Silty deposits with high moisture retention capabilities, supporting diverse hydrophytic species.
- Subsoil Layer: Sandy and gravel-based strata indicative of sediment transport influenced by the Hooghly River and related tributaries.

Sedimentological Classification and Lithological Characteristics of the Munia-2 Wetland

Alluvial Soil: This refers to soil brought down by rivers that you can often find in river deltas and floodplains) in the area, reflecting active deposition from riverine sources (Fig. 3.5). This organic-rich layer offers a nutrient-rich substrate for aquatic plants and microbial processes, which are crucial to the wetland ecosystem. The middle layer, as indicated by the silty deposits with high water-holding capacity, indicates a depositional setting controlled by flood and sedimentation cycles. The layer is instrumental in the maintenance of hydrophytic vegetation, which flourishes in water-saturated conditions. The occurrence of fluvial sediments in the area also supports the argument that this belt has been subjected to river dynamics, especially from the Hooghly River and its tributaries. The subsoil stratum of sandy and gravel-based strata shows sediment transport and deposition by fluvial and perhaps coastal processes. The classification of coastal glacial sediments by the map implies that past coastal or marine incursions could have added to the stratigraphy of the region. Moreover, the occurrence of aeolian sediments, which are usually characteristic of wind-driven deposition, suggests that parts of the landscape could have experienced episodic dryness or exposure to atmospheric processes, influencing the lithological nature over time.

Overall, the Munia-2 Wetland's lithological composition is in concordance with the sedimentological distribution depicted on the map, validating its origin from the combination of fluvial, coastal, and minor aeolian processes. All these have combined to

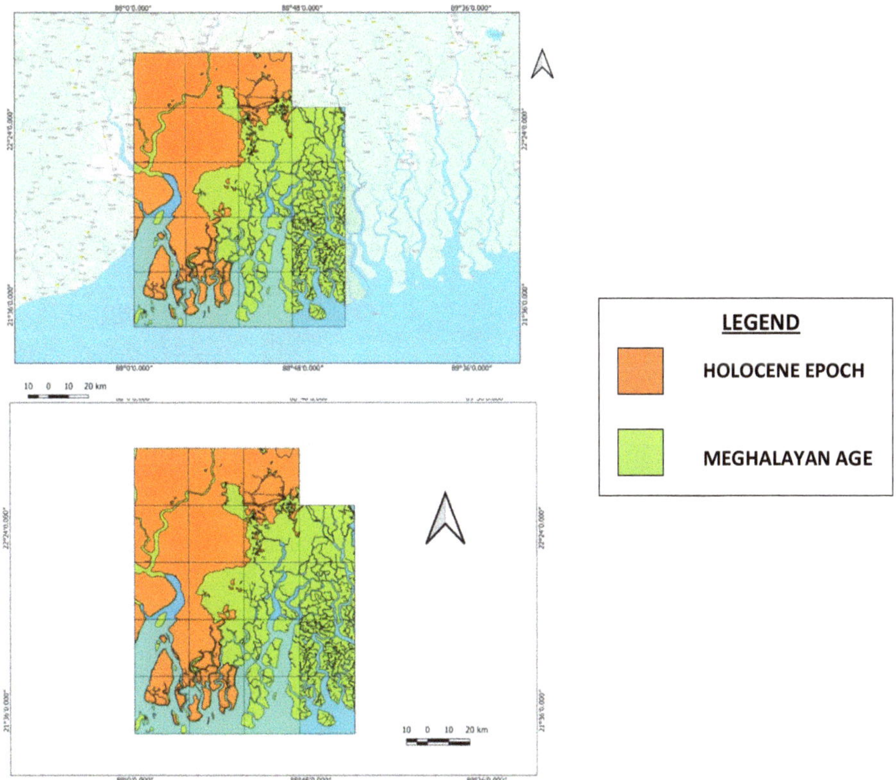

Fig. 3.4 The Munia-2 region of Rajarhat can be understood as a prolonged extension of the Sundarbans area as per geological and geomorphological data (Fig. 3.4). The occurrence of Holocene deposits at this location, as marked on the map, indicates that in the past it was under the control of deltaic processes like those of the Sundarbans. The Meghalayan Age (Around 4200 years ago, at the beginning of the Holocene's current age, global drought and changes among societies emerged) deposits, being considerably more stable landforms, stand in contrast to the actively dynamic tidal and fluvial processes typical of the Sundarbans delta. Hydrologically, the network of rivers and wetlands running from the South 24 Parganas into the Rajarhat area also points towards earlier tidal and estuarine environments. The low-lying character of this tract, in addition to proof of marine transgression and deltaic deposition, justifies the notion that Munia-2 was formerly a part of this developing coastal landscape. Even though urbanization and land reclamation have drastically altered the landscape, the underlying geology is still related to the Sundarbans ecosystem. *Source* Main data collected from Bhukosh and reproduced by QGIS 3.26

determine its hydrological holding capacity, ecological variety, and sediments, rendering it an important wetland system in the expanded Sundarbans deltaic complex.

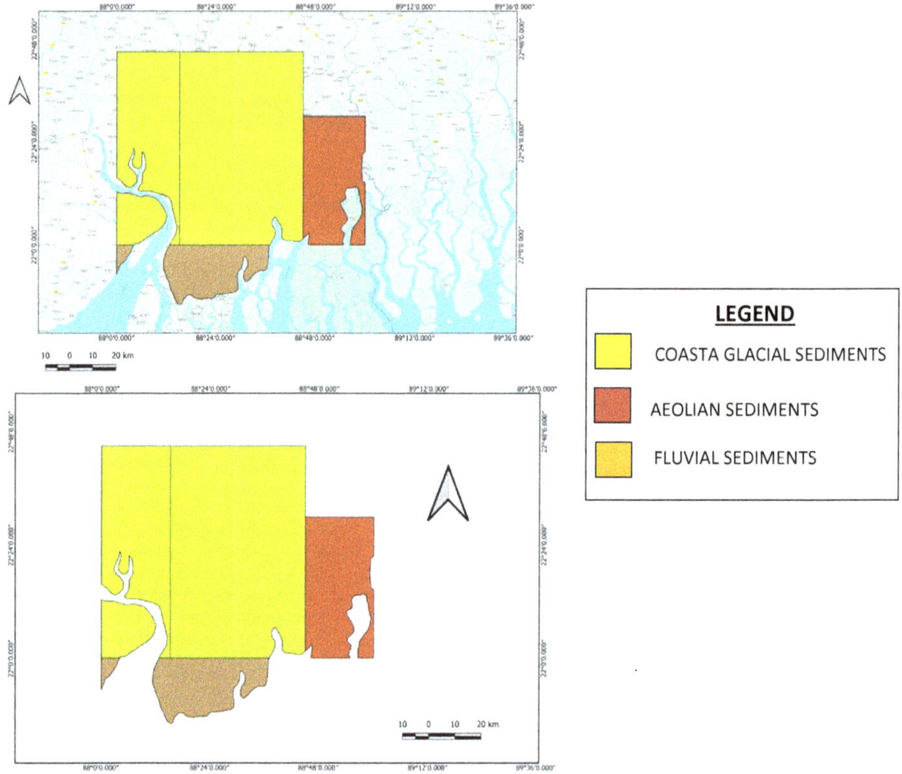

Fig. 3.5 The Munia-2 Wetland has a lithologically (Fig. 3.5) complex composition that is consistent with the sedimentological classification as seen in the given map. The topsoil layer, which is clayey loam-dominated, is consistent with the fluvial sediments (Sand, silt and clay) are brought and left by rivers and streams. *Source* Main data collected from Bhukosh and reproduced by QGIS 3.26

3.5.6 Sediment Composition

The sediment is quartz, feldspar, and mica-rich, characteristic of fluvial settings. Iron oxide levels cause intermittent reddish-brown spots, indicative of chemical weathering of minerals. Soil pH and Salinity Munia-2 soil profiles indicate a slightly acidic to neutral pH (5.8–7.1), which is typical of Sundarban soils. The occurrence of moderate salinity indicates periodic tidal influence, which further supports the deltaic nature of the region. The occurrence of heavy metals, including lead and arsenic, has been found in trace levels, which may indicate the possible impact of urban runoff. Hydrological Dynamics Seasonal fluctuations in the water table in Munia-2 are matched by monsoonal inflow that adds to the nutrient content. Water retention is ensured by clay layers, and drainage is promoted by sandy subsoil's to avoid stagnation.

Formation and Hydrological Dynamics of Munia-2 Wetland in the Expanding Sundarbans Delta

Munia-2, being a prolonged extension of the Sundarbans delta, is a testament to the dynamic interaction between tidal and fluvial sedimentation (Fig. 3.6). The perpetual supply of alluvial deposits has resulted in a progressive increase in the surface elevation of the wetland, thus shortening its depth over time. Such a growth in the deltaic landmass has modified the hydrological features of Munia-2, changing water retention capacity and drainage efficiency. The sedimentary composition, dominated by quartz, feldspar, and mica, testifies to its fluvial source. The iron oxide content in the deposits is responsible for the reddish-brown color in some spots, suggestive of mineral weathering processes. Further, the slightly acidic to neutral pH (5.8–7.1) of the soil falls in line with those of the general Sundarbans soils, while moderate salinity profiles indicate occasional tidal flooding, further supporting the deltaic character of the wetland.

The monsoonal inflows control the seasonal hydrological dynamics of Munia-2, which recharge the wetland's nutrient levels and maintain its ecological equilibrium. The clay bottom layers provide water retention, avoiding fast drainage, while the sandy subsoils allow percolation, minimizing the chances of extended stagnation. The occurrence of trace levels of heavy metals like lead and arsenic indicates possible anthropogenic factors, most probably due to urban runoff and industrial effluent discharge. As the Sundarbans delta extends further, the wetlands within this area, such as Munia-2, are also experiencing progressive geomorphological modifications. The decreasing depth of these wetlands could have consequences for aquatic biodiversity, water storage, and general ecosystem health. These current changes are important to understand to devise sustainable conservation and management strategies to ensure the wetland's ecological integrity.

Geomorphological Classification of the Bengal Delta

See Fig. 3.7

3.5.7 Ecological Importance

The clayey topsoil promotes nutrient retention, which favors rich aquatic vegetation and helps maintain the wetland as a carbon sink. The silty mid-section guarantees efficient water retention, which is crucial in maintaining wetland hydrology. Mangrove types like *Avicennia alba* and *Rhizophora mucronata* occur in sections of Munia-2, which suggests partial saline control. Economic Impact Agriculture is supported by the fertile soil, including aquaculture and paddy cultivation. The nutrient-rich silt deposited seasonally increases crop yields, enhancing local communities' economic stability. Sustainable fisheries associated with traditional systems of knowledge also increase income security. Potential for Eco-Tourism Munia-2's biodiversity and picturesque landscape are perfect for the development of eco-tourism. Nature walks, bird watching, and cultural visits can enhance local

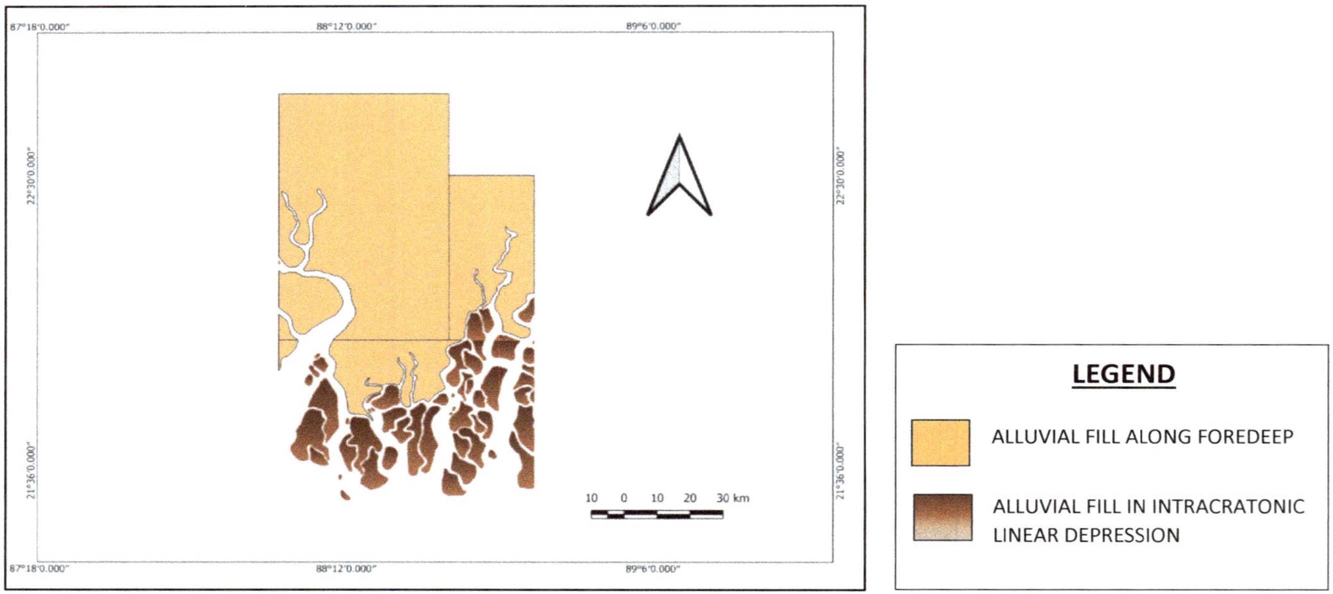

Fig. 3.6 The creation of wetlands within the Bengal Basin is a direct consequence of heavy sediment deposition in the intracratonic basin. During geological time, the Bengal Basin has served as a large sediment trap, building up deposits carried by the Ganges–Brahmaputra-Meghna river system. These Himalayan and Indian craton-sediment-laden riverine systems have filled the basin over time, resulting in the formation of several wetlands, such as Munia-2 (Fig. 3.6) *Source* Main data collected from Bhukosh and reproduced by QGIS 3.26.

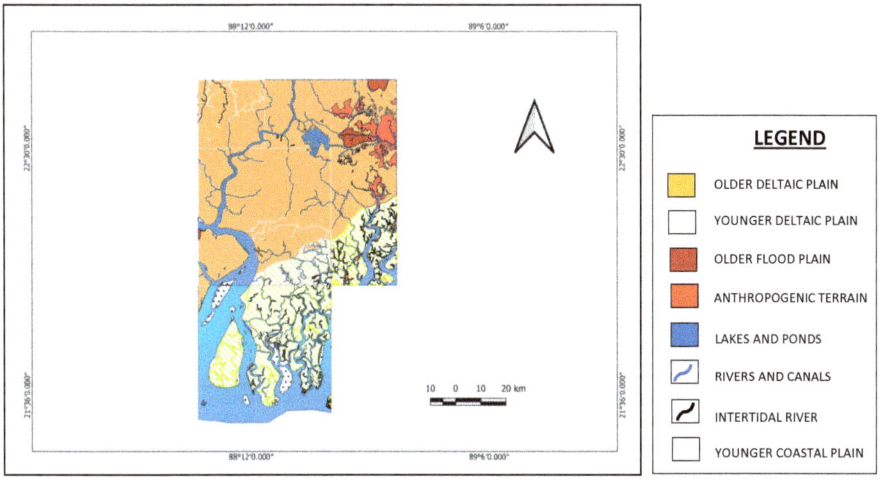

Fig. 3.7 The Bengal Basin has experienced extensive geomorphological change resulting from ongoing sediment deposition and fluvial processes. The process has molded the Older Deltaic Plain, which is an anciently deposited surface exposed to both fluvial and tidal forces for thousands of years. The progressive over-deposition of sediment in the intracratonic basin has resulted in the development of vast wetland ecosystems, one of which, Munia-2 (Fig. 3.7). Being a constituent of the Older Deltaic Plain, Munia-2's origin has a strong connection with the hydro-sedimentary processes of the area. The sediments are mostly composed of quartz, feldspar, and mica, typical of fluvial environments, and the occurrence of iron oxide deposits leaves with temporary reddish-brown patches, which reflect chemical weathering. Its slightly acidic to neutral soil pH (5.8–7.1) further indicates its correspondence with other Sundarban wetlands, and moderate salinity levels indicate exposure to tidal activities. These contribute to the implication that Munia-2 developed as a constituent of the wider deltaic system. Hydro regime of the wetland is controlled by the seasonal monsoon fluctuations, with effects on water table fluctuation. In monsoon season, high precipitation and riverine runoff lead to the retention of water, whereas, sandy sub-soils provide ready drainage so there is no chance of stagnation. Further, the occurrence of heavy metals, i.e., lead and arsenic, at trace levels, indicates probable human-induced influences possibly due to the urban runoff as well as discharges from the industries. One of the most remarkable changes in Munia-2 and other wetlands of the Older Deltaic Plain has been the progressive shallowing in depth because of deltaic progradation. Ongoing sediment influx from the Ganges–Brahmaputra-Meghna river system has increasingly raised the floor of the wetland, changing its hydro-morphological features. With newer deposits, the previously deeper areas of the wetland undergo shallowing, with implications for the ecological balance and water holding capacity of the area. This is a larger trend in deltaic settings in which landforms change due to sediment dynamics, sea-level fluctuations, and changes in hydrology. In short, Munia-2 wetland is a typical microcosm of the Older Deltaic Plain, moulded through centuries of sedimentation and hydrological alteration. Its present form is a product of the intricate interplay between natural processes and man-made factors, reflecting the dynamic character of the Bengal Delta and constantly changing wetland habitats within it. *Source* Main data collected from Bhukosh and reproduced by QGIS 3.26

sources of income while raising awareness for conservation. Challenges and Conservation Measures Despite its geologically rich base, Munia-2 is challenged by soil erosion, pollution, and altered hydrological regimes. Conservation measures need to incorporate soil stabilization, regulated land use, and low-impact eco-tourism development. Sediment Loss and Erosion Being near areas of urban growth, Munia-2 has a high risk of sediment runoff. Vegetation buffer zones and regulated water channel designs are advised for stabilization. Pollution Control Chemical runoff from adjacent farm fields and domestic waste is a challenge. Constructed wetlands along with integrated waste management solutions are suggested for enhanced water treatment. Community Involvement Enabling local communities to be trained in sustainable methods can guarantee equitable conservation practices. Participatory management (Managing and planning the use of natural resources with the help of the community and interested parties.) systems involving local fishers, farmers, and eco-tourism operators are crucial.

3.6 Results

The biodiversity survey of the Munia-2 wetland showed outstanding flora and fauna diversity. Some common species are Indian Major Carps, and a variety of catfishes, which should be there in local fisheries. Besides hosting several species of wetland-related organisms, aquatic macrophytes reside here, which also provides them shelter along with food. Second, as part of an important ecological value, wetland acts as an essential place to live by the passing or migration of birds. Indicators such as water quality, vegetation cover, and indices of biodiversity were used to determine the state of health of the wetland ecosystem. The water samples were good in terms of pH, dissolved oxygen, and nutrient levels, indicating the wetland is pretty healthy, but local pollution through agricultural runoff has been reported, hence adversely affecting the integrity of the ecosystem. The major economic activities around Munia-2 wetland are fishing, agriculture, and eco-tourism. This means that the ecological essence of the Munia Zone 2 region stands appealing mostly during the colder months when flora and fauna bloom because of perfect settings. In winter, the number of tourist visits peaks with about 800 visitors and this is due to the region's rich wildlife coupled with a more attractive climate that enhances activities such as wildlife visitation, trekking, and nature photography. There would be seasonal changes in the leaves and also the wildlife during the fall months which might increase the appeal of the area to nature lovers. The autumn and winter periods act as crucial opportunities for tourism for the year in a way that helps the development of Munia Zone 2 and demonstrates its ecosystem importance as well. On the other hand, spring experiences a relatively low number of arrivals with an average of about 400 tourists. Such splendor of the blooming flowers along with the spring season when most plants appear to be developing new growth could be less attractive for some fauna or tourism visitation. Very few tourists, only about 300, are recorded during summer this is because heat and humidity create an unpalatable condition, especially for outdoor sporting activities. Some other forest species may also be less noticeable to tourists since they are more active during the cool seasons. Conclusion The high and low seasons in the region Munia Zone 2 occur in Winter and Autumn with the highest and lowest numbers respectively with the given evidence that the region is attractive to tourists when the temperatures are moderate with high biodiversity. For this reason, too, these natural areas are worth conserving—not only due to the need to maintain the ecological equilibrium but also to contribute to the economic growth potential through tourism activities in the area (Fig. 3.7).

3.7 Discussion

The major economic activity is fishing activity, and about 60% of the households conducted in the survey depended on fishing as a source of income. Interviews with the locals have revealed the local fishermen catch an average of 10–15 kg a day during peak seasons which would bring in a decent income to sustain the family. The average monthly income generated from fishing is INR 8000. This mode of nutrition and earnings is precious for families. It is due to overfishing that these people complain, and they suggest taking necessary precautions to achieve sustainable practices for the future 4.2.2 Agriculture Another prominent economic activity is agriculture. A considerable percentage of families, among many activities, had participated in paddy cultivation, among others: other vegetables and pulses. Such land can yield prosperous agriculture only on the rich land with an abundant supply of water to create surplus foods at the family level or other related forms of extra family revenue. As said above, they reportedly make, per household, approximately a month an INR 10,000 (116.81USD) level. This is a new sector of eco-tourism emerging in the Munia-2 region, which would provide additional sources of income to the local people. The sector is still in the developing stages, but several initiatives among the locals are trying to attract visitors who might be interested in bird watching, nature walks, and experiences about culture. According to community members, eco-tourism could generate INR 5000–7000 (58.41–81.77) per month for the participants involved in guiding and hospitality services. It needs to be developed more for better awareness. There is a close tie between the locals and Munia-2 wetland as shown by interviews with community members. The wetland is seen to both generate income and serve as a cultural resource and spiritual resource. Traditional fishing and agriculture practices are mostly passed down from one generation to the next, which denotes a significant role that the wetland plays in defining the locals. However, the neighboring agriculture activities and climate change issues, such as uneven rainfall and increased temperatures, were cited most of the time. Better management practices would allow sustainable fishing and agricultural practices, thus securing future generations' continued benefits from the wetland. Direct use values come mainly from fishery and agriculture. The approximate annual direct economic value from fishing is about INR 1.2 million. Agriculture has the potential to produce around INR 1.5 million a year. All these are a testament to the potential of the wetland as an income source for the people nearby. The indirect use values include services of flood regulation, water purification, and carbon sequestration in the form of an ecosystem. Standard methods of valuation calculated that such indirect economic value of those services stood at around INR 2 million a year. This further assumes that the wetland supports both local economies as well as the greater ecological functions that will benefit the neighborhood environment. Munia-2 wetland analysis provides essential geological and ecological findings. Topsoil is clayey with dense organic matter content, which adds to nutrient holdup and encourages diversity. The intermediate silty stratum provides for good moisture storage, which is essential for supporting hydrophytic plant life. Sandy subsoil allows easy drainage, averting waterlogging. The pH levels of the wetland, which range from 5.8 to 7.1, are consistent with normal Sundarban delta soil conditions and allow for varied flora. Hydrological balance is facilitated by seasonal water inflow changes, while GIS analysis identifies areas prone to erosion that need to be conserved. The nutrient-rich soil of the wetland has promoted sustainable agriculture, particularly for paddy farming and aquaculture, which are key to the local economy. Moreover, the development of eco-tourism activities like guided tours and birdwatching has a potential income stream for the people. Nonetheless, the issues of pollution and soil erosion underscore the necessity for a holistic approach to conservation aimed at maintaining the region's geological stability and ecological equilibrium.

Our results fit with other research showing that wetlands can withstand pressure from climate change. In Munia-2, it was found that seasonal changes in water levels and variety of species are in agreement with Mitra (2019), who noticed that West Bengal's wetlands are able to change with the seasons, but can be damaged by human action. Unlike the steady environments which Ghosh et al. (2009) described in the East Kolkata Wetlands, Munia-2 now shows signs of damage because of increased waste around the nearby city. In addition, Sharma et al. (2024) found that after extreme climatic events, the ecology of West Asia's Ramsar wetlands worsened more rapidly. Interviews with local communities in Munia-2 indicate that issues with cyclonic windstorms (for example, Amphan in 2020) have pushed up the risk of flooding, land degradation and forest cover changes, affecting the wetland's ability to resist stresses. The results suggest that small, peri-urban wetlands like Munia-2 which are not protected by formal conservation, require well-planned methods to adapt to climate change and land changes.

3.8 Challenges and Opportunities

Despite these economic benefits, the Munia-2 wetland faces numerous problems that may make it fail to achieve its potential.

Environmental Degradation

The root problem is environmental degradation mainly through pollution and misuse. The quality of water has become poorer because agricultural runoff contains fertilizers and pesticides directly affecting aquatic life and human health. Interviews revealed that members of the community have gained an increased awareness of such problems but lack the resources and information to implement the solutions in the most effective way.

Resource Overexploitation

Another critical issue is overfishing, as an increase in fish demand may result in the depletion of fish stocks. There is a need for sustainable practices in fishing and community-based management strategies.

Climate Change

Climate change is one of the significant issues facing the Munia-2 wetland, where increased flooding and changed rainfall patterns negatively affect agricultural productivity. Adaptive strategies to cope with these changes in local communities need to be developed, and climate resilience should be integrated into wetland management. Despite these challenges, there are considerable opportunities for sustainable development in the Munia-2 wetland. Community Involvement: Involvement of the communities living around the wetlands will increase the sustainability of the wetland resources. Training and awareness programs will empower the communities to adopt sustainable fishing and agriculture practices. The development of eco-tourism can offer sources of alternative income and improve environmental stewardship in the communities. Improving road infrastructure and accommodation facilities can improve the number of visitors. Policies supported at the government level can protect the wetland along with economic gains by conserving natural resources through financial incentives and engaging the community.

Policy Recommendations and Suggestions

It is suggested that sustainable and resilient use of the Munia-2 wetland ecosystem can be maintained by introducing several policy rules and management approaches. First, a conservation zoning regulation should cover the wetland to forbid building on it, reclaiming the land and polluting. Munia-2 can be safeguarded by incorporating it into the Wetlands (Conservation and Management) Rules, 2017 used by state governments. Next, systems for local fisherman, farmers and eco-tourism groups to work together and manage their areas of interest should be made official and everyone should be taught sustainable actions and given the power to monitor the environment. It is also essential to consider climate resilience, so places are made with vegetative zones, structures for storing sediment and systems to inform people about cyclones and floods. Moreover, people and groups promoting eco-tourism need to pay attention to guidelines about who visits, how waste is kept and how wildlife habitat is considered. Sixth, existing laws should introduce required ecological monitoring with the use of remote sensing and GIS software. Lastly, classes for

Picture 3.1 Wet land area. *Source* P.C: Anushka Ghosh. Date: 22.10.2024.Time: 11.22 A.M. Place: Munia-II, Newtown

Picture 3.2 Waste dumping ground. *Source* P.C: Subhadip Mondal. Date: 22.10.2024. Time: 11.25 A.M. Place: Munia-II, Newtown

Picture 3.3 Land degradation due to waste disposal. *Source* P.C: Anushka Ghosh. Date: 22.10.2024. Time: 11.22 A.M. Place: Place: Munia-II, Newtown

schools and other institutions encourage students and adults to join in the care of wetland areas. All of these efforts have the potential to make Munia-2 an example of sustainable wetland conservation for nearby urban areas (Pictures 3.1, 3.2, 3.3, 3.4 and 3.5).

Picture 3.4 Land degradation by waste dumping. *Source* P.C: Anushka Ghosh. Date: 22.10.2024. Time: 11.42 A.M. Place: Munia-II, Newtown

Picture 3.5 Natural landscape. *Source* P.C: Anushka Ghosh. Date: 22.10.2024. Time: 11.56 A.M. Place: Munia-II, Newtown

3.9 Conclusion

Munia-2 wetlands are quite economically and ecologically significant wherein the economic scopes of the wetland are at the mercy of the environment as well as of the local peoples. The wetland is a valuable source of biodiversity and has huge tourist attractions primarily during winter as well as autumn when the wetland is in full bloom and diversified fauna is available. During such high periods of demand, the income that is derived from tourism provides not only income but also employment opportunities for the local people in terms of feeding the tourists, providing guided travel around the region, and selling local products. However, these benefits gained are only available if the people involved in these areas adopt means of creating wealth and simultaneously practicing sustainable development. The results of this study submit to me the relevance that social net intervention says the integration of sustainable practices can offer so that maximum economic gain is achieved from the wetland but in such a way that the ecological integrity is maintained. Integral to this is the role played by the community members as custodians of the wetland. By actively participating in decision-making concerning their wetland and receiving appropriate training aimed at promoting responsible and sustainable tourism practices, the communities will take an active role in resource management and conservation. This empowerment contributes to economic resilience as the local population acquires the know-how and the techniques to participate in the tourist economy more efficiently. Interaction among different sectors: such as authorities, environmental, and business sectors is seen as one other significant factor. In a cooperative mode of action, the stakeholders will be able to formulate policies and strategies that will go a long way in the sustainable management of wetland resources. For instance, controlled tourism, rehabilitation of affected resources, and measures against pollution will alleviate the negative effects caused by the increase of human activities in the wetlands and thus continue providing unique ecological characteristics of the wetlands. A fundamental principle in this case, therefore, is the integration of social and economic development with environmental protection in sustainable development. The incorporation of Munia-2 sustainable practices safeguards the environmental integrity of the region as well as the economic opportunities available to the local people attending the region. The focus on the environmental status of Munia-2 together with the need to generate income from the area ensures that Munia-2 will be economically viable, environmentally rich, and sustainable enhancing the well-being both of the ecosystems and the people within the region. Munia-2 wetland is a singular geological formation in the Sundarban deltaic environment with varied soil types that increase ecological stability and economic viability. The combination of clayey, silty, and sandy horizons produces a strong system for supporting agriculture, aquaculture, and eco-tourism. Increasing threats of erosion, pollution, and human disturbance, however, call for aggressive conservation measures. To ensure the long-term sustainability of the Munia-2 wetland, there needs to be a concerted effort integrating scientific studies, community participation, and government policy measures to conserve this critical ecosystem while fostering sustainable economic development.

This research prepares the way for further studies collaborating different areas of knowledge on intermediate wetlands close to urban development. From now on, using satellite pictures, models to forecast climate and automated sensors for water quality can help discover the effect of seasons and climates on wetlands. Similar research studies in other West Bengal and South Asian deltaic wetlands could reveal more about regional wetland resilience. This subject also gives us an opportunity to develop economic models that include cultural and ecosystem services. Moreover, developing an observatory and policy action toolkit led by the community supports the use of research findings in governance. Introducing education and co-management methods will maintain effective conservation efforts. Overall, Munia-2 can show the way for managing wetlands sustainably in other cities near the Global South.

References

Banerjee P, Chatterjee S (2022) Remote sensing-based analysis of deltaic expansion in eastern India. Earth Observ Environ Remote Sensing 10(2):88–102

Bera S, Ghosh S, Biswas R (2021) Sediment accumulation and landform changes in the Lower Ganga Delta. J Geomorphol Stud 12(3):56–69

Brander LM, Boisclair C, Jameus B (2023) The global economic value of the climate regulation of inland wetland ecosystems. Clim Econ. https://climateeconomics.pressbooks.tru.ca/chapter/the-global-economic-value-of-the-climate-regulation-of-inland-wetland-ecosystems-by-clara-boisclair-and-brook-jameus/

Das S, Sengupta S (2019) Sustainable wetland agriculture in India: a review. J Wetland Ecol 8(1):45–57

Garcia-Herrero L, Rodriguez-Sinobas L (2023) Cost-benefit analysis of constructed wetlands for sustainable water management in Mediterranean areas. arXiv https://arxiv.org/abs/2305.06284

Ghosh D, Das S, Mukherjee S (2009) Wastewater-fed aquaculture in East Kolkata Wetlands: resource recovery for sustainable urban ecosystem management. Aquacult Econ Manag 13(1):68–80. https://doi.org/10.1080/13657300902777595

Mitchell WJ, Gosselink JG (2015) Wetlands, 5th edn. Wiley, Hoboken

Mitra A (2019) Wetlands in West Bengal: conservation and management strategies. Wetland Sci Pract 36(2):24–32

Mukherjee N (2011) Wetland management and livelihood: a case study from Sundarbans. Int J Sustain Dev 8(2):89–102

Mulatu DW, Al Ghurair A, Adelphi R (2020) Valuing the ecosystem services of the Machar Marshes, South Sudan. Nile Basin Initiative. https://nilebasin.org/node/12023

Munia Wetland Research Committee (2023) Ecological and economic significance of Munia-2 Wetland in Rajarhat, West Bengal. Environ Conserv J 45(3):120–135

Prasad SN, Ramachandra TV, Ahalya N, Sengupta T, Kumar A (2002) Conservation of wetlands in India: a review. Trop Ecol 43(1):173–186

Roy K, Barman A, Sarkar P (2013) Sustainable fishery management in the wetlands of West Bengal. J Environ Manage 128:76–85

Sarkar T, Dey R, Mitra A (2023) Deltaic dynamics and lithological variations in the Sundarban region. Environ Earth Sci 82(1):101–114

Sharma P, Dar RA, Fawzi N (2024) Economic valuation of changes in ecosystem services of 77 Ramsar wetlands in West Asia over 37 years (1984–2021). Environ Conserv 51(1):1–15. https://doi.org/10.1017/S0376892924000029

Silvius MJ, Oneka M, Verhagen A (2000) Wetlands: lifeline for people at the edge. Phys Chem Earth Part B 25(7–8):645–652

4 Pressing Need to Protect Mangrove Diversity in the Deltas of the Indian Sundarbans

Subrata Mondal, Mousathi Tah, Sukanta Ankure, Ashok Kumar Murmu, and Saikat Naskar

Abstract

The Sundarbans, known as one of the most fertile mangrove ecosystems globally, consists of numerous islands within the estuaries of the Ganges-Meghna-Brahmaputra River system. The estuarine mangroves in the Indian part of the Sundarbans experience high siltation and relatively elevated salinity levels compared to those in Bangladesh. This disparity is primarily attributed to the gradual decline in freshwater flow through the Hooghly River and its tributaries. As a result, the Indian Sundarbans are now dominated by salt-tolerant species, while many less salt-tolerant mangroves have become increasingly rare. The hostile terrain of the Sundarbans renders it nearly impenetrable for plant exploration. Thus far, research on mangroves in the Indian Sundarbans has primarily been confined to riverine mudflat regions, where salt-tolerant species predominate. However, the elevated interior lands, which could potentially support other species, remain largely unexplored. Remote sensing techniques offer limited utility for the physical verification of plants in these interior regions. Nevertheless, the relatively flat terrain, extending over 80 km from Bay of Bengal in some parts of the Indian Sundarbans, provides a valuable opportunity to study less salt-tolerant species in inner estuaries where seawater intrusion is reduced. The mangroves of the Indian Sundarbans often form a dense, mixed flora. This association is further enriched and complicated by

S. Mondal (✉) · M. Tah · S. Naskar
Department of Botany, The University of Burdwan, Burdwan, West Bengal, India
e-mail: botsm95@gmail.com

S. Ankure
Department of Botany, S.B.S. Government College, Hili, Dakshin Dinajpur, West Bengal, India

A. K. Murmu
Department of Botany, Hooghly Women's College, Hooghly, West Bengal, India

C. Singha et al. (eds.), *Sustainable Management of Delta Ecosystems Resilience*, Deltas of the World, https://doi.org/10.1007/978-3-032-16261-8_4

mangrove-associated angiosperms, including herbs, vines, and parasitic species. Given the uniformly low elevation typical of the Indian Sundarbans, even a slight rise in sea levels—by just a few centimeters—could have a profound effect on the diversity and functioning of mangrove ecosystems in the foreseeable future. As well as this environmental challenge, the increasing human population around the mangrove ecosystem of Indian Sundarbans presents a significant and compounding threat. This review underscores the critical need for the protection of the mangroves in the Indian Sundarbans in light of increasing threats posed by climate change, pollution, and human activities. It examines the essential roles these mangroves play in coastal protection, biodiversity preservation, and climate mitigation. Furthermore, the review emphasizes the importance of implementing effective conservation and restoration strategies that are vital for the well-being of local communities.

4.1 Introduction

Mangroves, the tidal forest communities located in tropical and subtropical coastal regions, play an essential role in stabilizing shorelines, mitigating climate change, supporting rich biodiversity, and sustaining local livelihoods (Spadling et al. 2010; Tomlinson 2016). The Sundarbans, recognized as one of the largest and most ecologically complex mangrove ecosystems, exhibit remarkable resilience in confronting various environmental stressors (Ghosh et al. 2015). The Indian Sundarbans are intricately linked with rivers such as the Hooghly, Muriganga, Thakuran, Matla, Bidyadhari, Raimangal, and Harinbhanga, which present varying salinity conditions conducive to the growth of both low and high salinity-tolerant mangrove species (Mandal and Bar 2019). This unique hydrological environment fosters a diverse array of mangrove species that serve as nurseries for marine life, offering shelter and feeding grounds for juvenile fish, crustaceans, and mollusks. The complex root systems of mangroves create a structurally intricate habitat that enhances biodiversity while supporting the productivity of adjacent marine ecosystems (Tomlinson 2016).

Salinity gradients significantly influence species distribution throughout the Sundarbans (Mandal and Naskar 2008). The eastern region of the Indian Sundarbans, characterized by a greater influx of freshwater and narrower river mouths, exhibits lower salinity levels. As a result, species such as *Aglaia cuculata*, *Avicennia officinalis Brownlowia tersa, Bruguiera sexangula, Bruguiera parvifloraHeritiera fomes*, *Kandelia candel, Merope angulate, Nypa fruticans*, and *Sonneratia apetala* are more prevalent in this area (Mondal et al. 2025). Conversely, the western region, facing higher salinity due to diminished freshwater influx, is dominated by species such as *Avicennia alba, Avicennia marina*, *Bruguiera gymnorhiza*, *Bruguiera cylindrical, Lumnitzera racemosa, Rhizophora mucronata,* and *Sonneratia alba* (Mondal et al. 2025). These variations in species composition underscore the necessity of maintaining hydrological balance to sustain mangrove diversity.

In light of the accelerating climate change driven by greenhouse gas emissions, mangroves serve as vital natural climate solutions. In accordance with the objectives set forth by the Paris Climate Agreement to restrict the increase in global temperatures to less than 1.5 °C higher pre-industrial temperature, mangroves serve a vital function in the process of carbon sequestration (Tiwari et al. 2024). They sequester carbon at a rate 35% higher than terrestrial forests by trapping organic material in anoxic soils, thereby preventing decomposition and the subsequent release of carbon dioxide (Choudhary et al. 2024). Beyond their capacity for carbon storage, mangroves act as natural barriers against coastal hazards, including storm surges, cyclones, and erosion. The low-lying deltaic region of the Sundarbans is particularly vulnerable to submergence and land loss due to sea-level rise, with areas such as Sagar Island in the western Sundarbans facing severe threats from soil erosion (Fig. 4.1a–e). The degradation of these forests not only jeopardizes biodiversity but also threatens the livelihoods of millions who rely on them for sustenance and protection.

Anthropogenic pressures further exacerbate these threats. The overexploitation of mangrove resources for firewood, timber, and traditional medicine, coupled with unsustainable harvesting practices and inadequate conservation enforcement, accelerates habitat degradation (Mondal et al. 2025). Pollution from industrial and agricultural runoff, including heavy metals, pesticides, and untreated sewage, detrimentally affects estuarine waters, compromising mangrove growth and survival (Chowdhury and Maiti 2016). Additionally, plastic pollution disrupts the ecosystem by entangling roots and reducing soil oxygen availability, further undermining the resilience of mangrove forests (Bijsterveldt et al. 2021; Cappa et al. 2023).

In the aftermath of Cyclone ‘Amphan’, initiatives for mangrove restoration have been enhanced through active community engagement and the planting of mangrove seedlings (Bardhan 2021; Chowdhury et al. 2021). Effective restoration not only entails replanting key mangrove species but also restoring hydrological conditions that facilitate natural regeneration. Community-based afforestation initiatives in the Sundarbans have demonstrated promising results, restoring degraded areas while providing alternative livelihoods for local communities (Chowdhury and Maiti 2014). Nature-based solutions, such as utilizing mangroves as bio-shields to combat coastal erosion and mitigate severe storm surges, offer cost-effective and sustainable strategies to mitigate the impacts of climate change (Madramootoo and Virgil 2024).

Strengthening scientific research on mangrove ecology, carbon sequestration potential, and adaptive traits is crucial for the development of evidence-based conservation policies (Mondal et al. 2025). This review underscores the urgent need to protect the mangrove ecosystems of the Indian Sundarbans. With escalating threats from climate change, pollution, and human activities, it is imperative to emphasize the role of mangroves

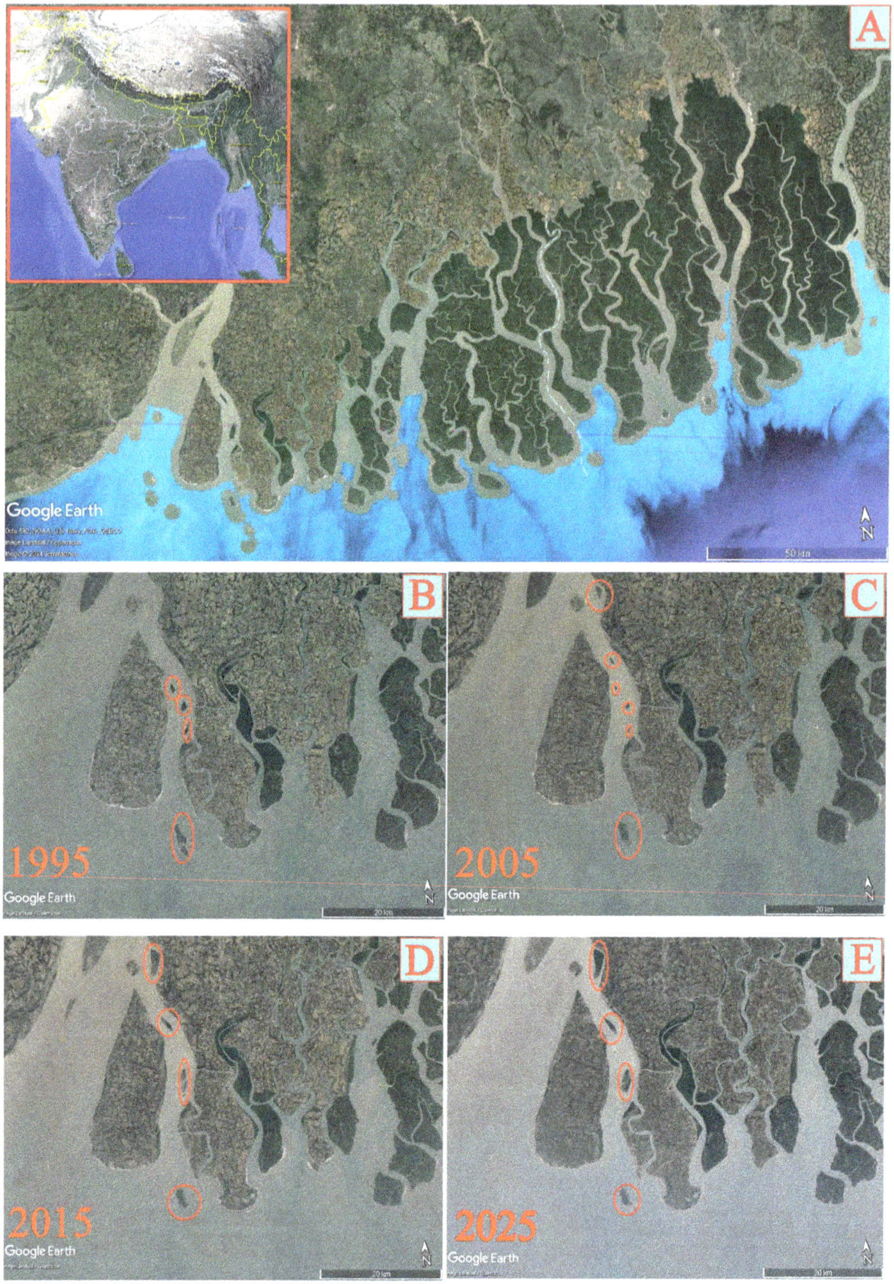
A
Google Earth
50 km
N
B
1995
Google Earth
20 km
C
2005
Google Earth
20 km
D
2015
Google Earth
20 km
E
2025
Google Earth
20 km

◂**Fig. 4.1** Map of the Sundarbans, illustrating that approximately 40% of the region is located within Indian territory (to the left of the blue dotted line), while the remaining area is situated in Bangladesh [A]. The western segment of the Indian Sundarbans is currently threatening significant environmental challenges, which encompass soil erosion, land subsidence, and sedimentation. These phenomena are primarily driven by changes in hydrology, climate change, and rising sea levels, resulting in dynamic alterations to island morphology. An analysis of Google Earth images from 1995, 2005, 2015, and projections for 2025 illustrates the changes of islands in the western Sundarbans, with small islands exhibiting observable shape changes marked by red circles [B-E]. The southernmost islands are experiencing the most severe impacts of soil erosion, while new land formations are emerging in the northernmost regions. This situation underscores the ongoing geomorphological and ecological transitions occurring within the Sundarbans landscape

in safeguarding coastal areas, supporting biodiversity, and mitigating climate change (Chowdhury et al. 2016). By examining the unique characteristics of the Indian Sundarbans, this review explores effective conservation strategies and restoration efforts while highlighting the importance of mangroves for local communities and their livelihoods, reinforcing the urgent need for their preservation.

4.2 Mangroves as a Carbon Sink: A Key to Climate Change Mitigation

In relation to climate change, blue carbon ecosystems—consisting of mangroves and their associated species (including tidal marshes and seagrasses)—are crucial for the process of capturing and storing carbon up to 35 times more quickly than tropical rainforests (Mcleod et al. 2011; Macreadie et al. 2021). Globally, these ecosystems are estimated to sequester between 8970 and 32,650 Tg of carbon, thereby preventing approximately 304 Tg of carbon dioxide emissions annually (Macreadie et al. 2019; Costa et al. 2023). The Sundarbans, the world's largest contiguous mangrove forest, is a significant carbon reservoir, storing approximately 26.5 Tg of blue carbon. A study conducted by Ray et al. (2011) in the Indian Sundarbans indicated that mangrove stands to sequester carbon across three key compartments: 39.9 ± 14.1 Mg C ha^{-1} in aboveground biomass (AGB), 9.6 ± 3.4 Mg C ha^{-1} in belowground biomass (BGB), and 17.4 ± 2.3 Mg C ha^{-1} within soil carbon reservoir (SCR). Among the various mangrove genera, *Avicennia* is particularly dominant in the Indian Sundarbans and contributes substantially to the region's blue carbon stocks (Chanda and Akhand 2023). In addition to their critical function in carbon sequestration, mangroves offer vital ecosystem services, particularly in the realm of coastal protection. The degradation of coastal ecosystems contributes to increased greenhouse gas emissions, compromising their effectiveness in climate mitigation (Choudhary et al. 2024). The protection and restoration of these ecosystems can assist nations in achieving their emission reduction objectives and Nationally Determined Contributions (NDCs) while also delivering essential benefits such as coastal protection and food security (Smith et al. 2020).

Local communities ascribe significant value to mangroves due to their ability to alleviate the effects of cyclones and storm surges. However, the awareness regarding payments for ecosystem services (PES) remains relatively limited. Government initiatives that aim to integrate PES mechanisms could not only enhance conservation efforts but also support India's aim to attain a 0.4% increase in carbon sinks, thereby aligning with broader climate mitigation strategies (Choudhary et al. 2024).

4.3 Mangroves as Natural Shields: Coastal Protection and Disaster Resilience

Mangroves serve as vital ecological buffers against climate change, safeguarding both coastal environments and the nearly 40% of the global population residing in these areas from the threats posed by tidal waves and cyclones (CIESIN 2007). The Sundarbans, situated at the convergence point of the Bay of Bengal, are instrumental in mitigating damage from cyclonic activities (Fig. 4.2a–d) by functioning as biological shields against powerful winds, tidal surges, and coastal erosion (Cornforth et al. 2013). Their intricate root systems stabilise shorelines, effectively reducing land loss due to erosion, while their canopies act as windbreaks, thereby diminishing the destructive potential of storms (Mondal et al. 2025).

Encompassing approximately 4260 km^2, the Indian Sundarbans are particularly crucial in protecting against the frequent occurrences of tropical cyclones, storm surge-induced flooding, breaches of embankments, and saline water intrusion—each posing substantial risks to human safety and livelihoods (Ghosh 2015). The mangrove forests create a protective barrier that shields not only local communities but also significant urban centers such as Kolkata from the devastating consequences of extreme weather events (Ghosh et al. 2015). The critical importance of mangrove conservation and restoration has become increasingly pronounced in light of escalating climate change challenges. Research employing satellite-based Earth observation and spatial data analysis tools has evaluated the impacts of tsunamis and storm surges on coastal regions with and without mangrove coverage (Marois and Mitsch 2015; Koshimura et al. 2020). These studies underscore the essential role of mangroves in mitigating damage, thus reinforcing the necessity for sustainable conservation initiatives. By preserving and expanding mangrove forest areas, we can bolster coastal resilience, ensuring the long-term security of ecological systems and human communities in the Indian Sundarbans.

Fig. 4.2 Mangroves act as a natural barrier, offering coastal protection and contributing to enhanced disaster resilience. However, the expansion of human settlements within the Indian Sundarbans has resulted in significant encroachment, with embankments replacing mangrove ecosystems. These embanked islands exhibit heightened vulnerability to environmental degradation; the removal of mangrove barriers increases their susceptibility to erosion and tidal influences [A-D]. In the locality of Pakhiraloy, South 24 Parganas, over one kilometer of land has been adversely affected by soil erosion attributed to the construction of embankments [A-B]. Following Cyclone Amphan, embankments in Jharkhali, South 24 Parganas, sustained considerable damage from elevated tidal waves, necessitating local communities to adopt interim protective measures, such as bamboo fencing [D]. This highlights the critical role of mangroves in coastal stability and the increasing challenges posed by anthropogenic modifications and extreme weather events

4.4 Sustaining Fisheries and Livelihoods: The Socio-economic Importance of Mangroves

The Sundarbans, one of the largest mangrove ecosystems, are increasingly utilised for human habitation and agricultural activities. Presently, the region is densely populated, with the human population of the Indian Sundarbans surpassing 5 million individuals (Chatterjee and Bhandari 2023). The livelihoods of the local communities are intricately linked to the resources derived from the Sundarbans, encompassing activities such as fishing, shrimp and crab harvesting, honey collection, agriculture, and tourism (Jana 2019).

The region's demographics are influenced by an extensive network of rivers and waterways, with fishing as the predominant income source for most residents (Fig. 4.3a–d). It is estimated that approximately 54.43% of the population is engaged in fishing, while 25.32% participates in shrimp collection (Chatterjee and Bhandari 2023). These economic activities not only underpin the local economy but also contribute to the complex socio-economic dynamics of the region.

Fig. 4.3 Mangroves function as fundamental nursery habitats for various aquatic species, including fish, crabs, prawns, and shrimp, thus supporting biodiversity and contributing to coastal food security. These ecosystems offer critical shelter and breeding grounds for marine life while playing a significant role in nutrient cycling and water filtration, which enhances overall ecosystem health. In Kaikhali, South 24 Parganas, local fishermen actively participate in their daily fishing practices, illustrating the profound interrelationship between mangroves and traditional livelihoods [A-D]. The sustainable management of these resources is imperative for ensuring long-term economic stability for coastal communities

4.5 Unique Features of Mangrove Diversity in the Indian Sundarbans

Mangrove forests, also known as mangal communities, represent highly specialised ecosystems that thrive in saline-rich, intertidal coastal, and estuarine zones of the tropical and subtropical regions (Spalding et al. 1997; McKee et al. 2007; Tomlinson 2016). These forests are integral to coastal stabilisation, biodiversity conservation, and carbon sequestration (Mandal and Naskar 2008). The Indian Sundarbans, recognised as one of the largest contiguous mangrove ecosystems globally, are influenced by three primary ecological factors: waterways, islands, and forests. India is home to a remarkable variety of mangrove flora (Table 4.1), which includes true mangroves, mangrove associates, halophytes, and back mangals (Naskar 1993; Naskar and Mondal 1999; Ghosh et al. 2008). According to Mandal and Bar (2019), India possesses 78 species of mangroves, categorised into 56 genera and 37 families. Of these, 32 species are classified as true mangroves, 14 as mangrove associates, 18 as mangrove halophytes, and 14 as back mangals. This diverse assortment of species showcases the complex ecological adaptations essential for survival in dynamic intertidal zones. Tomlinson (2016) identified five key characteristics of true mangroves: (1) complete fidelity to mangrove environments, (2) dominance within mangrove communities, (3) specialised morphological adaptations, (4) physiological tolerance to salinity and anoxia, and (5) systematic isolation from terrestrial relatives. True mangroves exhibit notable modifications in their roots, shoots, leaves, and reproductive structures to endure harsh saline conditions.

In contrast, although situated near intertidal zones, mangrove associates do not display such extensive structural or physiological adaptations. Some herbaceous and grass species within these environments, particularly halophytes, enhance leaf and stem succulence to manage high salinity levels (Fig. 4.4a–f). A global assessment of mangroves conducted by Polidoro et al. (2010) indicated that among the 77 mangrove species worldwide, 65 are classified as threatened according to the IUCN Red List. In the Indian Sundarbans, *Heritiera fomes* faces a significant threat from salinization, as salt levels often exceed its tolerance limit. *Heritiera fomes* have declined by 50–80% due to habitat degradation, climate change, and anthropogenic pressures (Mandal and Bar 2019). Owing their high species diversity and ecological importance, the Indian Sundarbans represent a crucial focal point for mangrove conservation efforts. Implementing effective conservation strategies, habitat restoration initiatives, and sustainable management practices is essential to protect this unique and vulnerable ecosystem from further decline.

Table 4.1 True mangroves (TM), mangrove associates (MA), back mangroves (BM), and salt marsh (SM) species in the Indian Sundarbans, along with their salinity tolerance levels and intertidal positions on the mudflat (LT = Low Tide, MT = Mid Tide, HT = High Tide, Highest abundance=*)

Sl. No.	Family	Scientific name	Type	Salinity tolerance			Intertidal position		
				High	Mid	Low	LT	MT	HT
1	Acanthaceae	*Acanthus ilicifolius* L.	MA		*				*
2	Acanthaceae	*Acanthus volubilis* Wall.	MA			*			*
3	Aizoaceae	*Sesuvium portulacastrum* (L.) L.	SM		*				*
4	Apocynaceae	*Cerbera odollam* Gaertn.	MA			*			*
5	Araceae	*Cryptocoryne ciliata* (Roxb.) Schott	SM			*			*
6	Arecaceae	*Nypa fruticans* Wurmb	TM			*			*
7	Arecaceae	*Phoenix paludosa* Roxb.	MA			*		*	
8	Apocynaceae	*Finlaysonia obovata* Wall.	MA		*			*	
9	Apocynaceae	*Pentatropis capensis* (L.f.) Bullock	MA			*			*
10	Apocynaceae	*Sarcolobus carinatus Wall.*	MA			*			*
11	Apocynaceae	*Sarcolobus globosus Wall.*	MA			*			*
12	Avicenniaceae	*Avicennia alba* Blume	TM		*		*		
13	Avicenniaceae	*Avicennia marina* (Forssk.) Vierh.	TM		*		*		
14	Avicenniaceae	*Avicennia officinalis* L.	TM			*		*	
15	Bignoniaceae	*Dolichandrone spathacea* (L.f.) K.Schum.	BM			*			*
16	Boraginaceae	*Heliotropium curassavicum* L.	MA			*			*
17	Amaranthaceae	*Suaeda pannonica* Beck	MA		*				*
18	Combretaceae	*Lumnitzera racemosa* Willd.	MA		*			*	
19	Euphorbiaceae	*Excoecaria agallocha* L.	TM			*			*
20	Fabaceae	*Cynometra iripa* Kostel.	BM			*		*	
21	Fabaceae	*Dalbergia horrida* var. horrida	BM		*			*	
22	Fabaceae	*Brachypterum scandens* (Roxb.) Wight & Arn. ex Miq.	BM			*			*
23	Fabaceae	*Derris trifoliata* Lour.	BM			*			*
24	Lamiaceae	*Volkameria inermis* L.	BM			*			*
25	Lythraceae	*Sonneratia alba* Sm	TM		*			*	
26	Lythraceae	*Sonneratia apetala* Banks	TM		*			*	
27	Lythraceae	*Sonneratia caseolaris* (L.) Engl.	TM			*		*	
28	Malvaceae	*Brownlowia tersa* (L.) Kosterm.	MA			*			*
29	Malvaceae	*Heritiera fomes* Banks	TM			*			*
30	Meliaceae	*Aglaia cucullata* (Roxb.) Pellegr.	TM			*			*
31	Meliaceae	*Xylocarpus granatum* J.Koenig	TM		*			*	
32	Meliaceae	*Xylocarpus moluccensis* (Lam.) M.Roem.	TM		*			*	
33	Plumbaginaceae	*Aegialitis rotundifolia* Roxb.	TM		*			*	
34	Poaceae	*Oryza coarctata* Roxb.	SM		*			*	
35	Poaceae	*Myriostachya wightiana* (Nees ex Steud.) Hook.f.	SM			*			*
36	Primulaceae	*Aegiceras corniculatum* (L.) Blanco	TM		*			*	
37	Rhizophoraceae	*Bruguiera cylindrica* (L.) Blume	TM		*			*	
38	Rhizophoraceae	*Bruguiera gymnorhiza* (L.) Lam. ex Savigny	TM		*			*	
39	Rhizophoraceae	*Bruguiera sexangula* (Lour.) Poir.	TM		*			*	
40	Rhizophoraceae	*Ceriops decandra* (Griff.) W.Theob.	TM		*			*	
41	Rhizophoraceae	*Ceriops tagal* (Perr.) C.B.Rob.	TM		*			*	
42	Rhizophoraceae	*Kandelia candel* (L.) Druce	TM			*		*	
43	Rhizophoraceae	*Rhizophora apiculata* Blume	TM		*			*	
44	Rhizophoraceae	*Rhizophora mucronata* Poir.	TM		*			*	
45	Rubiaceae	*Scyphiphora hydrophylacea* C.F.Gaertn.	TM		*			*	
46	Rutaceae	*Merope angulata* (Willd.) Swingle	TM			*		*	

Fig. 4.4 A top view of the dense mixed mangrove forest in Henry Island, South 24 Parganas, reveals a thick canopy where sunlight rarely reaches the forest floor. This indicates the high fertility of saline soils, which support diverse mangrove species. The dominant species in this region include Avicennia spp., Bruguiera gymnorhiza, and Xylocarpus granatum. In the side view, the climber vine Derris trifoliata is seen covering the true mangrove species Bruguiera gymnorhiza. The tallest true mangrove species, Sonneratia apetala, exhibits rapid growth, while the mangrove associate Acanthus ilicifolius thrives in the lower understory, providing additional protection to the forest. Some rare species, such as Sonneratia caseolaris [C], Scyphiphora hydrophylacea [D], and Sarcolobus globosus [E], are found sporadically in the Indian Sundarbans. Along coastal stretches, Ipomoea pes-caprae is commonly observed, stabilizing sandy shores and playing a crucial role in coastal dune ecosystems [F]. The presence of such diverse flora underscores the ecological richness and conservation significance of this mangrove habitat

4.6 Challenges Facing Mangrove Diversity in the Sundarbans: Freshwater Scarcity and Salinity Stress

The Sundarbans region, famous for its intricate network of creeks and rivulets, is facing serious challenges regarding freshwater availability (Manna et al. 2023). Despite substantial rainfall during the monsoon season, the freshwater supply remains inadequate. Silt accumulation gradually makes the water channels shallower, restricting water flow. Rising salinity levels and poor drainage systems worsen the situation, posing significant threats to the mangrove ecosystems vital to the region's ecological health (Das et al. 2023).

In the Indian Sundarbans, the freshwater crisis is particularly severe. Irregular monsoon patterns fail to provide a reliable water supply, while human interventions have disrupted the natural hydrology. Dams and barrages built along the Ganga River and its tributaries have drastically reduced the flow of freshwater reaching the delta. These changes have not only diminished the availability of freshwater but also altered soil salinity dynamics. This imbalance threatens the survival of mangroves and other plant species, making it increasingly difficult for this diverse ecosystem to thrive (Chanda and Akhand 2023).

4.7 Coastal Erosion in West Bengal: Impacts on Mangrove Ecosystems

West Bengal experiences the highest rate of coastal erosion in India, with approximately 70% of its shoreline affected (Das and Swain 2024). In the Indian Sundarbans, the southern and western portions of the islands are more significantly eroded than the northern and eastern sides. This phenomenon can be attributed mainly to the impact of strong tidal waves and the persistence of longshore currents. Hydrological alterations in the River Hooghly have resulted in the complete submersion of Lohachara Island, while Nayachar Island has expanded due to sediment accumulation (Chatterjee et al. 2015). The ongoing erosion along the shoreline in the Sundarbans is not solely a natural occurrence; anthropogenic activities, including the deforestation of mangroves, the expansion of agricultural land, the growth of fisheries, and the construction of infrastructure for tourism have notably exacerbated it. Geomorphological changes have further transformed the landscape, leading to a shift towards the eastern regions of the delta. Additionally, the region faces the pressing challenge of rising sea levels, currently observed at rates between 3.14 and 5.22 mm annually (Chatterjee et al. 2015). Sea-level rise and ongoing erosion pose significant threats to the ecosystem and the resident population (Fig. 4.5a–b).

Sea-level rise increases the mortality of mangroves along the seaward edges while pushing mangroves to migrate landward (Gilman et al. 2008). Such shifts can significantly affect mangrove species' distribution and abundance, thereby altering their habitats' ecology and structural integrity (Armitage et al. 2015). A field survey conducted by Ghosh et al. (2020) identified highly salt-tolerant true mangrove taxa, including *Avicennia* spp. (*A. alba*, *A. marina*, and *A. officinalis*), *Bruguiera gymnorhiza*, and *Sonneratia apetala,* within the lower stretches of the Hooghly River, near its confluence with the sea. Notably, *Sonneratia caseolaris* was absent from this region. However, the study revealed a northward migration of *Sonneratia caseolaris* upstream along the Hooghly River, with this species being observed in areas where mangroves had not been recorded for over a century. This phenomenon underscores the dynamic response of mangrove ecosystems to

Fig. 4.5 Hydrological alterations have profoundly affected the banks of Dakshin Chandanpiri in South 24 Parganas, resulting in significant erosion and habitat destabilization [A]. The highly exposed and interwoven root systems of the mangroves are prominently observable, indicating the severity of soil loss and root instability. In the Hingalganj area of North 24 Parganas, the construction of embankments encounters critical challenges due to the erosion of the natural barrier, which has rendered the region increasingly susceptible to tidal surges and flooding [B]. The ongoing depletion of these vital protective ecosystems highlights the urgent need for intervention

environmental changes, particularly concerning sea-level rise and variations in salinity gradients.

4.8 Anthropogenic Threats to Mangroves: Deforestation and Habitat Loss from Aquaculture, Agriculture, and Brickfield Expansion

Mangrove forests are rapidly declining, with 28% lost to commercial aquaculture across eight key mangrove-holding nations, including India (Hamilton 2013). In the Indian Sundarbans, numerous mudflats have been illegally transformed into agricultural or aquaculture land over the past two decades (Daw et al. 2021). Local fishers and farmers frequently clear mangrove areas for short-term economic benefits, often without recognising these forests' critical ecosystem services, such as coastal protection and carbon sequestration (Sinha et al. 2024). Aquaculture, particularly tiger shrimp farming, is expanding significantly in the Sundarbans region. This area produces approximately 50,000 tons of shrimp annually from 51,000 hectares of aquaculture farms, positioning it as the leading shrimp-producing region in India (Giri et al. 2022).

The destruction of mangrove ecosystems has a deep-rooted history. During the British colonial period, extensive portions of the Sundarbans were cleared for agricultural activities and settlement expansion (Ghosh 2015). Pramanik (2015) noted that mangrove forest cover in the Indian Sundarbans diminished from 44 to 31% due to anthropogenic interventions. The rapid establishment of brick kilns in the Ghusighata, Minakha, and Hemnagar

Fig. 4.6 A solitary *Avicennia officinalis* tree placed within a fishery at Ghusighata, North 24 Parganas, representing the prevailing encroachment of aquaculture into vital mangrove ecosystems [A & B]. The swift expansion of the fishery industry has resulted in extensive deforestation of mangrove forests, posing significant threats to biodiversity and diminishing coastal resilience. The degradation of these essential ecosystems not only disrupts habitat connectivity but also heightens susceptibility to erosion and extreme weather phenomena. The implementation of sustainable aquaculture practices and effective conservation strategies is imperative to reconcile economic development with ecological preservation in the region

regions has further exacerbated deforestation (Fig. 4.6a and b) and the extraction of riverine mud, leading to soil degradation that renders the environment unsuitable for the germination and establishment of mangroves (Mondal et al. 2025).

4.9 Pollution from Industrial Waste, Agricultural Runoff, and Domestic Sewage. In the Indian Sundarbans: A Rising Concern

Various anthropogenic activities significantly drive toxic metal pollution in mangrove ecosystems. Primary sources of this pollution include municipal waste disposal originating from large metropolitan areas, such as Kolkata, alongside mining operations, industrial processing, and the combustion of fossil fuels (Bakshi et al. 2018). Additionally, the issue is exacerbated by automobile emissions, agricultural runoff containing agrochemicals, discharge from paint industries, and untreated sewage (Alloway 2013). The drainage of fertilisers and pesticides from agricultural lands and the influx of untreated domestic wastewater further intensify the contamination problem. Mangroves serve a crucial ecological function as natural sinks for these pollutants, effectively trapping and accumulating toxic metals within their roots and sediments. However, the escalating levels of contamination pose serious threats to the health and biodiversity of these vital ecosystems. Non-essential elements like arsenic (As), lead (Pb), and cadmium (Cd) can infiltrate mangrove systems via membrane transporters (Verbruggen et al. 2009). Once absorbed, these elements may be transported into the xylem, sequestered within vacuoles, or stored in intercellular spaces. The increasing entrapment and accumulation of chemical pollutants in mangrove roots, along with the biotransformation of toxic metals, can significantly diminish mangrove biodiversity, reducing species richness and adaptability. Research by Maiti and Chowdhury (2013) indicates a linear increase in copper, zinc, and lead concentrations correlating with rising pollution. This trend is particularly pronounced in the cosmopolitan bioindicator species *Avicennia marina*. Furthermore, a study conducted by Sarangi et al. (2002) found that *Avicennia officinalis* exhibited the highest rates of metal accumulation—specifically for iron, copper, manganese, and zinc—when compared to other true mangrove species of Rhizophoraceae (*Bruguiera cylindrica*, *Ceriops decandra*, and *Rhizophora mucronata*) and Meliaceae (*Xylocarpus granatum*). The diversity and health of mangrove ecosystems in the Indian Sundarbans, particularly within the Chemaguri and Ghushighata regions, are experiencing heightened stress (Bakshi et al. 2018). This situation is primarily attributed to increased metal stress resulting from elevated levels of enrichment, geo-accumulation, and contamination stemming from anthropogenic activities, which significantly adversely affect the mangrove ecosystems. (Mondal et al. 2025).

4.10 Mangrove Restoration and Protection: Strategies for a Sustainable Future

The Indian Sundarbans, characterized by its complex tidal river network, extensive mudflats, and a diverse array of small islands and creeks, serves as an ideal habitat for mangrove ecosystems. However, this delicate environment is increasingly threatened by climate instability, ocean expansion, island subsidence, and intensifying meteorological phenomena, including tidal surges and severe cyclonic storms (Chowdhury and Maiti 2014). These factors significantly contribute to the decline of mangrove cover, jeopardizing both biodiversity and the livelihoods of local communities. Despite these challenges, the Sundarbans has demonstrated notable resilience through ongoing restoration efforts. Various mangrove species, such as *Avicennia alba* (Kala Baine), *Avicennia marina* (Peara Baine), *Avicennia officinalis* (Jat Baine), *Bruguiera gymnorhiza* (Kakra), *Ceriops tagal* (Garan), *Excoecaria agallocha* (Genwa), *Heritiera fomes* (Sundari), *Nypa fruticans* (Golpata), *Rhizophora apiculata* (Garjan), and *Sonneratia apetala* (Keora), have been successfully replanted on barren islands as part of community-driven initiatives. The participation of local stakeholders, particularly from marginalized groups, has proven essential in these conservation endeavors. In the wake of the widespread devastation inflicted by Cyclone Amphan in 2020, the West Bengal State Forest Department, in collaboration with various non-governmental organizations (NGOs), initiated a comprehensive mangrove restoration program (Bardhan 2021). From July 2020 to 2022, approximately 71.58 million true mangroves species and 52.19 million mangrove-associated species were established in the tidal mudflat (Banerjee et al. 2023). This large-scale afforestation initiative not only sought to restore ecological balance but also aimed to bolster coastal protection, enhance biodiversity, and secure local livelihoods. Notably, these efforts have significantly empowered women by actively involving them in plantation and conservation activities, thereby fostering socio-economic resilience within the region. Looking ahead, the long-term sustainability of mangrove restoration in the Sundarbans will necessitate ongoing monitoring, scientific intervention, and robust policy support. Strengthening natural regeneration processes, integrating traditional ecological knowledge with contemporary restoration techniques, and promoting alternative livelihood options can further improve the efficacy of these conservation strategies.

4.11 Advancing Mangrove Conservation: The Role of Research and Technology

The Indian Sundarbans Delta represents a dynamic and ecologically sensitive ecosystem that confronts substantial challenges stemming from both environmental changes and anthropogenic influences (Ghosh et al. 2015). Research within this region has predominantly concentrated on aspects of regional development, ecological sustainability, and the implications of shifting vegetation patterns and climate variability on hazard characteristics (Kumar 2023; Guha et al. 2020). The application of modern equipment, encompassing satellite imagery and spatial data systems, has been instrumental in mapping and analyzing coastal zones, assessing land utilization and surface characteristics changes, and monitoring hydrological alterations, soil erosion, and human encroachment (Hickey and Radford 2022). These investigations have yielded critical insights into the spatio-temporal dynamics of the Sundarbans, revealing significant transformations in mangrove coverage, salinity variations, and land use patterns over recent decades. A notable observation from these studies is the increasing prevalence of highly salt-tolerant species, which has resulted in the expansion of monospecific mangrove forests. Concurrently, several species, such as *Brownlowia tersa*, *Kandelia candel*, *Bruguiera sexangula*, *Bruguiera parviflora*, and *Rhizophora apiculata*, have become increasingly scarce within the transitional zones of the Indian Sundarbans. This decline can be attributed to habitat specificity, changing environmental conditions, tidal inundation, salinity fluctuations, and constraints in seed dispersal. Consequently, biodiversity in the region continues to adapt in response to climatic changes, thereby altering the ecological balance over time (Kumar 2023). In light of these pressing challenges, research highlights the urgent need for a comprehensive coastal governance strategy that incorporates both environmental and socio-economic factors (Mondal 2024). Sustainable conservation strategies should be designed to protect biodiversity while simultaneously supporting the livelihoods of local communities.

4.12 Mangrove Conservation Policies: A Global and Local Perspective

Global deterioration of mangrove health and biodiversity is occurring at an alarming rate. Strengthening conservation initiatives for these critical ecosystems necessitates a multi-tiered approach that incorporates policies at global, national, state, regional, and community levels (Mondal et al. 2025). In the Indian Sundarbans, where the delicate equilibrium between ecological sustainability and human livelihoods is persistently challenged, the establishment of a well-coordinated policy framework is imperative (Raha et al. 2013). At the international level, institutions such as UNESCO and the International Union for Conservation of Nature (IUCN) are instrumental in promoting for mangrove conservation, acknowledging their significance as carbon sinks and essential components of coastal resilience (Raha et al. 2013; Ghosh et al. 2015. International agreements and funding mechanisms facilitate large-scale restoration efforts, thereby reinforcing the ecological importance of these vital wetlands. On a national scale, India has implemented various policies, including the Coastal Regulation Zone (CRZ) regulations, the Integrated Coastal Zone Management (ICZM) and the National Biodiversity Action Plan (NBAC) program, to protect mangroves. However, effective enforcement and adaptive management strategies are critical for addressing site-specific challenges within the Sundarbans. At both the state and regional levels, conservation efforts are frequently influenced by governmental agencies, research institutions, and non-governmental organizations (NGOs). Joint Forest Management (JFM) programs and sustainable livelihood initiatives—such as permaculture and ecotourism—have been introduced to engage local stakeholders in conservation endeavors (DasGupta and Shaw 2017; Raha et al. 2013). By integrating sectors such as fisheries, agriculture, tourism, and forestry, a comprehensive approach to biodiversity conservation and resource management can be realized (Mondal 2024). Empowerment of local communities is fundamental to achieving long-term conservation success. Strengthening the roles of panchayats, self-help groups, and grassroots organizations can enhance capacity-building, disaster preparedness, and sustainable development (Mondal et al. 2025). Local populations, who rely on these forests for their livelihoods, must be actively engaged in decision-making processes to mitigate conflicts between conservation objectives and socio-economic needs.

4.13 Conclusions

Mangrove Diversity in the Deltas of the Indian Sundarbans, play a crucial role in climate change mitigation, coastal protection, biodiversity conservation, and sustaining local livelihoods. These tidal forests are recognized as some of the most effective blue carbon sinks, capable of capturing and storing significant quantities of carbon, thereby significantly contributing to mitigating greenhouse gas emissions. The prominence of genera such as *Avicennia* Spp. and *Sonneratia* Spp. underscores their ecological importance. Nevertheless, the escalation of human activities, including aquaculture expansion, agricultural encroachment, and infrastructure development, in conjunction with natural stressors such as rising sea levels and coastal erosion, poses substantial threats to their survival. These pressures contribute to alterations in species composition, thereby endangering rare mangroves such as *Brownlowia tersa* and *Heritiera fomes*.

In spite of their vital ecosystem services, pervasive habitat degradation—largely driven by shrimp farming, tourism development, and changes in land use—continues to hasten the loss of mangroves. The ecological uniqueness of the Sundarbans is manifest in its diverse mangrove species, which include true mangroves, mangrove associates, and salt-tolerant species. Effective conservation strategies must prioritize sustainable management, habitat restoration, and the implementation of Payment for Ecosystem Services frameworks to promote local engagement in mangrove preservation. Protecting these forests is not merely a regional issue but a global imperative. A comprehensive strategy that incorporates ecological restoration, measures for climate resilience, and socio-economic incentives will be essential for ensuring the long-term viability of these vital tidal mangal community.

References

Alloway BJ (2013) Sources of heavy metals and metalloids in soils. In: Alloway BJ (ed) Heavy metals in soils: trace metals and metalloids in soils and their bioavailability, vol 22. Springer Netherlands, pp. 11–50

Armitage AR, Highfield WE, Brody SD, Louchouarn P (2015) The contribution of mangrove expansion to salt marsh loss on the Texas Gulf Coast. PLoS ONE 10(5):e0125404. https://doi.org/10.1371/journal.pone.0125404

Bakshi M, Ghosh S, Chakraborty D, Hazra S, Chaudhuri P (2018) Assessment of potentially toxic metal (PTM) pollution in mangrove habitats using biochemical markers: a case study on *Avicennia officinalis* L. in and around Sundarban, India. Mar Pollut Bull 133:157–172. https://doi.org/10.1016/j.marpolbul.2018.05.030

Banerjee S, Ladd CJ, Chanda A, Shil S, Ghosh T, Large A, Balke T (2023) Securing the sustainable future of tropical deltas through mangrove restoration: Lessons from the Indian Sundarban. One Earth 6(3):190–194. https://doi.org/10.1016/j.oneear.2023.02.015

Bardhan M (2021) An empirical study on mangrove restoration in Indian Sundarbans—a community-based environmental approach. In: Modern cartography series, vol 10. Academic Press, pp 387–405

Bijsterveldt CEJV, Wesenbeeck BKV, Ramadhani S, Raven OV, van Gool FEV, Pribadi R, Bouma T (2021) Does plastic waste kill mangroves? A field experiment to assess the impact of macro plastics on mangrove growth, stress response and survival. Sci Total Environ 756:143826. https://doi.org/10.1016/j.scitotenv.2020.143826

Cappa P, Walton ME, Paler MKO, Taboada EB, Hiddink JG, Skov MW (2023) Impact of mangrove forest structure and landscape on macroplastics capture. Mar Pollut Bull 194:115434. https://doi.org/10.1016/j.marpolbul.2023.115434

Chanda A, Akhand A (2023) Challenges towards the sustainability and enhancement of the Indian Sundarban Mangrove's blue carbon stock. Life 13(8):1787. https://doi.org/10.3390/life13081787

Chatterjee N, Mukhopadhyay R, Mitra D (2015) Decadal changes in shoreline patterns in Sundarbans, India. J Coast Sci 2(2):54–64. https://doi.org/10.6084/m9.figshare.1578750

Chatterjee S, Bhandari G (2023) Study on development of sustainable livelihood framework approach at Indian part of Sundarbans by geospatial and geo-statistical analysis. J Environ Sci Econ 2(2):18–37. https://doi.org/10.56556/jescae.v2i2.508

Choudhary B, Dhar V, Pawase AS (2024) Blue carbon and the role of mangroves in carbon sequestration: Its mechanisms, estimation, human impacts and conservation strategies for economic incentives. J Sea Res 199:102504. https://doi.org/10.1016/j.seares.2024.102504

Chowdhury A, Maiti SK (2014) Mangrove reforestation through participation of vulnerable population: engineering a sustainable management solution for resource conservation. Int J Environ Res Dev 4(1):1–8

Chowdhury A, Maiti SK (2016) Assessing the ecological health risk in a conserved mangrove ecosystem due to heavy metal pollution: a case study from Sundarbans Biosphere Reserve, India. Hum Ecol Risk Assess Int J 22(7):1519–1541. https://doi.org/10.1080/10807039.2016.1190636

Chowdhury A, Sanyal P, Maiti S (2016) Dynamics of mangrove diversity influenced by climate change and consequent accelerated sea level rise at Indian Sundarbans. Int J Global Warm 09:486–506. https://doi.org/10.1504/IJGW.2016.076333

Chowdhury A, Naz A, Iyer AS, Bhattacharyya S (2021) Ecosystem based disaster risk reduction at Indian Sundarbans: a lesson learned from AMPHAN supercyclone. IOP Conf Ser Earth Environ Sci 796(1):012042. https://doi.org/10.1088/1755-1315/796/1/012042

CIESIN (2007) Center for international earth science information network-CIESIN-Columbia University. National aggregates of geospatial data collection: population, landscape, and climate estimates, version 2 (PLACE II). NASA Socioeconomic Data and Applications Center (SEDAC), Palisades, NY. https://doi.org/10.7927/H4JW8BSC

Cornforth W, Fatoyinbo L, Freemantle T, Pettorelli N (2013) Advanced land observing satellite phased array type L-Band SAR (ALOS PALSAR) to inform the conservation of mangroves: Sundarbans as a case study. Remote Sens 5(1):224–237. https://doi.org/10.3390/rs5010224

Costa M, Adame F, Bryant C, Hill J, Kelleway J, Lovelock C, Ola A, Rasheed M, Salinas C, Serrano O, Waltham APN, York P, Young M, Macreadie P (2023) Quantifying blue carbon stocks and the role of protected areas to conserve coastal wetlands. Sci Total Environ 874:162518. https://doi.org/10.1016/j.scitotenv.2023.162518

Das L, Bhowmick S, Goswami P, Choudhury RR, Akhter J (2023) Past and future freshwater availability scenarios and their impact on the Indian Sundarbans ecosystem and fisheries. Aquat Ecosyst Health Manag 26(3):53–61. https://doi.org/10.14321/6629.2022.05.19

Das A, Swain PK (2024) Navigating the sea level rise: Exploring the interplay of climate change, sea level rise, and coastal communities in India. Environ Monit Assess 196(11):1010. https://doi.org/10.1007/s10661-024-13191-z

DasGupta R, Shaw R (2017) Retrofitting joint forest management (JFM) in protected areas of Indian Sundarbans: how sustainable it is? In: Participatory mangrove management in a changing climate: perspectives from the Asia-Pacific. Disaster risk reduction. Springer, Tokyo, pp 59–71

Daw TM, Giri S, Protim Mondal P, Samanta S, Hazra S, Hutton C, Hornby DD, Harfoot A, Mukherjee K, Gupta R (2021) What is driving conversion of land to aquaculture in the Indian Sundarbans? EGU General Assembly Conference Abstracts. https://doi.org/10.5194/egusphere-egu21-16485

Ghosh P (2015) Conservation and conflicts in the Sundarban biosphere reserve, India. Geogr Rev 105(4):429–440. https://doi.org/10.1111/j.1931-0846.2015.12101.x

Ghosh A, Schmidt S, Fickert T, Nüsser M (2015) The Indian Sundarban mangrove forests: history, utilization, conservation strategies and local perception. Diversity 7(2):149–169. https://doi.org/10.3390/d7020149

Ghosh S, Bakshi M, Gupta K, Mahanty S, Bhattacharyya S, Chaudhuri P (2020) A preliminary study on upstream migration of mangroves in response to changing environment along River Hooghly, India. Mar Pollut Bull 151:110840. https://doi.org/10.1016/j.marpolbul.2019.110840

Ghosh A, Sarkar NS, Mukherjee S, Naskar KR (2008) Distribution of mangrove and mangrove associated plants in different blocks and comparatments of Sundarbans Tiger Reserve (STR) with special references to theirabundance. Phytogeogr Endemism RET Plants 541–553

Gilman EL, Ellison J, Duke NC, Field C (2008) Threats to mangroves from climate change and adaptation options: a review. Aquat Bot 89(2):237–250. https://doi.org/10.1016/j.aquabot.2007.12.009

Giri S, Daw TM, Hazra S, Troell M, Samanta S, Basu O, Marcinko CLJ, Chanda A (2022) Economic incentives drive the conversion of agriculture to aquaculture in the Indian Sundarbans: livelihood and environmental implications of different aquaculture types. Ambio 51(9):1963–1977. https://doi.org/10.1007/s13280-022-01720-4

Guha P, Aitch P, Bhandari G (2020) Effect of changing vegetation coverage and meteorological parameters on the hazard characteristics of Indian Sundarban region and its impact there on. In: An interdisciplinary approach for disaster resilience and sustainability. Disaster risk reduction. Springer, Singapore, pp 451–475

Hamilton S (2013) Assessing the role of commercial aquaculture in displacing mangrove forest. Bull Mar Sci 89(2):585–601. https://doi.org/10.5343/bms.2012.1069

Hickey SM, Radford B (2022) Turning the tide on mapping marginal mangroves with multi-dimensional space–time remote sensing. Remote Sens 14(14):3365. https://doi.org/10.3390/rs14143365

Jana R (2019) Climate related disaster vulnerability and livelihood security in Indian Sundarbans. University of Calcutta. https://shodhganga.inflibnet.ac.in:8443/jspui/handle/10603/366241

Koshimura S, Moya L, Mas E, Bai Y (2020) Tsunami damage detection with remote sensing: a review. Geosciences 10(5):177. https://doi.org/10.3390/geosciences10050177

Kumar S (2023) The regional development and planning in Sundarbans: a comprehensive analysis of south 24 Pargana's District; West Bengal, India. Int J Sci Res 12(12):1673–1679. https://doi.org/10.21275/SR231222133857

Macreadie PI, Anton A, Raven JA, Beaumont N, Connolly RM, Friess DA, Kelleway JJ, Kennedy H, Kuwae T, Lavery PS, Lovelock CE, Smale DA, Apostolaki ET, Atwood TB, Baldock J, Bianchi TS, Chmura GL, Eyre BD, Fourqurean JW, Hall-Spencer JM, Huxham M, Hendriks IE, Krause-Jensen D, Laffoley D, Luisetti T et al (2019) The future of blue carbon science. Nat Commun 10(1):3998. https://doi.org/10.1038/s41467-019-11693-w

Macreadie PI, Costa MDP, Atwood TW, Friess DA, Kelleway JJ, Kennedy H, Lovelock CE, Serrano O, Duarte CM (2021) Blue carbon as a natural climate solution. Nat Rev Earth Environ 2:826–839. https://doi.org/10.1038/s43017-021-00224-1

Madramootoo CA, Virgil C (2024) Nature-based solutions for coastal protection in the southern Caribbean. Blue-Green Syst 6(2):185–197. https://doi.org/10.2166/bgs.2024.013

Maiti S, Chowdhury A (2013) Effects of anthropogenic pollution on mangrove biodiversity: a review. J Environ Prot 4:1428–1434. https://doi.org/10.4236/jep.2013.412163

Mandal RN, Bar R (2019) Mangroves for building resilience to climate change. Apple Academic Press

Mandal RN, Naskar KR (2008) Diversity and classification of Indian mangroves: a review. Trop Ecol 49:131–146

Manna S, Mukherjee S, Das D, Saha A (2023) Sustainable management of aquaculture and water footprint analysis in Sunderban. J Clim Change 9(4):67–73. https://doi.org/10.3233/JCC230033

Marois DE, Mitsch WJ (2015) Coastal protection from tsunamis and cyclones provided by mangrove wetlands—a review. Int J Biodiv Sci Ecosyst Serv Manag 11(1):71–83. https://doi.org/10.1080/21513732.2014.997292

McKee KL, Cahoon DR, Feller IC (2007) Caribbean mangroves adjust to rising sea level through biotic controls on change in soil elevation. Glob Ecol Biogeogr 16(5):545–556. https://doi.org/10.1111/j.1466-8238.2007.00317.x

Mcleod E, Chmura G, Bouillon S, Salm R, Björk M, Duarte C, Lovelock C, Schlesinger W, Silliman B (2011) A blueprint for blue carbon: Toward an improved understanding of the role of vegetated coastal habitats in sequestering CO_2. Front Ecol Environ 9:552–560. https://doi.org/10.1890/110004

Mondal S, Tah M, Ankure S, Murmu AK, Naskar S (2025) Perils of the tidal forests: understanding threats to mangroves in the Indian Sundarbans. In: Tripathy B, Raha S, Mondal T (eds) Global warming, climate change & crisis of Sundarban, India. Evincepub Publishing House, Chhattisgarh, pp 62–103

Mondal J (2024) Sustainable development and ethno-inhabitant in mangrove ecology of India: a case study on governance policy in Sundarban. Proc Int Conf Polit Sci 2(1):1–18. https://doi.org/10.33422/politicalsciences.v2i1.593

Naskar K, Mandal R (1999) Ecology and biodiversity of Indian Mangroves. Daya Books

Naskar K (1993) Plant wealth of the lower Ganga delta: an eco-taxonomical approach. Daya Books

Polidoro BA, Carpenter KE, Collins L, Duke NC, Ellison AM, Ellison JC, Farnsworth EJ, Fernando ES, Kathiresan K, Koedam NE, Livingstone SR, Miyagi T, Moore GE, Nam VN, Ong JE, Primavera JH, Iii SGS, Sanciangco JC, Sukardjo S, Wang Y, Yong JWH (2010) The loss of species: mangrove extinction risk and geographic areas of global concern. PLOS One 5(4):1–10, e10095. https://doi.org/10.1371/journal.pone.0010095

Pramanik MK (2015) Changes and status of mangrove habitat in Ganges delta: case study in Indian part of Sundarbans. For Res 4(3):1–7. https://doi.org/10.4172/2168-9776.1000153

Raha AK, Zaman S, Sengupta K, Bhattacharyya SB, Raha S, Banerjee K, Mitra A (2013) Climate change and sustainable livelihood programmes: a case study from Indian Sundarbans. J Ecol Photon 107:335–348

Ray R, Ganguly D, Chowdhury C, Dey M, Das S, Dutta MK, Mandal SK, Majumder N, De TK, Mukhopadhyay SK, Jana TK (2011) Carbon sequestration and annual increase of carbon stock in a mangrove forest. Atmos Environ 45(28):5016–5024. https://doi.org/10.1016/j.atmosenv.2011.04.074

Sarangi R, Kandasamy K, Subramanian A (2002) Metal concentrations in five mangrove species of the Bhitarkanika, Orissa, east coast of India. Indian J Mar Sci 31(3):251–253

Sinha M, Bera N, Bhatta R, Geertz T, Chakraborty S, Dey A, Surathkal P, Majumder N (2024) Coastal aquaculture in the Indian Sundarban: evaluating production economics and farmers perception of mangrove ecosystem benefits. Int J Res Appl Sci Eng Technol 12:1257–1268. https://doi.org/10.22214/ijraset.2024.58535

Smith P, Soussana J-F, Angers D, Schipper L, Chenu C, Rasse DP, Batjes NH, van Egmond F, McNeill S, Kuhnert M, Arias-Navarro C, Olesen JE, Chirinda N, Fornara D, Wollenberg E, Álvaro-Fuentes J, Sanz-Cobena A, Klumpp K (2020) How to measure, report and verify soil carbon change to realize the potential of soil carbon sequestration for atmospheric greenhouse gas removal. Glob Change Biol 26(1):219–241. https://doi.org/10.1111/gcb.14815

Spalding MD, Blasco F, Field CD (eds) (1997) World mangrove Atlas. The International Society for Mangrove Ecosystems, Okinawa, Japan

Spalding MD, Kainuma M, Collins L (2010) World Atlas of mangroves. Routledge

Tiwari T, Kaur GA, Singh PK, Balayan S, Mishra A, Tiwari A (2024) Emerging bio-capture strategies for greenhouse gas reduction: navigating challenges towards carbon neutrality. Sci Total Environ 929:172433. https://doi.org/10.1016/j.scitotenv.2024.172433
Tomlinson PB (2016. The botany of mangroves, 2nd edn. Cambridge University Press
Verbruggen N, Hermans C, Schat H (2009) Molecular mechanisms of metal hyperaccumulation in plants. New Phytol 181(4):759–776. https://doi.org/10.1111/j.1469-8137.2008.02748.x

5 Livelihood Challenges and Governance Dynamics in the Indian Sundarbans Delta

Swagata Patra and Monirul Haque

Abstract

The Indian Sundarbans Delta (ISD), situated within the ecologically rich and geopolitically sensitive Ganges-Brahmaputra-Meghna basin, is home to nearly five million people whose lives are deeply entwined with a dynamic but increasingly fragile ecosystem. This region, known for its dense mangrove forests and complex tidal networks, faces mounting threats from climate change, including rising sea levels, frequent and severe cyclones, and increasing soil and water salinity. These environmental stressors have severely impacted traditional livelihoods, *viz.*, agriculture, artisanal fishing, and forest-based activities, forcing many communities into economic precarity. Vulnerability is especially acute in southwestern islands such as Sagar, Mousani, and Ghoramara, where land erosion, saline intrusion, and flooding are recurrent, and infrastructure and institutional support remain weak. While local communities have shown remarkable resilience through adaptive strategies like seasonal migration, crop diversification, and self-help groups, these responses often emerge without robust governance and long-term support. Fragmented institutional frameworks, inadequate coordination across agencies, and a lack of participatory governance have left many of the region's most vulnerable populations without adequate protection or opportunity for recovery. Despite growing recognition of these challenges, there remains a research gap

S. Patra
Department of Agricultural Extension, Palli-Siksha Bhavana (Institute of Agriculture), Visva-Bharati, Sriniketan 731236, West Bengal, India

M. Haque (✉)
Department of Biological Sciences (Agriculture), Midnapore City College, Midnapore 721129, West Bengal, India
e-mail: monirulhaque@mcconline.org.in

C. Singha et al. (eds.), *Sustainable Management of Delta Ecosystems Resilience*, Deltas of the World, https://doi.org/10.1007/978-3-032-16261-8_5

in understanding how integrated governance and community-driven approaches can be systematically implemented to enhance adaptive capacity. This chapter examines the intersecting challenges of climate vulnerability, livelihood insecurity, and governance inefficiency in the ISD. It also explores the potential for inclusive policy reforms such as decentralised planning, improved inter-agency collaboration, and ecosystem-based adaptation to support sustainable, community-driven development. In doing so, it advocates for a shift from reactive relief to proactive resilience, grounded in the lived experiences and agency of the people who call the Sundarbans home.

Nomenclature and Acronyms

AI	Artificial intelligence
BOB	Bay of Bengal
C.D.	Community development
CBMM	Community Based Co-management
CBOs	Community Based Organization
CRZ	Coastal regulation zone
FRA	The forest right act
GADF	Generic adoption decision framework
ICZM	Integrated Coastal Zone Management
IFS	Integrated farming system
IPCC	Intergovernmental panel on climate change
ISD	Indian Sundarbans Delta
ITKs	Indigenous technical knowledge
JFM	Joint Forest Management
JMM	Joint Mangrove Management
JMMCs	Joint Mangrove Management Committees
LVI	Livelihood vulnerability index
NBS	Nature-based solutions
NGOs	Non-Governmental Organizations
ppt	Parts per thousands
RBT	Royal Bengal Tiger
SHGs	Self help groups
SST	Sea surface temperatures
UNESCO	United Nations Educational, Scientific and Cultural Organization

5.1 Introduction

ISD is recognised as a UNESCO World Heritage Site and one of the world's most critical biodiversity hotspots, forming an integral socio-ecological region at the confluence of the lower Ganges-Brahmaputra-Meghna basin. This deltaic landscape, marked by a complex network of tidal rivers, estuaries, and dense mangrove forests, supports the livelihoods of approximately five million people. Most of this population inhabits low-lying islands and relies heavily on climate-sensitive economic activities such as agriculture, artisanal fishing, and forest-based resource collection.

ISD stands out as particularly vulnerable among the various segments of the larger Sundarbans region. Several studies, including those by Hajra et al. (2021), have employed composite indices to assess coastal vulnerability across blocks within the ISD, assigning vulnerability ranks based on exposure to environmental hazards. These assessments underscore the acute risks posed by soil salinisation, riverbank erosion, and loss of natural ecosystem factors that have repeatedly disrupted local livelihood structures and exacerbated food insecurity (Hajra and Ghosh 2018).

Study from a similar standpoint depicts that integrating scientific knowledge with indigenous and local practices significantly enhances climate resilience in wetland regions, particularly in South Asia's deltaic areas such as the Sundarbans (Haque et al. 2022). From a global perspective, ecosystem-based adaptation combined with local livelihood strategies is crucial to improve food security and reduce poverty under the pressures of climate change (Munang et al. 2014). Governance frameworks that promote community participation and empower local groups align conservation goals with livelihood needs, fostering transparency and inclusive decision-making, as demonstrated in coastal wetland management worldwide (Berkes 2017). The increasing impacts of sea-level rise and intensifying storms place immense pressure on deltaic wetlands, emphasising the need for governance reforms that ensure equitable resource access and bolster disaster preparedness (Reid et al. 2019). In the Ganges Delta, decentralised and community-driven resource management has enhanced institutional capacity and resilience, illustrating the critical role of local engagement in sustaining ecosystem services (Paul et al. 2021).

The challenges have intensified over recent decades, driven largely by climatic and anthropogenic pressures. Recurrent tropical cyclones, rising sea levels, saline water intrusion, and ongoing riverbank erosion have significantly undermined the agrarian and aquatic production systems that sustain the region. Particularly severe events such as Cyclones Sidr (2007) and Aila (2009) impacted around 2.3 million people in the ISD (Hazra et al. 2014). More recently, Super Cyclone Amphan caused widespread livelihood disruption across the region (Basu 2020; Pramanik et al. 2021; Mondal et al. 2022).

The southwestern islands of Sagar, Mousani, and Ghoramara have emerged as focal points of concern due to their high population densities and elevated levels of vulnerability (Hajra et al., 2021). These islands exhibit a complex geomorphological setup shaped by a dynamic tidal drainage system and the continuous processes of erosion and accretion.

The area is particularly susceptible to flooding and saline intrusion, situated on a flat alluvial plain with ground elevations ranging from 2.10 to 2.75 meters above mean sea level (Purkait 2008).

Climate change-induced phenomena such as rising sea levels, altered precipitation patterns, and the increasing frequency and intensity of extreme weather events have significantly impacted the socio-economic stability of island communities (Mullick et al. 2019). Cyclonic storm surges have frequently breached embankments, allowing saline water to inundate agricultural lands, leading to widespread crop failures and deepening rural poverty (Chand et al. 2012; Hajra and Ghosh 2018). In Ghoramara, land loss due to submergence has become a particularly severe threat, with significant tracts of agricultural land disappearing, compelling local populations to abandon traditional livelihoods and migrate (Mondal et al. 2022). These intersecting climatic and socio-economic stressors disproportionately affect economically marginalised groups, increasing their vulnerability to immediate disasters and long-term socio-ecological transformations (Jamal et al. 2022).

Compounding these environmental challenges are structural governance issues. The institutional framework managing the ISD is highly fragmented, with overlapping jurisdictions and poor inter-agency coordination. This fragmented governance structure often lacks adaptive capacity and participatory mechanisms, leading to insufficient support for sustainable livelihoods and local resilience. Critical gaps remain in disaster preparedness, climate-resilient agricultural extension, land-use planning, and natural resource management. Consequently, local communities remain exposed to various socio-ecological risks without adequate institutional support.

ISD presents notable strengths and challenges that shape livelihood and governance dynamics in the region. Substantial financial support from international sources and government initiatives such as the two trillion Indian rupee investment during India's twelfth five-year plan is crucial in advancing climate adaptation and sustainable development efforts (Mahadevia and Vikas 2012). Improved connectivity through affordable river transport and enhanced road networks facilitate access and communication, which is vital for local livelihoods and effective governance (Padhy et al. 2021). Furthermore, the region's rich biodiversity, including endangered species like the Royal Bengal tiger and saltwater crocodile, forms an ecological and cultural foundation that influences community practices and risk mitigation strategies (Jamal et al. 2022). However, the delta also faces significant vulnerabilities. Frequent cyclones and flash floods result in salinisation of land, displacement of residents, and damage to agriculture and ecosystems, while sudden onset of disasters limits effective relief operations (Jamal et al. 2022). Additionally, improper fertiliser use contributes to soil degradation and increased greenhouse gas emissions, threatening agricultural sustainability (Padhy et al. 2021). The overexploitation of prawn seeds disrupts aquatic biodiversity and compromises local food security (Chandra and Sagar 2003). Weak enforcement of environmental regulations and fragmented policy frameworks further hinder mangrove conservation and restoration, exacerbating the region's ecological fragility (Akram et al. 2023). Addressing these intertwined strengths

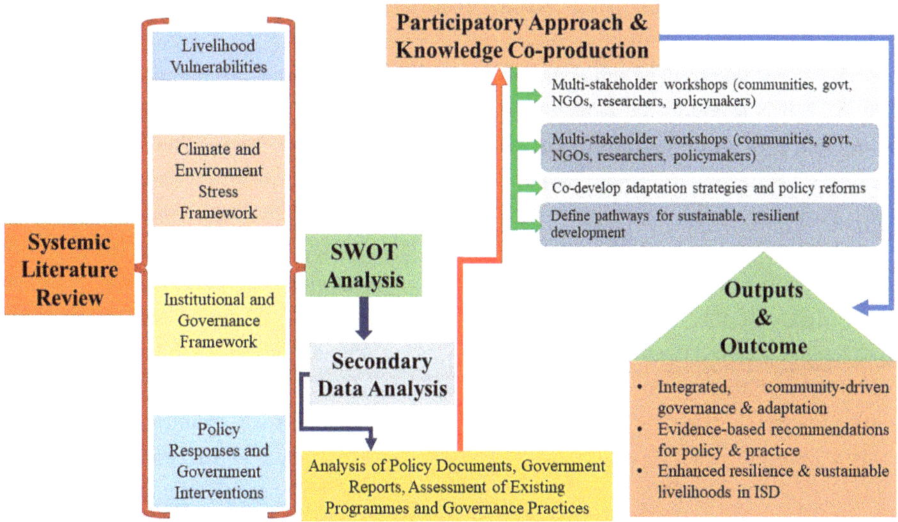

Fig. 5.1 Methodological framework of the study

and weaknesses is essential for developing robust, adaptive governance mechanisms that support resilient livelihoods in the Sundarbans.

This study aims to critically examine the complex interplay (Fig. 5.1) between climate change impacts, livelihood vulnerabilities, and governance shortcomings in ISD. It seeks to understand how the compounding effects of sea-level rise intensified institutional inefficiencies exacerbate cyclonic activity and land salinisation, ultimately deepening socio-economic marginalisation. Special attention is given to rural households whose economic security is intricately tied to natural resource-based livelihoods.

The research further evaluates existing governance frameworks for their capacity to support climate-resilient development, equitable resource access, and inclusive planning processes. By identifying institutional bottlenecks and policy deficits, the study proposes targeted reforms such as decentralised governance, improved inter-agency collaboration, enhanced community participation, and increased investment in ecosystem-based adaptation approaches.

Moreover, the study highlights local communities' adaptive strategies in response to escalating environmental stressors. These include seasonal and permanent migration, diversification of income sources, cultivation of saline-tolerant crops, and the formation of self-help and community support groups. While these grassroots adaptations underscore the resilience and agency of local populations, they also reveal the urgent need for systemic interventions in the form of supportive policies, infrastructural development, and robust institutional frameworks.

5.2 Geographical and Socio-Economic Context of the ISD

It is a unique geographical and ecological entity in the Ganges-Brahmaputra deltaic region (Fig. 5.2). It is home to the world's largest mangrove forest, which supports a rich biodiversity and provides critical ecosystem services (Bharati et al. 2016). The delta's dynamic riverine system, including the Ganges River's distributaries, shape its physical characteristics. However, these rivers have become increasingly disconnected from their upstream sources, resulting in acute freshwater scarcity during dry seasons and elevated surface water and soil salinity (Uddin et al. 2024). This hydrological alteration has had profound implications for the region's ecological integrity, adversely affecting mangrove health, biodiversity, and the overall ecosystem services on which millions of inhabitants depend (Akram et al. 2023).

The lower Gangetic Delta, often referred to as ISD, comprises an intricate archipelago of 103 low-lying, densely populated deltaic islands situated at the confluence of the Ganges-Brahmaputra-Meghna river system and the Bay of Bengal (BOB). Geographically, it is bounded between 21°32′ and 22°40′ N latitude and 88°05′ to 89°00′ E longitude, covering nearly 10,000 km^2 within the Indian state of West Bengal. Historically dominated by dense mangrove cover, the delta has witnessed extensive anthropogenic transformations, with over 54 islands reclaimed for agriculture by constructing approximately 3500 km of earthen embankments to resist daily tidal incursions (SBR 2008). Today, the region presents two distinct landscapes: a protected eastern sector dominated by mangrove forests and a western sector characterised by reclaimed agricultural land and rural settlements. About 2482.33 $km^2$of this reclaimed land is used for coastal agriculture, with rice as the principal crop. According to official estimates, 60–65% of the 4.37 million inhabitants are directly engaged in agricultural activities, while up to 95% may indirectly depend on it (Bagchi 2017). Nonetheless, landlessness is prevalent, with 54% of households lacking ownership, contributing to a substantial population of agricultural labourers. In addition to rice, cash crops such as chilli, cotton, watermelon, cucumber, and betel leaves are cultivated.

Demographically, the delta supports nearly five million people whose livelihoods are intricately linked to ecosystem services (Prasanya et al. 2024). These include artisanal fishing, crab and prawn seed collection, aquaculture, and the extraction of fuel, fodder, and honey from the forest (Mitra 2023). Despite their socio-economic importance, these services often remain underrepresented in official statistics and are declining due to population pressures and ecosystem degradation (Vishwapriya and Devaiah 2024). The rural economy relies heavily on natural resources, rendering it highly susceptible to environmental changes and governance deficits.

The broader Gangetic Delta region, which encompasses the ISD, is among the most densely populated areas in the world, with a population exceeding 400 million (Ghosh 2011). The ISD, home to approximately 4.5 million people, is particularly vulnerable to recurrent tropical cyclones, storm surges, and associated flooding. Rising sea surface

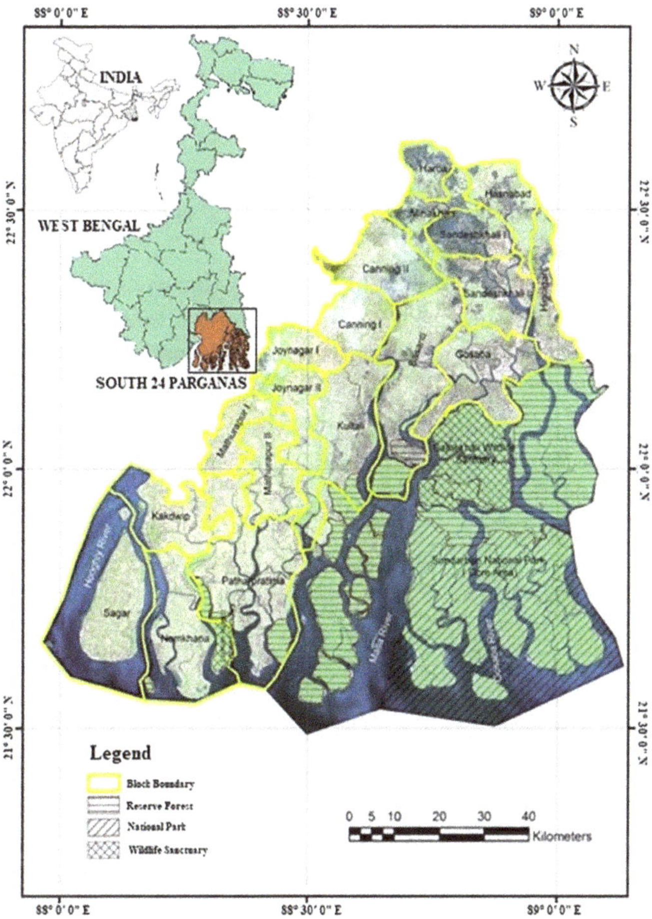

Fig. 5.2 Map of ISD

temperatures in the Bay of Bengal, increasing at 0.5 °C per decade, have created ideal conditions for cyclone formation (Yadav et al. 2024). There has been a 17% increase in the intensification of cyclonic disturbances into cyclones and a 25% rise in severe cyclonic storms over the northern Indian Ocean, especially during the pre-monsoon and post-monsoon periods (Deshpande et al. 2021). Recent cyclones, including the 2007 event named "Sidr," the 2009 cyclone "Aila," the 2013 cyclone "Phailin," the 2019 cyclone "Bulbul," the 2020 cyclone "Amphan," and the 2021 cyclone "Yaas," have severely disrupted livelihoods, damaging access to food, water, shelter, and health services. Rising soil salinity and waterlogging in several riverside Community Development (C.D.) Blocks have significantly affected agricultural productivity, resulting in a 30–40% decline in yield potential (Ghosh and Mistri 2020). These impacts have exacerbated food insecurity, unemployment, poverty, and migration (Mondal et al. 2022). The entrenched poverty cycle, strongly linked with agricultural decline and outmigration, has persisted for decades (Hajra and Ghosh 2018). Over 34% of the ISD's population still lives below the poverty line, and approximately 75% of households have at least one member who has migrated outside the region to sustain basic livelihoods (Hazra and Ghosh 2018). These socio-economic pressures have considerably undermined local coping capacities and increased vulnerability to environmental stressors.

Given the projected trajectory under a business-as-usual scenario, ecosystem provisioning and regulating services are expected to deteriorate further by 2030, intensifying rural poverty (Hussain et al. 2024). Addressing the sustainability of this complex socio-ecological system demands integrated strategies that target both environmental restoration and livelihood enhancement. Key biophysical constraints such as water stress, siltation, and drainage congestion aggravated by the disruption of freshwater flows from upstream must be mitigated (Haque and Nehar 2023).

5.3 Livelihood Vulnerabilities in the Sundarbans

Livelihoods in the ISD are highly vulnerable to various environmental stressors, significantly undermining local communities' resilience. Among the most pressing challenges are salinity intrusion, riverbank erosion, and frequent cyclones. Rising salinity levels driven largely by reduced freshwater inflow from upstream sources have led to the degradation of agricultural soils and a marked decline in productivity (Sarkar et al. 2024). This growing salinisation of soil and water resources has forced many farmers to either shift to salt-tolerant crops or adopt unsustainable practices that are often financially burdensome (Granados-Olivas et al. 2023). Erosion, particularly along the delta's riverbanks and coastlines, continues to reduce the available land for cultivation and habitation, intensifying economic insecurity and displacement risks (Barua et al. 2020). Recurrent cyclones amplify these vulnerabilities by causing widespread destruction of infrastructure, loss of standing crops, and prolonged disruptions to fishing and other livelihood activities (Mukherjee and Siddique 2024).

In the 1980s, livelihood vulnerability emerged to describe how people navigate and survive under adverse socio-environmental conditions (Scoones 2009). Livelihoods are the means and capabilities people rely on to satisfy essential needs such as food, shelter, clothing, health, and social belonging. Meeting these needs requires access to assets and forms of capital, both tangible and intangible (Gaillard et al. 2009). Vulnerability in this context refers to the degree to which individuals or systems cannot withstand and recover from the impacts of climate change and natural hazards (IPCC 2001). Coastal disasters often disrupt the foundations of livelihood systems, damaging income sources, personal property, community infrastructure, and social networks.

The IPCC framework disaggregates vulnerability into three core dimensions: exposure, sensitivity, and adaptive capacity (Yang et al. 2021). These dimensions are critical in evaluating the scope and intensity of damage and identifying intervention areas for achieving sustainable livelihoods. Based on this framework, tools such as the Livelihood Vulnerability Index (LVI) have been developed to assess the susceptibility of households to risks such as poverty, food insecurity, and environmental shocks (Hahn et al. 2009; Mukherjee et al. 2019). Households in similar geographical settings may display varying levels of resilience due to differences in resources, social capital, and institutional access (Cardona and Barbat 2000). These differences often manifest through distinct coping strategies, resistance thresholds, and adaptive capabilities (Buckle et al. 2001). Understanding how vulnerability is distributed across space and social groups is crucial before initiating any policy or planning interventions. It requires systematic documentation and household-level data collection to support local disaster risk reduction (Quader et al. 2021). Although government agencies often prioritise reconstruction and rehabilitation after disasters, the extent of recovery also depends on how effectively individuals and communities can adapt and respond to evolving risks. Strengthening these coping capacities remains essential for building resilience and ensuring more secure and sustainable livelihoods (Pomeroy et al. 2006).

Economic insecurity in the ISD is shaped by the dominance of informal, low-income livelihoods such as subsistence farming and artisanal fishing, which remain highly sensitive to climate variability and ecological stress (Kaschek 2021). Limited employment opportunities, poor access to financial resources, and weak service delivery systems contribute to persistent poverty and reduced well-being. As environmental degradation continues to erode traditional livelihood bases, migration has become a necessary strategy for survival. Many younger members of households migrate to urban centres or other regions in search of wage labour, often leaving behind dependents who face additional emotional and economic strain.

Human-wildlife conflict and restricted access to forest-based resources also impose significant livelihood limitations. Habitat loss from salinity and erosion has increased encounters with species such as tigers and crocodiles, often resulting in deadly conflict and long-term fear within communities. Policies aimed at biodiversity conservation frequently limit access to fuelwood, fodder, and other essential forest products, which forces households to seek alternative sources that may not be ecologically or economically viable. Managing ISD as a socio-ecological system involves difficult trade-offs between ecological integrity and human development. Achieving sustainability in this region depends on integrating conservation goals with the livelihood needs of its most vulnerable residents, recognising that resilience cannot be built without ensuring social and economic justice.

5.4 Climate Change and Environmental Stress in the Delta

Climate change poses significant environmental stress in ISD through several critical factors. One of the most pressing concerns is rising sea levels, which exacerbate tidal flooding and saltwater intrusion into freshwater sources and agricultural lands, threatening the region's biodiversity and affecting local communities' livelihoods by reducing the availability of arable land and freshwater resources. ISD, often termed the "biowall" of the coastal region, is particularly vulnerable to climate-induced changes. Recent studies show that South Asian countries, including India, are especially at risk from the impacts of climate change (Hijioka et al. 2014). India was ranked third in the Global Climate Risk Index in 2013, highlighting its population's vulnerability (Kreft et al. 2014).

Air temperatures over the Bay of Bengal have risen by 0.019 °C annually between 1985 and 2000, and projections indicate an increase of 1 °C by 2050 if current trends persist (Hazra et al. 2002). A more accelerated warming pattern was observed between 1993 and 2007 compared to earlier decades (Hazra et al. 2010). Sea surface temperatures (SST) in ISD have also increased notably, rising by 0.5 °C per decade during the pre-monsoon periods between 1980 and 2007 (Mitra et al. 2009). This warming trend has been linked with shifts in rainfall patterns, with monsoon and post-monsoon rainfall decreasing while pre-monsoon rainfall has increased slightly (Mandal et al. 2013). The changing rainfall regime has resulted in delayed monsoons, erratic rains during harvest periods, and rising vulnerability to crop damage and pest attacks (Rao and Filho 2015).

Salinity intrusion has become a significant threat across the delta, with the eastern Sundarbans recording an increase of 6 ppt over three decades, a rate much higher than the average across the Indian Ocean (Mitra et al. 2009). The northern areas, traditionally less saline, now report levels as high as eight ppt, while the southern parts experience salinity between 8 and 20 ppt. Soil salinity has sharply increased in many areas from 0.1 to 5.25 ppt, well beyond the safe thresholds for staple crops like rice (Haldar and Debnath 2014).

The increased frequency and intensity of cyclones further strain the region. The Bay of Bengal, though contributing to less than 6% of global tropical cyclones, accounts for the majority of the fatal ones, with ISD experiencing severe impacts (World Bank 2014). Cyclones like Aila in 2009 affected millions, caused over 300 deaths and led to massive economic losses exceeding $550 million in India alone (Rudra 2010). The rise of sea level in the region is also notable. From 1991 to 1999, ISD and adjoining Bay of Bengal saw a relative sea-level rise of 3.14 mm/year (Hazra et al. 2002). Tidal gauge data from the last 25 years across four locations in the delta suggest an even more alarming rate of 8 mm/year, with spikes of 12 mm/year between 2002 and 2009 (Hazra et al. 2010). This rise far exceeds the global average reported by the IPCC (2014), which estimated an increase of 3.2 mm/year from 1993 to 2010.

Erosion remains a significant concern. The region has experienced a land loss of 44 km^2 between 2001 and 2009, and about four islands, Bedford, Lohachara, Kabasgadi, and Suparibhanga, have been entirely submerged due to rising sea levels and erosion, displacing around 6000 families (Hazra and Bakshi 2003). Embankments, which play a crucial role in protecting against storm surges and sea level rise, are also under stress. A large proportion of the 3638 km-long embankment network in ISD is in a vulnerable state, with the length of such stretches increasing from 470 km before Cyclone Aila to around 1000 km in its aftermath (Hazra et al. 2010; World Bank 2014).

The impact of these environmental stressors on agriculture is particularly severe. Rising salinity levels and increased flooding make it difficult to maintain traditional farming practices, significantly reducing crop yields and agricultural productivity (Kumar et al. 2021). Farmers must adopt high-cost and often unsustainable strategies to sustain production. Fishery-based livelihoods are also disrupted due to water temperature and salinity changes, which alter fish populations and habitats (Okafor-Yarwood et al. 2023). Meanwhile, erosion and tidal flooding threaten residential areas, heightening displacement risks and social vulnerabilities (Harms 2024).

These cascading stressors position ISD as a climate hotspot with acute socio-ecological vulnerabilities. Despite the presence of adaptation practices, many of them fail to address the complex interplay of environmental and livelihood challenges in the region. Effective responses require systemic and inclusive adaptation strategies informed by local knowledge, adaptive governance, and sustainable environmental management (Kumar et al. 2021). Balancing ecological conservation with livelihood security is vital to safeguarding the future of the Sundarbans.

5.5 Institutional and Governance Framework

Institutional and governance frameworks are crucial in managing ISD, particularly in forest and coastal zone governance. The Sundarbans, a unique ecosystem with mangrove forests and diverse wildlife, require a governance structure that balances conservation with the needs of local communities. Forest governance involves managing mangrove forests, critical for biodiversity conservation, and providing ecosystem services such as coastal protection and carbon sequestration. However, the effectiveness of these regulations is often challenged by inadequate enforcement and conflicting interests between conservation and economic development (Kumar et al. 2021).

Coastal zone governance is equally important, as it involves managing the impacts of sea-level rise, erosion, and cyclones on coastal communities and ecosystems (Duxbury and Dickinson 2007). These require coordinated efforts to implement climate-resilient infrastructure and sustainable land-use practices. The governance of ISD involves multiple levels of authority, including state, central, and local governments. The central government plays a key role in setting national environmental conservation and disaster management policies, while state governments are responsible for implementing these policies at the regional level. Local authorities, such as panchayats and municipalities, are crucial for engaging with local communities and ensuring that governance decisions reflect local needs and priorities.

However, there is often a disconnect between policy intent and implementation, with local authorities facing challenges related to resource constraints and bureaucratic inefficiencies. Non-governmental organisations (NGOs) and civil society actors play a vital role in delta management by advocating for community rights, supporting sustainable livelihoods, and promoting environmental conservation. These organisations often work closely with local communities to develop and implement projects that enhance resilience to climate change impacts. Additionally, NGOs can facilitate stakeholder engagement and participatory governance processes, which are essential for ensuring that governance decisions are inclusive and responsive to local needs (Kumar et al. 2021).

The effectiveness of NGOs can be limited by funding constraints and the need for greater coordination with government agencies to ensure policy alignment and impact. Overall, effective governance in the ISD requires a collaborative approach that integrates the roles of state, central, and local authorities with those of NGOs and civil society actors. This collaboration is crucial for addressing the complex socio-ecological challenges faced by the region and ensuring sustainable development that balances human well-being with environmental conservation. By fostering such partnerships, it is possible to develop governance frameworks responsive to the needs of both the environment and local communities, ultimately enhancing the resilience of ISD in the face of climate change and other environmental stressors.

5.6 Livelihood Transitions and Emerging Opportunities

ISD, home to one of the World's largest mangrove ecosystems, has traditionally supported various livelihoods closely linked to its unique natural environment (Gopal and Chauhan 2006). For generations, communities in this region have depended on agriculture, fishing, and forest-based activities such as collecting fuelwood, honey, medicinal plants, and other forest products for survival. These activities have not only sustained local households but have also shaped the cultural and ecological identity of the region. However, the changing environmental and socio-economic conditions pose serious challenges to these traditional ways of living (Mahmood et al. 2021).

ISD has recently faced intensified environmental stress, including rising salinity, increasing coastal erosion, frequent cyclones, and tidal surges (Hasnine and Nagdeve 2024). These climatic pressures are steadily degrading the land and water resources on which local livelihoods depend. Due to saltwater intrusion, agricultural land is becoming less fertile, and freshwater sources are shrinking. At the same time, fishing and crab collection are becoming more unpredictable, often due to ecological degradation and overharvesting (Chandra and Sagar 2003). The region's limited infrastructure, poor road connectivity, irregular electricity, insufficient healthcare, and underdeveloped education facilities adds to the hardship. Most families have small landholdings or work as agricultural labourers, and stable, year-round employment opportunities are scarce. In this context, migration has emerged as a key livelihood strategy. People, particularly young men, increasingly move out of ISD temporarily or permanently to find work in other parts of the state or country. Many work as wage labourers in construction sites, brick kilns, or small factories. While these jobs are often low-paying and physically demanding, they provide a crucial source of income that many families in ISD have come to rely on. The money sent back by migrants to their family members plays a vital role in the local economy. These funds meet daily household expenses, support children's education, access better healthcare, and sometimes even invest in small businesses or housing improvements. Remittances often buffer against the financial uncertainty caused by crop failure or poor fishing seasons. As a result, migration is no longer just a response to crisis but becoming a deliberate strategy for survival and future planning. This shift towards migration has also brought about social changes. While traditional occupations still hold cultural importance, many younger people aspire to move beyond land and resource-based livelihood limitations. Migration has created new connections between rural and urban areas, exposing families to new ideas, opportunities, and ways of life. At the same time, it has placed emotional and social stress on families, especially those left behind. Nonetheless, migration and remittances have become essential to how households in ISD adapt to a changing environment (Moreda 2023).

Aquaculture has emerged as a viable substitute for conventional fishing, particularly in light of ecological degradation. Alongside this, eco-tourism creates job opportunities by leveraging the region's biodiversity. Wage labour in urban construction and manufacturing sectors is also expanding, especially among migrants, offering a new channel of income diversification. These shifts reduce reliance on ecologically vulnerable activities and lay the groundwork for more sustainable development. Migratory movement, particularly male labour outflow, remains a significant coping mechanism in response to declining rural opportunities. While this reduces local economic strain and brings in remittances that modernise agriculture, raise land value, and stimulate rural entrepreneurship, it also creates labour shortages and strains traditional occupations such as farming on saline lands.

Women's cooperatives and self-help groups are vital in this transformative phase. Traditionally confined to domestic roles, many women now participate in micro-enterprises like small-scale aquaculture, tourism, and value-added production. Cooperatives have facilitated access to training, financial credit, and market tools essential for empowering women and sustaining livelihood ventures. As male outmigration continues, women increasingly shoulder familial and financial responsibilities, gaining confidence and societal visibility. Recent trends also show a rise in female outmigration, driven by aspirations for higher education and formal employment, marking a shift in gender dynamics.

Several underutilised natural resources offer further potential. Honey and wax, harvested from wild and box bees, yield hundreds of quintals annually, yet collectors receive limited returns due to a lack of local processing. Establishing honey processing and candy-making units can enhance income by adding value locally. Similarly, mangrove plants like *Ceriops decandra* produce tannins in fisheries and leather industries. At the same time, various species offer natural dyes, medicinal properties, gums, resins, and edible products like wild fruits, tender leaves, and mushrooms. Reviving these traditional resources through controlled harvesting and local processing can revitalise grassroots economies.

5.7 Policy Responses and Development Interventions

The Sundarban Delta, a region of critical ecological significance, is increasingly vulnerable to the intersecting threats of climate change, environmental degradation, and socio-economic fragility. In response to these challenges, the Indian government has introduced several policy frameworks to safeguard the region's natural ecosystems while promoting sustainable development in coastal and deltaic zones. One key legislative tool is the Coastal Regulation Zone (CRZ) Notification of 2018, a revision of the earlier 1991 notification issued under the Environment Protection Act of 1986. This regulation governs developmental activities along the coastline by classifying areas into four categories: CRZ-I (ecologically sensitive regions), CRZ-II (urbanised coastal areas), CRZ-III (rural settlements), and CRZ-IV (islands) (Table 5.1). These zones are delineated based on environmental sensitivity, geomorphology, and demographic features, aiming to protect coastal biodiversity while permitting regulated development (Prasad 2020).

Table 5.1 Different level Governance body of ISD

Governance	Institutional body	Policy stakeholders
Global	International bodies	UNESCO International union for conservation of nature (IUCN)
National	Central government	Ministry of environment, forest and climate change national biodiversity authority
Regional	Statement government	Department of environment, government of West Bengal sundarban affairs department West Bengal biodiversity board
Territorial	Local bodies	Gram panchayats (village councils) Local forest committees—JFMC, FPC Irrigation department
Endemic/ tropical	Non-governmental organisations (NGOs)	Various national and international NGOs work on conservation, climate adaptation, and community development

Complementing the CRZ framework, the National Conservation Strategy and Policy Statement on Environment and Development of 1992 emphasises the importance of conserving wetlands, particularly mangrove ecosystems (Mohanty et al. 2024). It recommends decentralised planning mechanisms to prevent resource depletion, contamination, and the adverse impacts of natural disasters. Similarly, the National Environment Policy of 2006 calls for an Integrated Coastal Zone Management (ICZM) approach that encourages collaborative governance models to align ecological preservation with economic advancement (Frost, 2009).

The Forest Rights Act (FRA) of 2006 is pivotal in empowering forest-dependent communities, especially indigenous groups. It acknowledges their traditional rights over forest lands and resources, promoting their active participation in conservation and resource stewardship even within protected areas like wildlife sanctuaries and national parks. This legislation fosters a more equitable framework for environmental governance by addressing historical exclusions (Das 2024).

Mangrove restoration initiatives in ISD have gained momentum under these policies, especially in light of increasing climate-related threats. Between 2020 and 2022, the Indian Sundarban initiative facilitated the plantation of more than 71 million mangrove species and 52 million mangrove associates. These efforts were intended to reinforce coastal embankments and improve the region's resilience to extreme weather events. Despite a substantial financial investment of approximately 1.23 billion Indian rupees, the program encountered operational setbacks, including delayed wage disbursements and insufficient community involvement. These issues led to reduced local engagement and diminished trust. Moreover, although the mangrove ecosystem provides critical services such as carbon storage, erosion control, and biodiversity conservation valued at nearly USD 19 billion annually, the lack of communication about these benefits has further limited grassroots participation.

To enhance the long-term viability of such restoration projects, linking them with carbon credit markets could generate financial incentives for local stakeholders. Equally important is the need for effective dialogue between scientists, policymakers, and communities to set pragmatic expectations around ecological recovery timelines. Scientific input during the planning stages is vital for outlining realistic benchmarks for the growth and environmental functionality of mangrove plantations. The misalignment between scientific timelines and public expectations was evident during Cyclone Yaas when newly planted mangroves were prematurely expected to serve as effective natural buffers.

The success of future restoration programs will also depend on sustained resource planning and inter-agency cooperation. Engaging stakeholders, including community members, throughout all project development and implementation stages, is crucial. Recognising this, the Government of West Bengal has recently engaged academic institutions to conduct independent assessments and facilitate participatory research, signalling a more inclusive and evidence-based approach to coastal management. A collaborative framework that brings together local knowledge, scientific expertise, and policy instruments is essential to ensure the durability and effectiveness of ecological interventions.

Nevertheless, challenges persist, particularly at the implementation level. One of the major obstacles is the persistent gap between policy design and on the ground execution. While national policies often present well-structured solutions for disaster resilience and environmental protection, their actual application is hindered by logistical constraints and bureaucratic inefficiencies. Moreover, insufficient public engagement in the planning phases leads to interventions that often fail to align with the specific needs of local populations, exacerbating existing vulnerabilities.

Emergency relief efforts following natural disasters like cyclones are indispensable for short-term recovery. However, they rarely translate into strategies for building long-term resilience. Likewise, development schemes intended to improve livelihoods—such as aquaculture and eco-tourism face constraints in terms of funding, infrastructure, and access to markets. Without sustained support, these initiatives may fall short of their intended goals.

Emerging technologies and innovative governance practices offer promising avenues to address these gaps. Artificial intelligence (AI) and remote sensing can significantly enhance disaster preparedness and environmental monitoring. Yet, these technologies require reliable infrastructure and capable institutions, often underdeveloped in remote regions like ISD. Additionally, technological adoption must be inclusive and reflective of community needs, thereby underscoring the importance of participatory governance frameworks.

The history of forest policy in ISD offers insights into the evolution of conservation approaches. During colonial rule, sections of the delta were declared "protected forests" in 1878 to maximise timber extraction, significantly curtailing the rights of local inhabitants. This designation facilitated the conversion of forest land for agricultural and commercial purposes. By 1943, these areas were upgraded to "reserved forests," with even more stringent restrictions on resource use.

Following India's independence, the emphasis on conservation intensified. The Sundarbans National Park was declared a UNESCO World Heritage Site in 1987, reinforcing its ecological significance. The National Forest Policy of 1988 introduced a shift toward community participation, laying the groundwork for Joint Forest Management (JFM) programs launched in 1990. These initiatives attempted to balance conservation objectives and local livelihood needs.

Further progress was made through enacting the FRA in 2006, which sought to rectify historical exclusions by formally recognising the rights of tribal and forest-dwelling communities. Despite this, many beneficiaries remain unaware of their entitlements, and bureaucratic red tape impedes meaningful participation.

In recent years, community-led organisations have actively bridged these governance gaps. Projects like "The Community Forest," spearheaded by the Kolkata Society for Cultural Heritage, involve over 3500 families in mangrove restoration and sustainable livelihood generation. Similarly, Mukti's "Green Defence" program empowers women through self-help groups to take part in mangrove plantation activities, thereby enhancing both environmental and socio-economic resilience.

Education and outreach also play a key role in fostering environmental stewardship. Initiatives like the Environment Academy incorporate creative methods such as music, drama, and visual arts to instil ecological awareness among youth. These programs emphasise the value of combining indigenous knowledge systems with scientific understanding to promote sustainable management of fragile ecosystems.

5.8 Adaptive Strategies and Local Resilience

Adaptation strategies are defined as purposeful actions that assist human and ecological systems in coping with actual or anticipated changes, especially those induced by climate variability and extremes (Lawler et al. 2013). In the ISD, where rising sea levels, frequent cyclones, saline intrusion, and livelihood disruptions are increasingly common, formulating adaptive strategies has become essential. In light of these challenges, policymakers are exploring the Generic Adaptation Decision Framework (GADF) to rationalise decisions between localised resilience measures and managed retreats (Danda et al. 2019). The GADF guides choices based on a comparative economic viability analysis, environmental impact, and socio-political acceptability (Fig. 5.3).

The decision to adopt localised resilience measures is supported when socio-economic well-being can be sustained at a reasonable cost. If, however, the cost of protection outweighs the value of what is being preserved, and if managed retreat offers better long-term returns (net benefits of ex-situ adaptation), it may justify population resettlement

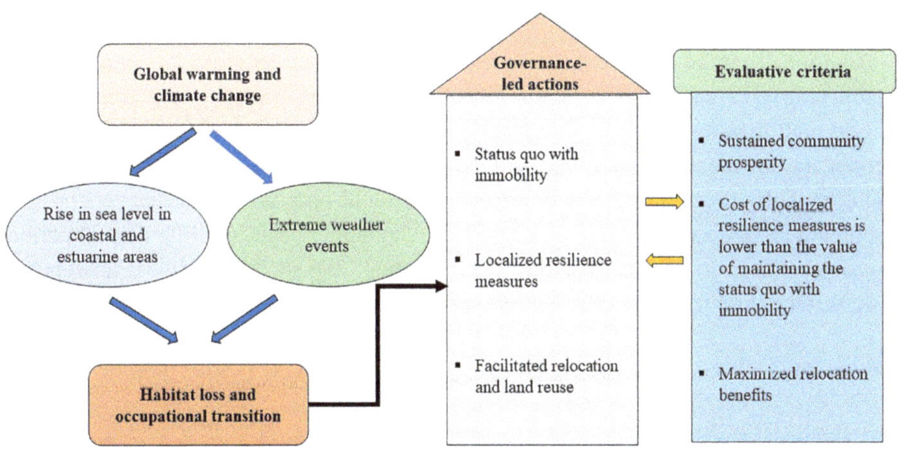

Fig. 5.3 Generic adaptation decision framework

and policy-driven relocation. This process involves a comprehensive valuation of ecosystem services, costs, benefits, and possible scenarios through comparative static analysis. Within this broader framework, two pillars of resilience stand out in the Sundarbans context, i.e. community based adaptation initiatives and strengthening of social capital and institutions.

5.8.1 Community-Based Adaptation Initiatives

Community-based adaptation initiatives play a central role in enhancing climate resilience across the ISD by actively engaging local stakeholders in the restoration, protection, and sustainable management of natural and agricultural systems. A prominent strategy is mangrove restoration, which supports biodiversity, enhances coastal protection, and sequesters carbon, thus contributing to both environmental and livelihood goals. Such initiatives have improved soil blue carbon pools and increased humic and fulvic acid concentrations, though sustained external support remains crucial for their continuity.

Water management forms another critical component. Given the region's vulnerability to saltwater intrusion and freshwater scarcity, local solutions, such as rainwater harvesting through pond and canal renovation and community-managed irrigation systems, help maintain freshwater sources, support post-monsoon crops, and generate seasonal employment (Padhy et al. 2021).

Alternative farming practices are equally vital. Traditional agriculture in the delta faces threats from salinity and flooding, prompting a shift toward salt-tolerant crop varieties (e.g., Luna Swarna, Luni Shree), organic farming, and integrated crop-livestock-fish farming systems. Land-shaping technologies, particularly farm ponds and paddy-cum-fish models, are financially viable interventions that address salinity, drainage congestion, and irrigation challenges. These improve productivity and income and reduce pressure on fragile ecosystems.

Additionally, promoting less perishable cash crops such as chilli and sunflower and establishing community seed banks for climate-resilient varieties can improve food security and disaster readiness. Freshwater aquaculture in community ponds offers supplementary income, particularly for landless and marginal farmers. Physical infrastructure also contributes to resilience. The construction of multi-purpose cyclone shelters provides emergency refuge and can serve as schools or community centres during normal periods. Improved village roads, pucca jetties, solar lighting, and mobile health camps enhance basic services and reinforce community engagement in environmental stewardship.

5.8.2 Strengthening Social Capital and Institutions

A robust foundation of social capital and local institutions is indispensable for effectively implementing and sustaining adaptive strategies in ISD, including investments in capacity building, governance reforms, and empowerment of grassroots organisations like self-help groups (SHGs).

Capacity building entails equipping communities with knowledge and skills in climate adaptation, disaster preparedness, and sustainable resource management. Training programs focused on sustainable agriculture, mangrove conservation, and alternative livelihoods can empower local actors to lead initiatives and reduce dependency on external interventions (Kumar et al. 2021).

Local governance reforms aimed at decentralisation and participatory planning must ensure adaptation strategies reflect community needs and are implemented effectively. Strengthening institutions like panchayats and municipalities enables more responsive and inclusive governance, allowing communities to better navigate environmental and socio-economic challenges.

Self-help groups (SHGs) catalyse collective action, especially among women, by facilitating access to microfinance, training, and market opportunities. SHGs can potentially drive the adoption of eco-friendly practices such as organic farming, small-scale aquaculture, mushroom cultivation, floriculture, and apiculture. Additionally, promoting village-based enterprises such as sewing, piggery, duckery, and seedling nurseries diversifies income sources and reduces dependency on fragile mangrove ecosystems. These institutions also function as intermediaries between communities and funding agencies, enhancing access to resources and ensuring that adaptation actions are well-supported and context-specific. Empowering these social structures enhances resilience and reinforces social cohesion and collective well-being (Pandhare et al. 2024).

5.9 Socio-Economic Transformation and Pathways Forward

The socio-economic transformation of ISD is a complex but necessary shift from vulnerability to resilience, aimed at improving the lives of local communities and ensuring sustainable development in the face of environmental challenges. This change requires a mix of economic growth, social support, and care for the environment. Adaptation strategies in the region depend on factors like history, community behaviour, gender roles, social connections, and the fragile ecosystem, making each community's needs and recovery process unique. Therefore, it is important to include local perspectives and evolving ideas about adaptation that focus on people and nature. Looking at the situation through social and ecological lenses highlights key issues such as land ownership disputes, loss of traditional knowledge, misuse of funds, and weak public health systems (Bhowmick 2023).

Another critical step toward building resilience in ISD is promoting economic diversification, especially as traditional livelihoods like fishing and farming face increasing risks from climate-induced challenges such as rising salinity and frequent cyclones. Transitioning into alternative income-generating sectors such as aquaculture, eco-tourism, and wage labour offers a pathway to economic stability and reduced vulnerability. However, this shift demands structured support through skill development, credit access, and strong market linkages to ensure local communities can adapt effectively. The Sundarban mangrove ecosystem is central to diversification as it offers various ecosystem services. These include provisioning goods like food, timber, medicinal plants, and bio-resources used for renewable energy; critical ecological functions such as nutrient cycling, habitat provision, and water transport; and vital regulating services like air and water purification, erosion control, storm buffering, and flood mitigation (Walters et al. 2008). The mangroves also provide breeding and nursery grounds for commercially important fish and shellfish species, making them indispensable to local fisheries. Juvenile fish benefit from the mangrove's structural complexity, which offers protection from predators and ideal feeding conditions. In addition, non-timber forest products such as honey, wax, and medicinal plants contribute significantly to household incomes (Saifullah and Jewel 2021). With its unique biodiversity, including the Royal Bengal Tiger, estuarine crocodiles, and migratory birds, ISD also holds immense potential for eco-tourism, which can further strengthen local economies while promoting conservation.

Recurrent natural disasters in ISD have significantly altered traditional livelihoods, particularly agriculture, by narrowing the range of viable livelihood options. Once practised biannually, paddy cultivation has largely shifted to a monocrop system due to increasing soil salinity, unpredictable rainfall, and cyclone induced disruptions. This change reflects a broader adaptation strategy where livelihood practices are increasingly intertwined with educational attainment and local cultural context. In rural areas like the ISD, education and culture are deeply interrelated; cultural norms shape educational systems and, in turn, influence livelihood choices. As a result, education plays a transformative role in rural development by enhancing individual capabilities and fostering a mindset of innovation and adaptability essential for coping with environmental and socio-economic stresses. Improving access to quality education, particularly in subjects like environmental science, technology, and entrepreneurship, equips communities with practical skills to pursue sustainable livelihoods and navigate emerging challenges. Additionally, education empowers local populations to engage meaningfully in participatory governance and decision-making processes, ensuring that development initiatives align better with the community's specific needs, values, and aspirations (Mousazadeh 2024).

A comprehensive analysis of healthcare distribution in ISD, conducted using primary and secondary datasets, highlights significant disparities in health services across different areas and population groups. Approximately 4.3 million residents in the region face persistent health challenges, including malnutrition, reduced life expectancy, and widespread prevalence of waterborne and vector-borne diseases. Women, in particular, experience irregular and inadequate dietary intake, contributing to high rates of maternal, infant, and child mortality, along with a large proportion of underweight children (Ghosh and Mistri 2020). In this context, healthcare plays a crucial role in facilitating socio-economic development. Ensuring access to quality healthcare services is fundamental to improving public health, especially in environmentally vulnerable regions like ISD, where ecological degradation exacerbates health risks. Enhancing healthcare infrastructure and increasing awareness through health education can significantly reduce disease susceptibility, improve living standards, and bolster economic resilience. Healthier populations are better equipped to engage in livelihood activities and adapt to socio-environmental challenges.

Technology plays a transformative role in rural development by offering innovative solutions to socio-economic challenges. It can be leveraged to enhance climate resilience through early warning systems for cyclones and floods, significantly reducing the risk of loss of life and property. Additionally, digital technologies can facilitate access to markets, improve agricultural productivity through precision farming, and support the development of sustainable livelihoods such as eco-tourism. However, adopting technology requires robust infrastructure and institutional capacity, which may be lacking in remote regions like ISD. Therefore, investments in digital infrastructure and capacity building are essential for harnessing the potential of technology in rural transformation.

Despite India's longstanding commitment to mangrove conservation through stringent environmental legislation, top-down protectionist strategies have often fallen short of ensuring long-term ecological sustainability. Recognising these limitations, the government introduced a more participatory model in the early 1990s under the Joint Forest Management (JFM) program, which aimed to empower local communities and stakeholders, including NGOs and Community-based Organisations (CBOs), to actively engage in the management of forest resources (Selvam 2003; DasGupta and Shaw 2013). This community-based co-management (CBMM) model marked a significant shift in mangrove governance in India, offering a collaborative framework that integrates traditional knowledge, local needs, and conservation priorities. NGOs, in particular, have played a crucial role as intermediaries, facilitating dialogue between forest departments and resource-dependent communities, forming local user groups, and mobilising self-help initiatives (Sundar 2000). However, the effectiveness of NGOs remains uneven. While they have contributed significantly to awareness building and community mobilisation, their long-term impact is often constrained by the project-based nature of their interventions, lack of sustained funding, and absence of formal authority in dispute resolution, especially in heterogeneous communities divided along caste or livelihood lines. Furthermore, the discretion to include NGOs in JFM initiatives lies with the forest department, leading to inconsistent application of co-management strategies across regions. Despite these challenges, successful case studies such as NGO-led mangrove restoration in Pondicherry demonstrate the potential for civil society engagement to achieve superior conservation outcomes compared to solely government-led initiatives (Datta et al. 2012).

The collaborative model is particularly critical due to the region's complex socio-ecological dynamics and vulnerability to climate-induced disasters. With over 4.3 million residents dependent on mangrove ecosystems for livelihood and protection, a decentralised and inclusive approach to mangrove management is essential for long-term resilience. The Joint Mangrove Management (JMM) model typically involves three key stakeholders: local forest departments, mangrove-dependent communities, and facilitating NGOs. However, these user groups are further differentiated into resource-dependent and resource-independent communities, complicating coordination and equitable participation (Kathiresan 2011). NGOs and CBOs operate with limited legal and tenurial rights despite their central role, often undermining their authority and continuity. The current governance structure also allows forest departments full discretion over the formation and dissolution of Joint Mangrove Management Committees (JMMCs), leaving little room for institutional stability or community ownership. To address these limitations, a growing consensus exists on integrating JMM holistically with national climate change adaptation strategies and disaster risk reduction frameworks. Doing so could enhance technical and financial capacities, mobilise larger resource pools, and motivate vulnerable coastal populations to actively engage in mangrove conservation and restoration (Datta et al. 2012; DasGupta and Shaw 2013). Mangroves are ecological assets and function as natural buffers against storms, tidal surges, and erosion, reducing disaster vulnerability in coastal rural areas. Therefore, any future direction in mangrove governance must prioritise statutory recognition of community rights, institutionalise participatory governance, and promote integrated frameworks that link environmental sustainability with economic empowerment and disaster resilience. This approach would improve conservation outcomes and strengthen coastal communities' socio-economic fabric, making them active stakeholders in the stewardship of critical ecosystems.

5.10 Conclusion

5.10.1 Synthesis

ISD exemplifies a fragile balance between human communities and their natural environment, where livelihoods have traditionally depended on agriculture, fishing, and forest resources supported by mangrove ecosystems and tidal rivers. However, this equilibrium is increasingly disrupted by environmental stressors such as salinity intrusion, coastal erosion, frequent cyclones, and broader climate change impacts, including sea-level rise and rising temperatures (Sarkar et al. 2024; Hazra et al. 2010; Mukherjee and Siddique 2024). These pressures degrade traditional livelihoods, compelling residents to adopt coping mechanisms like migration, livelihood diversification, and community-driven initiatives, including mangrove restoration and sustainable water management (Chandra and

Sagar 2003; Padhy et al. 2021). Governance structures are multilayered but often hampered by weak enforcement, bureaucratic inefficiencies, and limited participation from local communities (Kumar et al. 2021; Das 2024). Building resilience requires strengthening social capital, decentralising governance, and empowering local groups, particularly women's self-help organisations (Pandhare et al. 2024). Existing policies such as the Coastal Regulation Zone Notification, Forest Rights Act, and community co-management frameworks provide important foundations but face implementation and integration challenges with climate adaptation strategies (Prasad 2020; Mohanty et al. 2024). Adopting emerging technologies and participatory governance approaches offers promising avenues for sustainable development (Dasgupta et al. 2020; Banerjee and Dutta 2023). Achieving sustainable livelihoods in this climate-sensitive delta ultimately depends on balancing ecological conservation with socio-economic empowerment.

5.11 Recommendations

To address the complex vulnerabilities of the Sundarbans, livelihood diversification and emphasis on ecological sustainability are essential. Introducing salt-tolerant crops and integrated aquaculture systems can stabilise food production and incomes while alleviating pressure on wild fisheries and forests (Mahmood et al. 2021). Eco-tourism and community-based conservation offer alternative livelihood opportunities that foster biodiversity conservation and economic benefits (Saifullah and Jewel 2021). Empowering marginalised groups, especially women, through skill development and micro-enterprises strengthens social inclusion and resilience (Moreda 2023; Pandhare et al. 2024). Renewable energy technologies like solar and biogas help mitigate energy scarcity, supporting value addition and market access. Enhancing community-based disaster preparedness and natural resource co-management improves adaptive capacities (Paul et al. 2021). Digital tools that link local producers with wider markets can increase profitability and reduce exploitation. These strategies collectively promote an integrated approach grounded in ecological care, social justice, and technological innovation to build a sustainable future for the Sundarbans.

5.11.1 Limitations and Future Strategies

Nature-based solutions (NBS), such as mangrove restoration, offer substantial ecological benefits but encounter barriers, including socio-economic inequalities, insufficient institutional frameworks, and socio-political complexities like land tenure disputes and low community engagement (Chowdhury et al. 2021; Ghosh and Jana 2019; Mukhopadhyay and Saha 2020). Fragmented governance and lack of coordinated policy enforcement hinder the scaling and sustainability of these solutions (Roy and Pal 2022). Moving forward, participatory governance models that empower local stakeholders, foster knowledge co-production and ensure equitable benefit-sharing are imperative (Dasgupta et al. 2020). Strengthening governance across multiple levels and embedding climate resilience into NBS frameworks are crucial for tackling the Sundarbans' dynamic environmental and livelihood challenges (Banerjee and Dutta 2023). These comprehensive strategies can create adaptive, inclusive, and sustainable outcomes consistent with the socio-ecological complexity of the region.

Acknowledgments The authors gratefully acknowledge Prof. S K Acharya, Department of Agricultural Extension, Bidhan Chandra Krishi Viswavidyalaya, Mohanpur, India and Dr. Darshan N. P., Department of Agricultural Extension, Visva-Bharati, Sriniketan, India for their continuous support in formulating and conceptualizing this study. Their guidance was invaluable to this work.

Competing Interests Both authors declare that they have no competing interests related to this chapter.

References

Akram H, Hussain S, Mazumdar P, Chua KO, Butt TE, Harikrishna JA (2023) Mangrove health: a review of functions, threats, and challenges associated with mangrove management practices. Forests 14(9):1698. https://doi.org/10.3390/f14091698

Bagchi E (2017) Development of basic infrastructure: an analysis of south 24 Parganas district in West Bengal, India. Bull Geogr Socioecon Ser 36:33–60

Banerjee S, Dutta A (2023) Enhancing climate resilience through multi-level governance in the Sundarbans Delta. Environ Policy Gov 33(2):112–127. https://doi.org/10.1002/eet.1950

Barua P, Rahman SH, Mitra A, Zaman S (2020) Review on coastal erosion, displacement and resettlement strategies of south Asian countries. Glob Environ Eng 7:52–72

Basu J (2020) Cyclone Amphan and high tide: a 1 hour gap saved Sundarbans from an even bigger disaster. Down Earth 23. https://www.downtoearth.org.in/news/natural-disasters/cyclone-amphan-and-high-tide-a-1-hour-gap-saved-sundarbans-from-an-even-bigger-disaster-71317

Berkes F (2017) Environmental governance for the anthropocene? Sustainability 9(7):1232. https://doi.org/10.3390/su9071232

Bharati L, Sharma BR, Smakhtin V (2016) Ganges River Basin. Taylor and Francis, London

Bhowmick D (2023) Political ecology of climate change in Sundarbans, India: understanding Well-being, social vulnerabilities, and community perception. Environ Qual Manag 33(3):371–382. https://doi.org/10.1002/tqem.22125

Buckle P, Marsh G, Smale S (2001) Assessing resilience and vulnerability: rinciples, strategies and actions. In: Strengthening resilience in post-disaster situations, p 245

Cardona OD, Barbat AH (2000) El riesgo sísmico y su prevención. In: El riesgo sísmico y su prevención, pp 190–190

Chand BK, Trivedi RK, Biswas A, Dubey SK, Beg MM (2012) Study on impact of saline water inundation on freshwater aquaculture in Sundarban using risk analysis tools. Explor Anim Med Res 2(2):170–178

Chandra G, Sagar R (2003) Fisheries in Sundarbans: problems and prospects. SSRN Electron J. https://doi.org/10.2139/ssrn2084014

Chowdhury S, Roy S, Bhattacharya B (2021) Socio-economic vulnerabilities and adaptation in the Indian Sundarbans: challenges for nature-based solutions. Ecol Econ 182:106952. https://doi.org/10.1016/j.ecolecon.2020.106952

Danda AA, Ghosh N, Bandyopadhyay J, Hazra S (2019) Managed retreat: adaptation to climate change in the Sundarbans ecoregion in the Bengal Delta. J Indian Ocean Reg 15(3):317–335. https://doi.org/10.1080/19480881.2019.1652974

Das BK (2024) Beyond the 'protected area' paradigm in conservation: exploring India's forest legislation as a new conservation model for developing countries. Environ Manag 74(6):1223–1238

Dasgupta R, Mukherjee S, Singh P (2020) Participatory governance and livelihood security in coastal ecosystems: lessons from the Sundarbans. Sustain Sci 15(5):1361–1375. https://doi.org/10.1007/s11625-020-00824-1

DasGupta R, Shaw R (2013) Cumulative impacts of human interventions and climate change on mangrove ecosystems of south and Southeast Asia: an overview. J Ecosyst 2013:1–15. https://doi.org/10.1155/2013/379429

Datta D, Chattopadhyay R, Guha P (2012) Community based mangrove management: a review on status and sustainability. J Environ Manag 107:84–95. https://doi.org/10.1016/j.jenvman.2012.04.013

Deshpande M, Singh VK, Ganadhi MK, Roxy MK, Emmanuel R, Kumar U (2021) Changing status of tropical cyclones over the north Indian Ocean. Clim Dynam 57(11–12):3545–3567. https://doi.org/10.1007/s00382-021-05880-z

Duxbury J, Dickinson S (2007) Principles for sustainable governance of the coastal zone: in the context of coastal disasters. Ecol Econ 63(2–3):319–330

Forst MF (2009) The convergence of integrated coastal zone management and the ecosystems approach. Ocean Coastal Manag 52(6):294–306. https://doi.org/10.1016/j.ocecoaman.2009.03.007

Gaillard JC, Maceda EA, Stasiak E, Le Berre I, Espaldon MVO (2009) Sustainable livelihoods and people's vulnerability in the face of coastal hazards. J Coast Conserv 13:119–129

Ghosh A, Jana S (2019) Mangrove restoration and community engagement: socio-political constraints in the Sundarbans. Ocean Coast Manag 176:74–83. https://doi.org/10.1016/j.ocecoaman.2019.04.006

Ghosh N (2011) Indian Sundarbans Delta (ISD) in 2050: a vision—economic valuation of costs and benefits of the adaptation options. WWF-India, Kolkata

Ghosh S, Mistri B (2020) Coastal agriculture and its challenges: a case study in Gosaba Island, Sundarban. India Space Cult India 8(2):140–154. https://doi.org/10.20896/saci.vi0.747

Gopal B, Chauhan M (2006) Biodiversity and its conservation in the Sundarban mangrove ecosystem. Aquat Sci 68(3):338–354. https://doi.org/10.1007/s00027-006-0868-8

Granados-Olivas A, Hargrove HL, Heyman JM, Mayer A, Mirchi A, Ganjegunte G, Walker WS (2023) The future of water in a desert river basin facing climate change and competing demands: a holistic approach to water sustainability in arid and semi-arid regions. Instituto de Ingeniería y Tecnología

Hahn MB, Riederer AM, Foster SO (2009) The livelihood vulnerability index: a pragmatic approach to assessing risks from climate variability and change—a case study in Mozambique. Glob Environ Chang 19(1):74–88

Hajra R, Ghosh A, Ghosh T (2021) Micro level vulnerability assessment of estuarine islands: a case study from Indian Sundarban. Springer Water, pp 135–153. https://doi.org/10.1007/978-3-319-54612-4_11

Hajra R, Ghosh T (2018) Agricultural productivity, household poverty and migration in the Indian Sundarban Delta. Elem Sci Anth 6. https://doi.org/10.1525/elementa.196

Haldar A, Debnath A (2014) Assessment of climate induced soil salinity conditions of Gosaba Island, West Bengal and its influence on local livelihood. In: Climate change and biodiversity: proceedings of IGU Rohtak conference, vol 1. Springer Japan, pp 27–44

Haque CE, Prowse M, Alam M, Kuriakose A, Aggarwal P, Ahsan R (2022) Indigenous knowledge and climate resilience: the case of coastal Bangladesh. Environ Dev 41:100675. https://doi.org/10.1016/j.envdev.2021.100675

Haque SE, Nahar N (2023) Bangladesh: climate change issues, mitigation, and adaptation in the water sector. ACS ESandT Water 3(6):1484–1501

Harms A (2024) Enduring erosions: environmental displacement and relocation on India's sinking coasts. University of Hawai'i Press

Hasnine M, Nagdeve DA (2024) Natural hazard induced coastal vulnerability in Indian Sundarbans: a village-level study by using geospatial and statistical techniques. Nat Hazards Res. https://doi.org/10.1016/jnhres202410004

Hazra S, Bakshi A (2003) Environmental refugees from vanishing islands. In: Environment and human security. Lancers' Books, pp. 219–227

Hazra S, Das I, Samanta K, Bhadra T (2014) Impact of climate change in Sundarban area West Bengal, India. Earth Sci Clim Book 9326(17.02):00

Hazra S, Ghosh T, DasGupta R, Sen G (2002) Sea level and associated changes in the Sundarbans. Sci Cult 68(9/12):309–321

Hazra S, Samanta K, Mukhopadhyay A, Akhand A (2010) Temporal change detection (2001–2008) of the Sundarban. Unpublished report, WWF-India

Hijioka Y, Lin E, Pereira JJ, Corlett RT, Cui X, Insarov G, Wang W (2014) IPCC fifth assessment report: chapter 24 Asia. In: climate change 2014—impacts, adaptation and vulnerability: part B: regional aspects. Working group II contribution to the IPCC fifth assessment report, pp 1327–1370

Hussain S, Hussain E, Saxena P, Sharma A, Thathola P, Sonwani S (2024) Navigating the impact of climate change in India: a perspective on climate action (SDG13) and sustainable cities and communities (SDG11). Front Sustain Cities 5. https://doi.org/10.3389/frsc.2023.1308684

IPCC (2001) Climate change 2001: synthesis report. A contribution of working groups I, II, and III to the third assessment report of the intergovernmental panel on climate change. In: Watson RT (ed) Core writing team. Cambridge University Press, Cambridge, UK, New York, NY, USA, p 398

IPCC (2014) Climate change 2014: impacts, adaptation, and vulnerability. Part B: regional aspects. Contribution of working group II to the fifth assessment report of the intergovernmental panel on climate change. Cambridge University Press, Cambridge, United Kingdom; New York, NY, USA, p 688

Jamal S, Ghosh A, Hazarika R, Sen A (2022) Livelihood, conflict and tourism: an assessment of livelihood impact in Sundarbans, West Bengal. Int J Geoherit Park 10(3):383–399. https://doi.org/10.1016/j.ijgeop.2022.07.004

Kaschek TS (2021) Exploring the experiences of female small-scale organic cocoa farmers about gender-based inequality in agency and empowerment in light of the sustainable development goal 5: a case study from rural Ghana

Kathiresan K (2011) United nation university chapters. Training course on mangroves and biodiversity

Kreft S, Eckstein D, Junghans L, Kerestan C, Hagen U (2014) Global climate risk index 2015: who suffers most from extreme weather events? Weather-related loss events in 2013 and 1994 to (2013)

Kumar N, Chhokar RS, Meena RP, Kharub AS, Gill SC, Tripathi SC, Gupta OP, Mangrauthia SK, Sundaram RM, Sawant CP, Gupta A, Naorem A, Kumar M, Singh GP (2021) Challenges and opportunities in productivity and sustainability of rice cultivation system: a critical review in Indian perspective. Cereal Res Commun 50(4):573–601. https://doi.org/10.1007/s42976-021-00214-5

Lawler J, Spencer B, Olden J, Kim S, Lowe C, Bolton S, Beamon B, Thompson L, Voss J (2013) Mitigation and adaptation strategies to reduce climate vulnerabilities and maintain ecosystem services. Elsevier eBooks, pp 315–335. https://doi.org/10.1016/b978-0-12-384703-4.00436-6

Mahadevia K, Vikas M (2012) Climate change–impact on the Sundarbans, a case study. Int Sci J Environ Sci 2(1):7–15

Mahmood H, Ahmed M, Islam T, Uddin MZ, Ahmed ZU, Saha C (2021) Paradigm shift in the management of the Sundarbans mangrove forest of Bangladesh: issues and challenges. Trees For People 5:100094. https://doi.org/10.1016/j.tfp.2021.100094

Mandal S, Choudhury BU, Mondal M, Bej S (2013) Trend analysis of weather variables in Sagar Island, West Bengal, India: a long-term perspective (1982–2010). Curr Sci 104:947–953

Mitra A (2023) Impact of COVID-19 on livelihoods of lower Gangetic Delta. Springer eBooks, pp 217–308. https://doi.org/10.1007/978-3-031-27242-4_5

Mitra A, Gangopadhyay A, Dube A, Schmidt AC, Banerjee K (2009) Observed changes in water mass properties in the Indian Sundarbans (northwestern bay of Bengal) during 1980–2007. Curr Sci 97:1445–1452

Mohanty S, Pandey PC, Pandey M, Srivastava PK, Dwivedi CS (2024) Wetlands contribution and linkage to support SDGs, its indicators and targets—a critical review. Sustain Dev 32(5):5348–5392

Mondal M, Biswas A, Haldar S, Mandal S, Mandal P, Bhattacharya S, Paul S (2022) Climate change, multi-hazards and society: an empirical study on the coastal community of Indian Sundarban. Nat Hazards Res 2(2):84–96. https://doi.org/10.1016/j.nhres.2022.04.002

Moreda T (2023) The social dynamics of access to land, livelihoods and the rural youth in an era of rapid rural change: evidence from Ethiopia. Land Use Policy 128:106616. https://doi.org/10.1016/j.landusepol.2023.106616

Mousazadeh H (2024) Unraveling the nexus between community development and sustainable development goals: a comprehensive mapping. Community Dev:1–27. https://doi.org/10.1080/15575330.2024.2388097

Mukherjee N, Siddique G (2024) Impact of climate change in the Indian Sundarbans region. https://doi.org/10.1007/978-3-031-54238-1

Mukherjee N, Siddique G, Basak A, Roy A, Mandal MH (2019) Climate change and livelihood vulnerability of the local population on Sagar Island, India. Chin Geogr Sci 29:417–436

Mukhopadhyay D, Saha S (2020) Institutional challenges in implementing nature-based solutions in the Indian Sundarbans. Environ Manag 66(3):397–410. https://doi.org/10.1007/s00267-020-01295-z

Mullick MRA, Tanim A, Islam SMS (2019) Coastal vulnerability analysis of Bangladesh coast using fuzzy logic based geospatial techniques. Ocean Coast Manag 174:154–169. https://doi.org/10.1016/j.ocecoaman.2019.03.010

Munang RM, Thiaw I, Alverson K, Liu J, Han Z, Lwasa S (2014) Climate change and ecosystem-based adaptation: a new pragmatic approach to buffering climate change impacts. Curr Opin Environ Sustain 7:1–7. https://doi.org/10.1016/j.cosust.2013.11.014

Okafor-Yarwood I, Nyangoko BP, Andriesse E, de la Torre-Castro M, Lindström L, Max A (2023) African ocean stewardship: navigating ocean conservation and sustainable marine and coastal resource management in Africa. 136

Padhy SR, Dash PK, Bhattacharyya P (2021) Challenges, opportunities, and climate change adaptation strategies of mangrove-agriculture ecosystem in the Sundarbans, India: a review. Wetl Ecol Manag 30(1):191–206. https://doi.org/10.1007/s11273-021-09844-2

Pandhare A, Bellampalli PN, Yadava N (2024) Transforming rural women's lives in India: the impact of microfinance and entrepreneurship on empowerment in self-help groups. J Innov Entrep 13(1). https://doi.org/10.1186/s13731-024-00419-y

Paul BK, Routray JK, Saha SK (2021) Community-based disaster risk reduction and institutional capacity building in the Ganges delta. Int J Disaster Risk Reduct 58:102193. https://doi.org/10.1016/j.ijdrr.2021.102193

Pramanik M, Szabo S, Pal I, Udmale P, Pongsiri M, Chilton S (2021) Population health risks in multi-hazard environments: action needed in the cyclone Amphan and COVID-19–hit Sundarbans region, India. Clim Dev 14(2):99–104. https://doi.org/10.1080/17565529.2021.1889948

Prasad N (2020) Policy interventions in India to improve coastal risk reduction in the face of climate change. University of Washington

Prasanya J, Kanmani S, Kumar PS (2024) A review of the wetland's restoration mechanisms and its economic and social benefits. Water Prac Technol. https://doi.org/10.2166/wpt.2024.241

Pomeroy RS, Ratner BD, Hall SJ, Pimoljinda J, Vivekanandan V (2006) Coping with disaster: rehabilitating coastal livelihoods and communities. Marine Policy 30(6):786–793. https://doi.org/10.1016/j.marpol.2006.02.003

Purkait B (2008) Coastal erosion in response to wave dynamics operative in Sagar Island, Sundarban delta, India. Front Earth Sci China 3(1):21–33. https://doi.org/10.1007/s11707-009-0001-0

Quader MA, Khan AU, Kervyn M (2021) Spatial variation in household-level risk to natural hazards across the coast of Bangladesh. GeomatS Nat Hazards Risk 12(1):1532–1559

Rao P, Filho WL (2015) Local community perception of climate change and scientific validation: a review of initiatives and perspectives in the Indian region. In: Climate change in the Asia-Pacific Region, pp 89–101

Reid H, Seddon N, Stewart F, MacGregor J, Béné C, Magnan A, Hutton C, Neufeldt H, Cajani L, Henly-Shepard S (2019) Ecosystem-based adaptation: theory, practice and future prospects. Annu Rev Environ Resour 44:1–27. https://doi.org/10.1146/annurev-environ-101718-033144

Roy K, Pal D (2022) Governance fragmentation and policy enforcement: barriers to effective nature-based solutions in the Sundarbans Delta. J Environ Plan Manag 65(4):695–712. https://doi.org/10.1080/09640568.2021.1907685

Rudra K (2010) Aila: unpublished records of the government of West Bengal. South Asian J forced Migr. MCRG, Kolkata. pp 86–93

Saifullah M, Jewel KNA (2021) Sustainable management of non-wood forest products for rural livelihoods in Bangladesh. In: Nonwood and livelihood: sustainable Management of non-wood Forest Products for rural livelihoods in South Asia. SAARC Agriculture Centre, SAARC, Dhaka, Bangladesh, pp 166

Sarkar SK, Haydar M, Rudra RR, Mazumder T, Nur MS, Islam MS, Sany SM, Noor TA, Ahmed S, Ahmad M, Sakib A, Ravela S (2024) A root-zone soil salinity observatory for coastal Southwest Bangladesh. *arXiv* (Cornell University). https://doi.org/10.48550/arxiv.2412.19740

SBR (2008) Sundarban atlas. In: Sundarban biosphere reserve

Scoones I (2009) Livelihoods perspectives and rural development. In: Critical perspectives in rural development studies. Routledge, pp 159–184

Selvam V (2003) Environmental classification of mangrove wetlands of India. Curr Sci 84(6):757–765

Sundar N (2000) Unpacking the 'joint' in joint forest management. Dev Change 31(1):255–279. https://doi.org/10.1111/1467-7660.00154

Uddin SA, He L, Hossain MJ, Nusrat N, Debi M (2024) Ganges-Brahmaputra-Meghna River Delta. In: Delta sustainability: a report to the Mega-Delta programme of the UN Ocean decade. Springer Nature Singapore, Singapore, pp 89–116

Vishwapriya C, Devaiah NG (2024) Conservation priorities for Indian biodiversity: spatiotemporal patterns, policy efficacy, and future outlook. Biodivers Conserv. https://doi.org/10.1007/s10531-024-02924-8

Walters BB, Rönnbäck P, Kovacs JM, Crona B, Hussain SA, Badola R, Primavera JH, Barbier E, Dahdouh-Guebas F (2008) Ethnobiology, socio-economics and management of mangrove forests: a review. Aquat Bot 89(2):220–236. https://doi.org/10.1016/j.aquabot.2008.02.009

World Bank (2014) Building resilience for sustainable development of the Sundarbans: through estuary management, poverty reduction, and biodiversity conservation strategy report (Report No. 88061–IN). South Asia region sustainable development department, environment and water resources management unit. The World Bank, Washington, DC

Yadav V, Dey S, Pan S (2024) Regional planning framework for addressing flood vulnerability: a comprehensive case study of south 24 Parganas, West Bengal, India. Springer Natural Hazards, pp 425–456. https://doi.org/10.1007/978-981-97-2688-2_20

Yang X, Guo S, Deng X, Wang W, Xu D (2021) Study on livelihood vulnerability and adaptation strategies of farmers in areas threatened by different disaster types under climate change. Agriculture 11(11):1088. https://doi.org/10.3390/agriculture11111088

Exploring Wastewater Management in Deltas 6

Annisa Utami Rauf and Ari Prayogo Pribadi

Abstract

Delta regions face significant wastewater management challenges due to rapid urbanization, industrial expansion, and climate change impacts. Inadequate wastewater treatment, poor sanitation infrastructure, and the direct discharge of pollutants into water bodies contribute to severe environmental degradation, including water eutrophication and biodiversity loss. Despite the growing urgency, there remains a critical gap in the development of integrated, climate-sensitive, and context-specific wastewater management strategies for delta environments. To address these challenges, innovative technologies such as decentralized wastewater treatment, multi-level reuse systems, and nature-based solutions are being explored to enhance efficiency and sustainability. Findings suggest that effective wastewater management in delta regions requires more than just infrastructure development—emphasizing adaptive governance, ecological protection, and inclusive community participation. Case studies from the Niger Delta, Yangtze River Delta, Mekong Delta, and others highlight both shared and site-specific issues, reinforcing the importance of localized, context-driven solutions. Policy instruments, such as integrated water resource management, regulatory enforcement, and community engagement, play a crucial role in mitigating wastewater-related risks. This chapter examines key wastewater challenges in delta regions worldwide,

A. U. Rauf (✉) · A. P. Pribadi
Department of Health Behavior, Environment, and Social Medicine, Faculty of Medicine, Public Health, and Nursing, Universitas Gadjah Mada, Yogyakarta, Indonesia
e-mail: annisaur@ugm.ac.id

A. P. Pribadi
e-mail: ari.prayogo.p@ugm.ac.id

C. Singha et al. (eds.), *Sustainable Management of Delta Ecosystems Resilience*,
Deltas of the World, https://doi.org/10.1007/978-3-032-16261-8_6

highlights the need for holistic, multi-scalar responses, and the underscores the technological advancements and governance innovations essential for achieving sustainable water management.

Abbreviations

ADM	Adaptive Delta Management
BOD	Biochemical Oxygen Demand
CO_2	Carbon Dioxide
COD	Chemical Oxygen Demand
CP	Cyclopentane
ERT	Electrical Resistivity Tomography
EU	European Union
HWTP	Hawai‘i Wastewater Treatment Plant
LDH	Layered Double Hydroxide
Lac	Laccase
MNB	Micro-Nano Bubbles
MPs	Microplastics
MQR	Minimum Quality Requirements
NH_3-N	Ammonia Nitrogen
OSDS	Onsite Sewage Disposal Systems
PAHs	Polycyclic Aromatic Hydrocarbons
PCBs	Polychlorinated Biphenyls
PGs	Progestins
$PM_{2.5}$	Particulate Matter with a diameter of 2.5 micrometers or less
POP	Persistent Organic Pollutants
PPCPs	Pharmaceuticals and Personal Care Products
PRD	Pearl River Delta
RQ	Risk Quotient
SGW	Shale Gas Wastewater
SSA	Strategic Stepwise Approach
SURE	Seemingly Unrelated Regression Estimator
TDS	Total Dissolved Solids
TMW	Treated Municipal Wastewater
UHI	Urban Heat Island
WWTPs	Wastewater Treatment Plants
YRD	Yangtze River Delta

6.1 Introduction

Delta areas are among the world's most densely populated and ecologically sensitive zones, yet they face growing threats from wastewater pollution (Mansur et al. 2016; Safra De Campos et al. 2020). These low-lying, water-rich regions serve as centers for agriculture, industry, and urban development, making them particularly susceptible to contaminants from multiple sources—ranging from fossil fuel combustion and agricultural runoff to urban wastewater and industrial effluents (Maliga et al. 2025). While increased energy use supports development in these regions, higher energy intensity has been linked to greater emissions of CO_2 and fine particulates ($PM_{2.5}$) (He et al. 2022), often diminishing the effectiveness of pollution and carbon reduction efforts. Moreover, growing population density and human mobility within delta urban clusters often lead to spillover effects that exacerbate environmental degradation in adjacent areas (Small et al. 2018; Xu et al. 2024).

Ecological transformations in delta landscapes, such as peatland drainage and mangrove deforestation, further intensify wastewater-related challenges (Hoang et al. 2025). These changes reduce the ability of delta ecosystems to absorb carbon and filter pollutants, increasing vulnerability to climate-related stressors like sea level rise and saltwater intrusion. Industrial activities, particularly extractive operations, have also contributed to the loss of biodiversity and the deterioration of aquatic and coastal ecosystems (Aransiola et al. 2024). To address these interconnected pressures, wastewater management in delta areas must adopt an integrated approach that accounts for ecological integrity, demographic shifts, and spatial interdependence—ensuring not only environmental sustainability but also the protection of human health and livelihoods in these fragile regions.

One frequently discussed example is pollution in the Niger Delta, which faces severe environmental challenges due to oil spills. This pollution degrades water quality, damages ecosystems, and threatens biodiversity. Contaminated soil loses fertility, disrupts agriculture, and worsens food security (Ewim et al. 2023). Several delta regions also facing these challenges include the Yangtze River Delta, Pearl River Delta, and Yellow River Delta in China. These areas are significantly affected by factors such as low residue levels and small debris, which have demonstrated a scale effect in mitigating mulch residue pollution. These regions often carry more sediment than water, resulting in the transport of substantial amounts of silt and sediments, further impacting the ecological stability and water quality of the deltas (Deng and Liu 2023; He et al. 2022; Lin et al. 2018; Shi et al. 2024; Sun et al. 2023; Xu et al. 2024). In States, The Sacramento-San Joaquin Delta faces growing environmental concerns due to wastewater treatment plant discharges, which introduce pharmaceutical and personal care pollutants into surface waters, endangering aquatic ecosystems (Shaefer et al. 2013). In Mississippi River Delta, nutrient runoff from agricultural activities leads to eutrophication, prompting the implementation of solutions like on-farm water retention systems (Brock et al. 2019). The Danube Delta in Europe experiences challenges from inadequate water and sanitation infrastructure, exacerbating social and economic vulnerabilities, particularly during periods of drought (Damian et al. 2023). Across various other deltas—such as the Indus, Mekong, and Ganges—salinity intrusion caused by sea-level rise and reduced freshwater inflows continues to undermine agricultural sustainability and weaken local community resilience (Abedin et al. 2013; Day et al. 2021). These diverse yet pressing issues highlight the urgent need for integrated, context-specific wastewater and resource management in delta environments.

There is an urgent need to prioritize wastewater management in delta regions due to their heightened vulnerability to environmental and health-related threats. Without adequate systems for wastewater collection and treatment, these areas face escalating risks—ranging from contamination of water sources and spread of waterborne diseases to long-term degradation of ecosystems (Hung et al. 2012; Tripathi et al. 2021). The challenges are intensified by poor infrastructure, rapid urban growth, and limited access to clean water, particularly in low-income communities. This situation underscores the importance of integrated and adaptive wastewater strategies that not only address technical gaps but also consider the socio-economic conditions and ecological fragility of delta environments. Strengthening wastewater management in these regions is essential to protect human well-being, preserve biodiversity, and build resilience against the growing pressures of climate change and urbanization.

This review offers a comprehensive and interdisciplinary assessment of wastewater management in delta regions by integrating ecological vulnerability, demographic dynamics, and governance challenges into a unified analytical perspective. Moving beyond the limitations of prior studies that often address technical or regional issues in isolation, it draws upon case studies from major deltas to capture both shared pressures and region-specific complexities. Through the synthesis of technological, socio-political, and environmental dimensions, this work contributes a valuable and timely perspective to the discourse on adaptive and context-sensitive wastewater management approaches in deltaic environments.

6.2 Major Wastewater Management Challenges in Delta Regions

6.2.1 Urbanization and Industrial Expansion

The swift pace of industrialization and urbanization has significantly increased energy consumption across manufacturing, transportation, and residential sectors, leading to elevated emissions of CO_2 and other greenhouse gases. Delta regions such as the Pearl River Delta (PRD) and the Yangtze River Delta (YRD) have undergone rapid economic expansion and population growth, elevating them into prominent global economic centers (Gu et al. 2011; Song et al. 2021). In the PRD, the sharp rise in industrialization and urban development has significantly contributed to its economic prominence (Peng et al. 2024). However, this swift advancement has come at an environmental cost, leading to increased flood risks, water contamination, and ecological decline. Urbanization in the YRD, for instance, has caused widespread flooding and reduced production efficiency, while the PRD's urban sprawl has intensified air and water pollution issues, posing major sustainability challenges. Urban expansion has been accompanied by a rise in water pollutants such as chemical oxygen demand (COD) and ammonia nitrogen (NH_3-N) (Kyere-Yeboah and Qiao 2024; Meng et al. 2023), worsening existing water quality problems.

In South Asia, most cities have expanded as isolated hubs of growth, often driven by unplanned urbanization. Constrained financial resources and a lack of coordinated planning have left municipal authorities facing significant challenges including environmental issues (Jha et al. 2019). The rise in vehicular traffic within urban centers and the operation of unregulated industries discharging untreated waste have led to worsening environmental conditions in South Asia's urban clusters, particularly through heightened air and water pollution (Liu and Li 2021). The expansion of industrial activities in densely populated urban centers contributes significantly to localized temperature rises through the amplification of the urban heat island (UHI) effect as happened in Germany, China, India and Mexico (Benz et al. 2017; Dwivedi and Khire 2018; Ramírez Hernández 2024; Zhao et al. 2024). This phenomenon is driven by direct heat emissions from industrial processes and the widespread reduction of vegetative cover. Quantitative analyses have established a positive correlation between industrial zone expansion and land surface temperature increases, such as the observed 0.39–0.48 °C rise associated with industrial growth in Linyi City, China, and similar patterns documented in Mexico (Ramírez Hernández 2024). Moreover, specific industries, notably coal and steel manufacturing, disproportionately elevate local temperatures due to their intensive energy demands and emission profiles. The intersection of accelerated urbanization, industrial proliferation, and rising temperatures has profound implications, including increased energy consumption for cooling, deteriorating air and water quality, heightened public health risks, and the destabilization of sensitive deltaic ecosystems (Rao et al. 2018).

Delta regions represent some of the most climate-sensitive areas globally due to their low elevations and exposure to multiple stressors, including sea-level rise, coastal flooding, storm surges, and land subsidence (Amin et al. 2025; Döll and Zhang 2010). Climate-driven alterations in precipitation and the frequency of extreme events are intensifying river discharge patterns and reshaping riverine landscapes, placing ecosystems and human settlements at heightened risk, as seen in the Mississippi River Delta (Day et al. 2024). These hydrological changes directly impact wastewater management, with rising floodwaters and encroaching seas increasing the likelihood of system failures and the release of untreated or partially treated wastewater into natural water bodies. This not only deteriorates water quality but also amplifies threats to ecosystem health and human well-being. Moreover, wastewater treatment processes contribute to climate change through the release of greenhouse gases such as methane and nitrous oxide, creating a reinforcing cycle of environmental degradation. Comprehensive strategies, including the adoption of adaptive delta management (ADM), the strengthening of wastewater infrastructure, wetland conservation, and enhanced institutional governance, are critical to building resilience against these interconnected climate risks (Amin et al. 2025; Dewulf and Termeer 2015). Ho Chi Minh City, Vietnam's largest metropolis, is predicted to be among the ten most at-risk cities globally due to climate change. At the same time, the Mekong Delta is considered one of the three most susceptible deltas worldwide to rising sea levels (Green Denmark in Southeast Asia 2021).

6.2.2 Direct Discharge of Pollutions

In several developing nations, effluents from industrial, household and agricultural activities are frequently released directly into river systems or deltaic waterways, often with insufficient filtration or untreated. For instance, between 4.83 and 28.34% of households in Jakarta and West Java lack access to proper sanitation. Only a small proportion of households manage their greywater, with the prevailing practice being the direct discharge of wastewater into drainage systems, which ultimately flow into water bodies. This widespread lack of adequate sanitation in certain areas of Jakarta and West Java, where direct discharge into water systems is common, exacerbates water quality issues (Pratama and Constantine 2023). Another case in Niger Delta, high concentrations of petroleum and related chemicals, which can jeopardize human health and disrupt the biodiversity arises from pipeline leaks due to industrial activities, which have a severe impact on nutrition (Babatunde 2020). Notably, oil spills, such as the notorious Bodo oil spill of 2008–2009, have led to significant declines in fish populations throughout the delta, a crucial protein source for the local community (Fentiman and Zabbey 2015; Ola et al. 2024).

Evidence of direct discharge of pollutants into rivers or deltas is reflected in the accumulation of heavy metals in sediments from rivers receiving shale gas wastewater (SGW) and hydrophobic compounds like progestins (PGs) suggesting inadequate management practices and environmental risks in China (Li et al. 2023; Shen et al. 2024). Pollutants in delta regions, particularly those with bioaccumulative properties, pose significant environmental threats due to their persistence and ability to move through various natural media such as soil and rock (Astuti et al. 2025; Chojnacka and Mikulewicz 2014; Garg et al. 2022; Liu and Guo 2023). These substances gradually accumulate in living organisms, often leading to toxic effects within ecosystems (Li et al. 2015; Van Cauwenberghe and Janssen 2014). What makes them especially hazardous is their capacity to migrate across multiple environmental layers—including air, water, soil, and geological substrates—highlighting the importance of understanding their pathways and long-term behavior (Jiang 2021).

Several types of pollutants are commonly found in delta areas. Heavy metals such as mercury, cadmium, arsenic, lead, and chromium are often introduced through industrial activities, agricultural runoff, and atmospheric deposition (Astuti et al. 2025; Rusprayunita et al. 2025). Persistent organic pollutants (POPs), including organochlorine pesticides, polycyclic aromatic hydrocarbons (PAHs), and polychlorinated biphenyls (PCBs), also contribute to environmental contamination, particularly in regions with intensive petrochemical and agricultural activity (Gan et al. 2019; Li et al. 2016). The migration and distribution of these pollutants are influenced by both environmental conditions and the physical characteristics of the landscape. In areas with high-risk contours—combinations of geographic and chemical factors—pollutants can spread more rapidly or become concentrated. These contours are useful for mapping environmental risks and guiding land-use planning and pollution control strategies. Pollutants may travel through soil and groundwater systems, especially in porous or permeable areas, increasing the risk of groundwater contamination (Klimkowicz-Pawlas et al. 2023; Liu and Guo 2023; Pobi et al. 2020). As observed in wastewater from four hospitals in Bamako, Mali—sampled at both exit points (Po) and discharge sites (Pf) into the Niger River—in addition to various chemical pollutants, antibiotics from nine different classes, mainly fluoroquinolones, macrolides, and sulfonamides, were also detected (Coulibaly et al. 2025).

6.3 Environmental Consequences of Poor Management

6.3.1 Biodiversity Loss and Ecosystem Disruption

Discharges of wastewater from industrial and urban areas bring heavy metals, excess nutrients, and micropollutants into aquatic environments, disrupting natural biogeochemical cycles and harming aquatic biodiversity, including microbial life (Coulibaly et al. 2025; Kyere-Yeboah and Qiao 2024; Li et al. 2025; Śniatała et al. 2024). In the Sacramento-San Joaquin Delta, for example, contaminants like pharmaceuticals and personal care products (PPCPs) from treatment plants have been associated with declines in pelagic fish populations, signaling serious ecosystem damage (Shaefer et al. 2013). Wastewater also alters the structure of microbial communities, diminishing their diversity and richness, which affects essential processes such as organic matter breakdown and nutrient recycling. In Swiss streams, while wastewater-derived nutrients stimulated microbial activity, the concurrent introduction of micropollutants had a detrimental effect on microbial community balance (Burdon et al. 2019).

The impact of wastewater pollution extends to plant and animal populations, leading to broader ecological disturbances. In riparian areas, deteriorating water quality shifts plant composition, reducing overall species richness and enabling a few species to dominate (Musisi et al. 2025). In the YRD, for instance, declining water flow combined with pollution has caused biodiversity loss and ecosystem deterioration, endangering ecological sustainability (Deng and Liu 2023; Gan et al. 2019; Li et al. 2016; Shen et al. 2024). Observations from different regions further highlight these threats: in Oklahoma's Red

River, proximity to wastewater outfalls increased macroinvertebrate diversity but did not significantly affect fish populations, while industrial discharges and shrimp farm effluents in the Persian Gulf and Gulf of Oman sharply reduced biodiversity in estuarine zones (Bakshi et al. 2023; Sharifinia et al. 2025). Similarly, in China's Liao River Delta, pollution and land use changes—especially cadmium contamination—have led to shrinking wetlands and loss of bird habitats (Yin et al. 2021).

6.3.2 Public Health Risks and Waterborne Diseases

Many wastewater treatment plants suffer from poor maintenance and frequent overloading, resulting in untreated wastewater being widely used for irrigation. Pathogenic viruses like SARS-CoV-2, norovirus, rotavirus, adenovirus, and enterovirus are commonly detected in wastewater treatment plants (WWTPs) worldwide (Ma et al. 2024). Their persistence in water—from days to over a year—heightens public health risks, yet global-scale studies on viral presence in reclaimed water remain limited. Farmers commonly apply this water to grow vegetables such as lettuce, cabbage, peppers, tomatoes, and beans, as well as non-food crops like alfalfa, maize, cotton, and sunflowers (Emmanouil et al. 2024; Ibrahimi et al. 2025; Todd 2024). In Iraq, the biological quality of water in the Al-Dewanyia River often fails to meet WHO standards, with reported health issues among people who rely on the river. The agricultural lands near wastewater discharge points show significantly elevated nitrate levels, adding to public health concerns (Hamad et al. 2022).

The link between water quality and public health is especially evident in dry regions, where outbreaks are growing more frequent and health risks are rising. Weather events often trigger parasitic infections, placing vulnerable groups at greater risk (Ismail et al. 2021; Ogata et al. 2021). In regions like the Mekong Delta, recurring climate extremes and changes in water flow are intensifying health problems related to unsafe water (Phung et al. 2015). Diseases spread through contaminated water remain a major health challenge worldwide, driven by a variety of pathogens. Factors such as weather patterns, water infrastructure, and local hydrology all influence how these illnesses are transmitted (Nichols et al. 2018; Todd 2024). People living in rural areas and low-income communities are disproportionately affected by waterborne diseases. Infants, pregnant women, the elderly, and those with weakened immune systems are especially vulnerable. Preventing disease outbreaks depends on effective controls to keep water safe, regular water quality checks, and strong infection prevention efforts (Levallois 2006; Yin et al. 2022).

6.4 Innovative Ways and Policy Strategies

With increasing environmental pressures and public health challenges in delta regions, innovation in wastewater treatment technologies has become critically important. Densely populated and pollution-prone, deltas require adaptive and sustainable approaches to safeguard water quality and minimize health risks. Table 6.1 presents several innovations implemented in various countries as efforts to address these challenges.

Starting in 2024, 297 cities throughout China will carry out urban evaluations to identify barriers to sustainable development, respond to community concerns, and enhance the coordinated management of urban challenges (Song et al. 2021). Earlier, in April 2015, China's State Council introduced the "Action Plan for the Prevention and Control of Water Pollution," designating the cleanup of black-odorous water bodies as a key national priority. The plan set a target to eliminate at least 90% of these polluted waters in built-up areas of prefecture-level cities and above by 2020. This initiative represents an integrated strategy to tackle urban water pollution, focusing on large-scale environmental rehabilitation, upgrades to wastewater treatment facilities, and the mitigation of ecological and public health threats linked to deteriorated aquatic systems (Xue et al. 2023). Another study mentioned that industrial activities often shift to areas near administrative boundaries where environmental oversight is weaker, leading to cross-boundary pollution. In delta regions, this creates significant challenges for wastewater management, as pollution can easily spread across jurisdictions. To effectively address this issue, stronger, coordinated policies and regulations are needed to control wastewater discharge and protect fragile delta ecosystems (Wang et al. 2024).

In Europe, countries such as Spain, Cyprus, France, and Italy comply with regulations governing wastewater management, specifically the European Regulation on Minimum Quality Requirements (MQR) for Water Reuse (EU 2020/741), issued in May 2020. This directive sets the standards for using reclaimed water in agricultural irrigation and became directly applicable across all Member States on June 26, 2023 (Moghimi Dehkordi et al. 2024). Following Cyclone Amphan in 2020, over 56 tons of plastic waste accumulated in the Sundarban, largely from unregulated relief materials. Each year, about four million tonnes of microplastics flow from Indian and Bangladeshi rivers into the Sundarban and Bay of Bengal, posing risks to biodiversity through trophic transfer and bioaccumulation. While both countries have existing mangrove protection policies, the scale of microplastic pollution calls for enhanced bilateral policy coordination, stricter transboundary river waste regulations, and targeted action plans to curb plastic discharge (Iqbal 2020; Kumar et al. 2022). A regional optimization model assessed the economic impacts of reusing treated municipal wastewater (TMW) in New Mexico's Middle Rio Grande Basin. The study found that water holds higher economic value in urban greenspace ($205.26/acre-foot) and environmental river flows ($220.39/acre-foot), while agriculture has a lower value ($28.30/acre-foot) (Olofinsao et al. 2025).

Table 6.1 Effective wastewater treatment innovations reported in various studies

No	Location	Methods/innovation	Result	Source
1	Bamako, Niger River, Mali	Providing baseline data is essential and supports the adoption of cost-effective, nature-based solutions, such as constructed wetlands, along with regulatory measures to reduce antibiotic pollution	Despite treatment, high ecological and microbial risks (RQ > 1) persisted at all sites, though three hospitals reduced these risks by over 50%, showing potential to mitigate environmental and health threats	Coulibaly et al. (2025)
2	China	The super-stable mineralization process significantly reduces chromium levels in wastewater	Using NiFe-LDH electrodes, wastewater treatment achieved over 99% removal of citric acid within 12 h at just 2.52 V and 50 mA, whereas bare carbon paper electrodes removed only ~50%, highlighting the critical role of LDH in enhancing treatment efficiency	Li et al. (2025)
3	China	Laccase (lac) effectively removed BPA from wastewater, especially in the presence of microplastics (MPs) like polyethylene (PE)	Laccase (lac) effectively removed BPA from wastewater, with removal rates reaching up to 96.1% in the presence of MPs such as PE. MPs enhanced both BPA adsorption and lac-catalyzed degradation, while ultrasound and micro-nano bubbles (MNB) further increased efficiency by 20–46%	Zhang et al. (2025)

(continued)

Table 6.1 (continued)

No	Location	Methods/innovation	Result	Source
4	Makassar, Indonesia	Lowering mercury concentrations in coastal regions is crucial. The use of polymer sulfur effectively reduces mercury in water, with reduction rates ranging from 39 to 100%.	The Monte Carlo simulation indicated that mercury exposure presents health risks for both adults and children (Total Hazard index >1). Mercury levels (23.3%) and exposure duration (21.3%) were identified as the key contributors to non-carcinogenic risk	Mallongi et al. (2023)
5	Keaukaha, Hawai'i island	Dye tracer tests, water quality analysis, $\delta^{15}N$ and $\delta^{18}O$ isotope measurements, and stable isotope mixing models were used to trace OSDS (onsite sewage disposal systems) connections to coastal waters	Sewage was detected in coastal waters and reefs. OSDS reached springs within 20 h to 3 days, with faster flow rates than expected. Springs showed higher bacteria and nutrients than the HWTP outfall. $\delta^{15}N$ confirmed sewage along the coast. Odor reports matched high bacteria and HWTP discharges	Waiki et al. (2025)
6	Nile Delta, Egypt	A pilot system combining facultative lagoons and hybrid constructed wetlands treated 250 m^3/day of drainage canal water in the Nile Delta over 530 days, using a surface loading rate of 200 L/day/m^2	The system removed 90% solids, 84% nitrogen, 80% phosphate, 67% COD, and reduced faecal coliforms by 2.2 log. The treated water met EU reuse standards and is suitable for crop irrigation, with low energy use, no chemicals, and minimal land needs (1.1–1.5%)	Frascari et al. (2024)

(continued)

Table 6.1 (continued)

No	Location	Methods/innovation	Result	Source
7	Mekong Delta, Vietnam	This study evaluates the impact of shrimp farming models on the Mekong Delta's socio-ecological systems, using risk assessments and community input to analyze aquaculture's effects on livelihoods and mangrove ecosystems.	Shrimp farming poses risks to mangroves and local communities. The study emphasizes the need for better wastewater management and stakeholder collaboration to promote sustainability in the Mekong Delta	Hua et al. (2024)
8	Nile Delta, Egypt	A multidisciplinary approach using hydrogeochemical, microbiological, and geophysical (ERT) data was applied to assess groundwater pollution and contamination pathways near an unlined wastewater drain in the eastern Nile Delta	High levels of pollutants (Pb, Cd, Cr) and bacteria were found in groundwater. ERT identified flow paths and lateral flow zones, aiding in the understanding of groundwater contamination. This approach helps improve monitoring and water quality management	Abu Salem et al. (2021)
9	Capitanata area, Italy	An econometric model (SURE) was used to analyze irrigation practices in southern Italy, assessing whether farms use collective water and groundwater as substitutes or complements, based on data from national agricultural databases	Farms using both sources show a substitution pattern—using more collective water (including treated wastewater) reduces groundwater withdrawals. Thus, providing treated wastewater to these farms can help lower aquifer exploitation without causing rebound effects	Manganiello et al. (2024)
10	Mekong Delta, Vietnam	Cyclopentane (CP) and R134a hydrates were tested for desalinating saline water from Vietnam's Mekong Delta, examining their thermodynamic and kinetic properties	Both hydrate methods achieved over 70% removal of salts and solids, offering an efficient and sustainable solution for saline water treatment with minimal waste, suitable for addressing saltwater intrusion in the region	Tran et al. (2024)

(continued)

Table 6.1 (continued)

No	Location	Methods/innovation	Result	Source
11	Córdoba department, in the Colombian Caribbean	This first study tracked MP transfer from a Colombian wastewater pond system to a marine protected area, analyzing MP levels in wastewater and sludge	While 67% of MPs were removed, about 1.3 million items per day still enter nearby waters, highlighting the need for better maintenance and efficiency in wastewater treatment to protect coastal ecosystems	Arregocés-Garcés et al. (2024)
12	Poland	This study analyzed stable carbon (δ13C) and nitrogen (δ15N) isotopes in aquatic macrophytes from 46 sites in northern Poland to assess nutrient pollution linked to land use	Isotopic differences among species and catchment types showed that δ13C and δ15N in macrophytes reflect human impacts, with urban areas displaying the lowest δ13C and highest δ15N values, confirming their use as indicators for river pollution monitoring	Pronin et al. (2025)
13	Amazon River Delta	This study analyzed soil cores from seven sites in the Amazon River delta to assess MPs distribution, burial rates, and potential ecological and economic impacts in mangrove ecosystems	MPs were widespread, with highest concentrations near the surface and evidence of downward migration; mangroves sequester 0.55 ± 0.28 tons/year of MPs, valued at up to 1.1 million USD annually, with current MP levels posing low ecotoxicological risks to benthic organisms	Rico et al. (2024)
14	Slave River Delta, Canada	Due to the declining capacity of the existing sewage lagoon system, a comprehensive assessment was conducted. Based on engineering analysis, cost considerations, site conditions, and regulatory requirements, a two-cell lined retention lagoon with an aligned primary cell was identified as the most suitable and sustainable solution	The chosen two-cell lagoon system is projected to meet effluent quality standards set in the community's water license for the next 20 years. The combination of engineered treatment and wetland polishing is expected to significantly reduce pollutants such as BOD, TSS, ammonia, and pathogens	Johnson (2019)

Current policy approaches to wastewater pollution primarily rely on high-tech treatment technologies, which often prove incomplete, inefficient, and costly. These limitations highlight the necessity for a more integrated and comprehensive strategy, such as the three-step strategic approach (3-SSA), which encompasses minimization and prevention, treatment for reuse, and stimulated natural self-purification (Galvis et al. 2018; Hotchkies 2018). Research has demonstrated that the 3-SSA is more effective than conventional methods, yielding higher pollutant load removal and more favorable cost-benefit outcomes. Recently, there is ongoing debate regarding the use of environmental policy instruments, with taxes and command-and-control approaches being prevalent in developed urban wastewater systems. These observations underscore the need for more effective policy frameworks to address the rapidly growing global wastewater challenge.

The 3-SSA offers a holistic solution by combining technological advancements with natural purification processes, emphasizing wastewater minimization, recycling of treated water, and the enhancement of natural self-purification. This integrated approach has been shown to provide superior economic outcomes when compared to traditional strategies. The cost-benefit analysis of the 3-SSA suggests that it is a more economically efficient option for wastewater management (Xenarios and Bithas 2012). Given the significant environmental and public health risks associated with wastewater pollution, the adoption of the 3-SSA and refinement of policy instruments are essential to improving wastewater management practices and mitigating the negative impacts on water resources.

Figure 6.1 illustrates not only the current conditions but also presents several suggestions and viewpoints regarding pollution control challenges in industry. These include recommended pathways for improvement such as enhancing regulatory supervision, promoting inter-agency cooperation, encouraging demonstration projects, and strengthening economic and environmental policy instruments. The issues depicted such as inadequate energy benchmarks, lack of comprehensive environmental and technical standards, a fragmented standardization system, and inappropriate saline wastewater treatment methods are not unique to a single nation. Rather, they reflect broader systemic challenges faced by many countries, particularly those with rapidly expanding industrial sectors. These unresolved issues place immense pressure on delta regions, which are already ecologically fragile and socioeconomically complex. In such vulnerable landscapes, pollution control failures can trigger cascading effects on ecosystems, human health, and livelihoods. Addressing these pressures demands integrated solutions that transcend sectoral silos. The involvement of multiple stakeholders is essential—ranging from government supervisors and industry demonstrators to transboundary cooperatives and economic policymakers. Their roles in shaping effective environmental management, enforcing regulations, promoting innovation, and harmonizing technical standards are crucial to overcoming persistent gaps in wastewater governance.

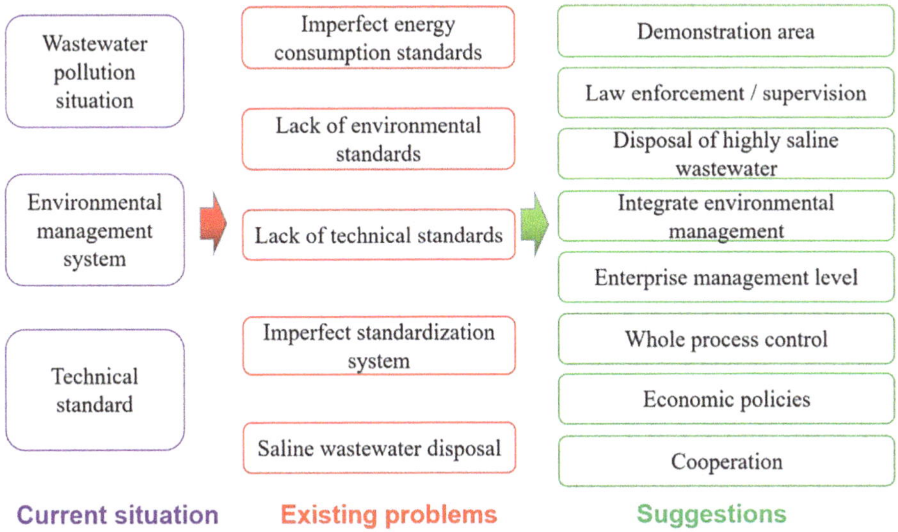

Fig. 6.1 Current challenges and strategic recommendations for wastewater pollution (Shi et al. 2021)

6.5 Conclusion and Recommendations

Ensuring proper wastewater treatment and sanitation is essential for protecting public health and preventing environmental contamination. Inadequate treatment of wastewater can lead to the spread of disease, degradation of ecosystems, and increased pressure on public health systems. Beyond immediate health risks, poor wastewater management undermines long term sustainability and community resilience. Addressing these issues requires advances in treatment technologies such as lagoon systems, filtration, and natural wetland based polishing processes, along with strong regulatory frameworks and community led planning. A systems based approach that emphasizes long term health and environmental outcomes, while integrating technical innovation and cross sector collaboration, provides a promising path forward.

References

Abedin MA, Habiba U, Shaw R (2013) Salinity scenario in Mekong, Ganges, and Indus River deltas. In: Abedin MA, Habiba U, Shaw R (eds) Community, environment and disaster risk management, vol 13. Emerald Group Publishing Limited, pp 115–138. https://doi.org/10.1108/S2040-7262(2013)0000013012

Abu Salem H, Gemail KS, Nosair AM (2021) A multidisciplinary approach for delineating wastewater flow paths in shallow groundwater aquifers: a case study in the southeastern part of the Nile Delta, Egypt. J Contam Hydrol 236:103701. https://doi.org/10.1016/j.jconhyd.2020.103701
Amin F, Gupta AK, Katyal D (2025) Impact assessment of climate change on lakes and Rivers: a case study of dal Lake and Jhelum River. In: Das J, Umamahesh NV, Pu JH, Pandey M (eds) Climate change impact on water resources, vol 561. Springer Nature Singapore, pp 75–89. https://doi.org/10.1007/978-981-97-9180-4_6
Aransiola SA, Zobeashia SSL-T, Ikhumetse AA, Musa OI, Abioye OP, Ijah UJJ, Maddela NR (2024) Niger Delta mangrove ecosystem: biodiversity, past and present pollution, threat and mitigation. Reg Stud Mar Sci 75:103568. https://doi.org/10.1016/j.rsma.2024.103568
Arregocés-Garcés R, Garcés-Ordóñez O, Vivas-Aguas L-J, Canals M (2024) Microplastics transfer from a malfunctioning municipal wastewater oxidation pond into a marine protected area in the Colombian Caribbean. Reg Stud Mar Sci 69:103361. https://doi.org/10.1016/j.rsma.2023.103361
Astuti RDP, Mallongi A, Hatta M, Rauf AU, Amiruddin R, Shoedarto RM (2025) Occurrence, distribution, and ecological risk of heavy metals using species sensitivity distribution model in Pangkajene river, Indonesia. Int J Environ Sci Technol 22(5):3691–3706. https://doi.org/10.1007/s13762-024-05834-y
Babatunde AO (2020) Local perspectives on food security in Nigeria's Niger delta. Extr Ind Soc 7(3):931–939. https://doi.org/10.1016/j.exis.2020.07.011
Bakshi B, Bouchard RW, Dietz R, Hornbach D, Monson P, Sietman B, Wasley D (2023) Freshwater mussels, ecosystem services, and clean water regulation in Minnesota: formulating an effective conservation strategy. Water 15(14):2560. https://doi.org/10.3390/w15142560
Benz SA, Bayer P, Blum P (2017) Identifying anthropogenic anomalies in air, surface and groundwater temperatures in Germany. Sci Total Environ 584–585:145–153. https://doi.org/10.1016/j.scitotenv.2017.01.139
Brock, M. L., Tagert, M. L. M., Paz, J. O., & Yasarer L (2019) Characterizing surface water use for irrigation in the Mississippi delta. 2019 Boston, Massachusetts July 7—July 10, 2019. https://doi.org/10.13031/aim.201901506
Burdon D, Potts T, McKinley E, Lew S, Shilland R, Gormley K, Thomson S, Forster R (2019) Expanding the role of participatory mapping to assess ecosystem service provision in local coastal environments. Ecosyst Serv 39:101009. https://doi.org/10.1016/j.ecoser.2019.101009
Chojnacka K, Mikulewicz M (2014) Bioaccumulation. In: Encyclopedia of toxicology. Elsevier, pp 456–460. https://doi.org/10.1016/B978-0-12-386454-3.01039-3
Coulibaly B, Pastor-López EJ, Diawara A, Savane FB, Escolà-Casas M, Matamoros V, Ba S (2025) Occurrence of antibiotics in hospital wastewater effluents discharged into The Niger River in Bamako, Mali. Risk assessment and solutions. Environ Pollut 371:125912. https://doi.org/10.1016/j.envpol.2025.125912
Damian N, Mitrică B, Mocanu I, Grigorescu I, Dumitraşcu M (2023) An index-based approach to assess the vulnerability of socio-ecological systems to aridity and drought in the Danube Delta, Romania. Environ Dev 45:100799. https://doi.org/10.1016/j.envdev.2022.100799
Day J, Goodman R, Chen Z, Hunter R, Giosan L, Wang Y (2021) Deltas in Arid Environments. Water 13(12):1677. https://doi.org/10.3390/w13121677
Day JW, Xu YJ, Keim BD, Brown VM, Giosan L, Mann ME, Stephens JR (2024) Emerging climate threats to the Mississippi River Delta: moving from restoration to adaptation. One Earth 7(4):558–571. https://doi.org/10.1016/j.oneear.2024.03.001
Deng M, Liu H (2023) Driving factors and potential offshore pollution of plastic mulch residue in farmland in the Yellow River Delta, China. https://doi.org/10.2139/ssrn.4603666

Dewulf A, Termeer C (2015) Governing the future? The potential of adaptive delta management to contribute to governance capabilities for dealing with the wicked problem of climate change adaptation. J Water Clim Change 6(4):759–771. https://doi.org/10.2166/wcc.2015.117

Döll P, Zhang J (2010) Impact of climate change on freshwater ecosystems: a global-scale analysis of ecologically relevant river flow alterations. Hydrol Earth Syst Sci 14(5):783–799. https://doi.org/10.5194/hess-14-783-2010

Dwivedi A, Khire MV (2018) Application of split-window algorithm to study urban Heat Island effect in Mumbai through land surface temperature approach. Sustain Cities Soc 41:865–877. https://doi.org/10.1016/j.scs.2018.02.030

Emmanouil C, Giannakis I, Kyzas GZ (2024) Terrestrial bioassays for assessing the biochemical and toxicological impact of biosolids application derived from wastewater treatment plants. Sci Total Environ 931:172718. https://doi.org/10.1016/j.scitotenv.2024.172718

Ewim DRE, Orikpete OF, Scott TO, Onyebuchi CN, Onukogu AO, Uzougbo CG, Onunka C (2023) Survey of wastewater issues due to oil spills and pollution in The Niger Delta area of Nigeria: a secondary data analysis. Bull Natl Res Cent 47(1):116. https://doi.org/10.1186/s42269-023-01090-1

Fentiman A, Zabbey N (2015) Environmental degradation and cultural erosion in Ogoniland: a case study of the oil spills in Bodo. Extr Ind Soc 2(4):615–624. https://doi.org/10.1016/j.exis.2015.05.008

Frascari D, Rashed A, Girometti E, Pinelli D, Toscano A, Lavrnić S (2024) Demonstration scale treatment of drainage canal water in the Nile Delta through a combination of facultative lagoons and hybrid constructed wetlands. J Environ Manag 370:122663. https://doi.org/10.1016/j.jenvman.2024.122663

Galvis A, Van Der Steen P, Gijzen H (2018) Validation of the three-step strategic approach for improving urban water management and water resource quality improvement. Water 10(2):188. https://doi.org/10.3390/w10020188

Gan Y, Huang X, Li S, Liu N, Li YC, Freidenreich A, Wang W, Wang R, Dai J (2019) Source quantification and potential risk of mercury, cadmium, arsenic, lead, and chromium in farmland soils of Yellow River Delta. J Clean Prod 221:98–107. https://doi.org/10.1016/j.jclepro.2019.02.157

Garg A, Gautamb P, Salunkeb D (2022) Advanced treatment Technologies for Industrial Wastewater. In: Roy S, Garg A, Garg S, Tran TA (eds) Advanced industrial wastewater treatment and reclamation of water. Springer International Publishing, pp 25–44. https://doi.org/10.1007/978-3-030-83811-9_2

Green Denmark in Southeast Asia. (2021). Flood control, wastewater management and climate change adaption in Ho chi Minh City & Mekong Delta. Green Denmark in Southeast Asia. https://www.greendkinsea.com/post/flood-control-wastewater-management-and-climate-change-adaption-in-ho-chi-minh-city-mekong-delta

Gu C, Hu L, Zhang X, Wang X, Guo J (2011) Climate change and urbanization in the Yangtze River Delta. Habitat Int 35(4):544–552. https://doi.org/10.1016/j.habitatint.2011.03.002

Hamad AY, Mohammed NS, Hasan AS (2022) Molecular detection and phylogenetic analysis of Astrovirus and noroviruses from sewage water in Diyala-Iraq. J Pharm Negative Results 13(S01) https://doi.org/10.47750/pnr.2022.13.s01.42

He Y, Lin K, Liao N, Chen Z, Rao J (2022) Exploring the spatial effects and influencing factors of PM2.5 concentration in the Yangtze River Delta urban agglomerations of China. Atmos Environ 268:118805. https://doi.org/10.1016/j.atmosenv.2021.118805

Hoang HG, Hadi M, Nguyen MK, Hai Nguyen NS, Huy Le PQ, Nguyen KN, Tran H-T, Mishra U (2025) Assessing heavy metal pollution levels and associated ecological risks in peatland areas in the Mekong Delta region. Environ Res 274:121319. https://doi.org/10.1016/j.envres.2025.121319

Hotchkies JW (2018) Future of water resource recovery. In: Biswas AK, Tortajada C, Rohner P (eds) Assessing global water megatrends. Springer Singapore, pp 77–85. https://doi.org/10.1007/978-981-10-6695-5_5

Hua HH, Cremin E, Huynh DV, Long G, Renaud FG (2024) Impacts of aquaculture practices on the sustainability of social-ecological systems in coastal zones of the Mekong Delta. Ocean Coast Manag 258:107392. https://doi.org/10.1016/j.ocecoaman.2024.107392

Hung Y-T, Wang LK, Shammas NK (2012) Handbook of environment and waste management: air and water pollution control. World Scientific Publishing. https://doi.org/10.1142/7971

Ibrahimi S, Gasmi A, Elboughdiri N, Aydi SS, Ghernaout D, Hannachi A (2025) Nanofiltration and electrocoagulation treatment of Moorish Bath wastewater for irrigation purposes: case study of El-Hamma, southeastern Tunisia. J Water Process Eng 70:106955. https://doi.org/10.1016/j.jwpe.2025.106955

Iqbal MH (2020) Valuing ecosystem services of Sundarbans mangrove forest: approach of choice experiment. Glob Ecol Conserv 24:e01273. https://doi.org/10.1016/j.gecco.2020.e01273

Ismail EM, Kadry M, Elshafiee EA, Ragab E, Morsy EA, Rizk O, Zaki MM (2021) Ecoepidemiology and potential transmission of vibrio cholerae among different environmental niches: an upcoming threat in Egypt. Pathogens 10(2):190. https://doi.org/10.3390/pathogens10020190

Jha, S., Raghuram, S., & Siddhant, A. (2019). Exploring strategies for planned urban cluster development in South Asia. ADB South Asia Working Paper Series. https://www.adb.org/sites/default/files/publication/496556/swp-064-urban-cluster-development-south-asia.pdf

Jiang X (2021) Future directions of environmental chemistry. Pure Appl Chem 93(12):1403–1409. https://doi.org/10.1515/pac-2020-0806

Johnson K (ed) (2019) Process options for a sewage lagoon in sub Arctic Canada. Curran Associates, Inc

Klimkowicz-Pawlas A, Smreczak B, Maliszewska-Kordybach B (2023) Integrated ecological risk assessment of the agricultural area under a high Anthropopressure based on chemical, Ecotoxicological and ecological indicators. Agriculture 13(7):1353. https://doi.org/10.3390/agriculture13071353

Kumar R, Sinha R, Rakib MRJ, Padha S, Ivy N, Bhattacharya S, Dhar A, Sharma P (2022) Microplastics pollution load in Sundarban delta of bay of Bengal. J Hazard Mater Adv 7:100099. https://doi.org/10.1016/j.hazadv.2022.100099

Kyere-Yeboah K, Qiao X (2024) Ammonia nitrogen removal from coal gasification wastewater using a falling-film dielectric barrier discharge reactor: process optimization and toxicity analysis. J Water Process Eng 62:105358. https://doi.org/10.1016/j.jwpe.2024.105358

Levallois P (2006) Eau potable et santé publique: Défis actuels et futurs. Revue Des Sciences de l'eau 19(2):127–135. https://doi.org/10.7202/013046ar

Li J, Chen C, Li F (2016) Status of POPs accumulation in the Yellow River Delta: from distribution to risk assessment. Mar Pollut Bull 107(1):370–378. https://doi.org/10.1016/j.marpolbul.2016.03.063

Li J, Yang D, Li L, Jabeen K, Shi H (2015) Microplastics in commercial bivalves from China. Environ Pollut 207:190–195. https://doi.org/10.1016/j.envpol.2015.09.018

Li Y, Bai H, Li Y, Zhang X, Zhang L, Zhang D, Xu M, Zhang H, Lu P (2023) An integrated approach to identify the source apportionment of potentially toxic metals in shale gas exploitation area soil, and the associated ecological and human health risks. J Hazard Mater 458:132006. https://doi.org/10.1016/j.jhazmat.2023.132006

Li Z, Dong W, Wang Y, Wang R, Liu S, Ren J, Xu N, Jin X, Peng Y, Zhao Y (2025) Super-stable mineralization coupled with electrolysis towards the treatment of leather wastewater containing high-concentration organic pollutants. Chem Eng Sci 310:121545. https://doi.org/10.1016/j.ces.2025.121545

Lin C, Li Y, Lau AKH, Li C, Fung JCH (2018) 15-year PM2.5 trends in the Pearl River Delta region and Hong Kong from satellite observation. Aerosol Air Qual Res 18(9):2355–2362. https://doi.org/10.4209/aaqr.2017.11.0437

Liu T, Li Y (2021) Green development of China's Pan-Pearl River Delta mega-urban agglomeration. Sci Rep 11(1):15717. https://doi.org/10.1038/s41598-021-95312-z

Liu X, Guo W (2023) Dynamic nonlinear effects of urbanization on wastewater discharge based on inertial characteristics of wastewater discharge. Sci Total Environ 904:166514. https://doi.org/10.1016/j.scitotenv.2023.166514

Ma J-X, Wang X, Pan Y-R, Wang Z-Y, Guo X, Liu J, Ren N-Q, Butler D (2024) Data-driven systematic analysis of waterborne viruses and health risks during the wastewater reclamation process. Environ Sci Ecotechnol 19:100328. https://doi.org/10.1016/j.ese.2023.100328

Maliga I, Purwono S, Harini R (2025) Marine pollution in small Island ecosystems and the impact of domestic wastewater. Glob J Environ Sci Manag 11(1) https://doi.org/10.22034/gjesm.2025.01.11

Mallongi A, Rauf AU, Astuti RDP, Palutturi S, Ishak H (2023) Ecological and human health implications of mercury contamination in the coastal water. Glob J Environ Sci Manag 9(2) https://doi.org/10.22034/gjesm.2023.02.06

Manganiello V, Zucaro R, Dono G (2024) The use of purified wastewater for irrigation: possible strategies in the Capitanata area (Apulia, Italy). Agric Water Manag 291:108614. https://doi.org/10.1016/j.agwat.2023.108614

Mansur AV, Brondízio ES, Roy S, Hetrick S, Vogt ND, Newton A (2016) An assessment of urban vulnerability in the Amazon Delta and estuary: a multi-criterion index of flood exposure, socio-economic conditions and infrastructure. Sustain Sci 11(4):625–643. https://doi.org/10.1007/s11625-016-0355-7

Meng X, Zeng P, Lin S, Bao H, Wu M, Yang L, Jing G, Han H, Zhang C, Jiang X, Wang S, Ding W, Sun W (2023) Removal of chemical oxygen demand and ammonia nitrogen from high salinity tungsten smelting wastewater by one-step electrochemical oxidation: from bench-scale test, pilot-scale test, to industrial test. J Environ Manag 340:117983. https://doi.org/10.1016/j.jenvman.2023.117983

Moghimi Dehkordi M, Pournuroz Nodeh Z, Soleimani Dehkordi K, Salmanvandi H, Rasouli Khorjestan R, Ghaffarzadeh M (2024) Soil, air, and water pollution from mining and industrial activities: sources of pollution, environmental impacts, and prevention and control methods. Results Eng 23:102729. https://doi.org/10.1016/j.rineng.2024.102729

Musisi M, Quadros C, Sellappan K (2025) Impact of anthropogenic pollution on the plant species diversity and composition along the riparian ecotones of Goa's Sal and Zuari Rivers. Plant-Environ Interact 6:e70037. https://doi.org/10.1002/pei3.70037

Nichols G, Lake I, Heaviside C (2018) Climate change and water-related infectious diseases. Atmos 9(10):385. https://doi.org/10.3390/atmos9100385

Ogata T, Racault M-F, Nonaka M, Behera S (2021) Climate precursors of satellite water marker index for spring cholera outbreak in Northern Bay of Bengal coastal regions. Int J Environ Res Public Health 18(19):10201. https://doi.org/10.3390/ijerph181910201

Ola I, Drebenstedt C, Burgess RM, Mensah M, Hoth N, Külls C (2024) Remediating oil contamination in The Niger Delta region of Nigeria: technical options and monitoring strategies. Extr Ind Soc 17:101405. https://doi.org/10.1016/j.exis.2024.101405

Olofinsao OA, Wang J, Berrens RP (2025) Reuse of treated municipal wastewater in drylands: multi-sector optimization analysis for middle Rio Grande case study. Util Policy 95:101941. https://doi.org/10.1016/j.jup.2025.101941

Peng X, Liao L, Tan X, Yu R, Zhang K (2024) City health assessment: urbanization and eco-environment dynamics using coupling coordination analysis and FLUS model—a case study of the Pearl River Delta urban agglomeration. Land 14(1):46. https://doi.org/10.3390/land14010046
Phung D, Huang C, Rutherford S, Chu C, Wang X, Nguyen M (2015) Climate change, water quality, and water-related diseases in the Mekong Delta basin: a systematic review. Asia Pac J Public Health 27(3):265–276. https://doi.org/10.1177/1010539514565448
Pobi KK, Nayek S, Gope M, Rai AK, Saha R (2020) Sources evaluation, ecological and health risk assessment of potential toxic metals (PTMs) in surface soils of an industrial area, India. Environ Geochem Health 42(12):4159–4180. https://doi.org/10.1007/s10653-020-00517-2
Pratama MA, Constantine DM (2023) A spatiotemporal analysis of organic pollutants contamination in Ciliwung River, Indonesia. CSID J Infrastruct Dev 6(2). https://doi.org/10.7454/jid.v6.i2.1119
Pronin E, Wrosz Z, Banaś K, Merdalski M (2025) Following the footsteps of macrophytes: potential application of isotope signals in pollution monitoring: a case study of northern polish rivers. Ecohydrol Hydrobiol 100650. https://doi.org/10.1016/j.ecohyd.2025.100650
Ramírez Hernández R (2024) Expansión urbana y clima en las ciudades mexicanas. Un análisis del crecimiento, usos de suelo y clima urbano en México. Investigaciones Geográficas 114. https://doi.org/10.14350/rig.60852
Rao Y, Xu Y, Zhang J, Guo Y, Fu M (2018) Does subclassified industrial land have a characteristic impact on land surface temperatures? Evidence for and implications of coal and steel processing industries in a Chinese mining city. Ecol Indic 89:22–34. https://doi.org/10.1016/j.ecolind.2018.01.058
Rico A, Redondo-Hasselerharm PE, Schell T, Sanders CJ, Bernardino AF (2024) Microplastic burial potential and ecological risks in mangrove forests of the Amazon River delta. Sci Total Environ 957:177666. https://doi.org/10.1016/j.scitotenv.2024.177666
Rusprayunita NRA, Puspitasari S, Mahiroh H, Setyani ER, Citraningtyas VI, Wahyuningsih WS, Rauf AU (2025) Water pollution reduction for sustainable urban development. In: Parray JA, Shameem N, Haghi AK (eds) Sustainable urban environment and waste management. Springer Nature Singapore, pp 1–21. https://doi.org/10.1007/978-981-96-1140-9_1
Safra De Campos R, Codjoe SNA, Adger WN, Mortreux C, Hazra S, Siddiqui T, Das S, Atiglo DY, Bhuiyan MRA, Rocky MH, Abu M (2020) Where people live and move in deltas. In: Nicholls RJ, Adger WN, Hutton CW, Hanson SE (eds) Deltas in the Anthropocene. Springer International Publishing, pp 153–177. https://doi.org/10.1007/978-3-030-23517-8_7
Shaefer M, Doyle L, Fleenor W, Johnson M (2013) Fate and transport of three Pharmaceuticals in the Sacramento-san Joaquin Delta. San Franc Estuary Watershed Sci 11(2) https://doi.org/10.15447/sfews.2013v11iss2art3
Sharifinia M, Taherizadeh M, Namin JI, Kamrani E (2025) How do macrobenthic-based indices respond to anthropogenic pressures? Insights from estuaries of the Persian Gulf and Gulf of Oman. Deep-Sea Res II Top Stud Oceanogr 221:105481. https://doi.org/10.1016/j.dsr2.2025.105481
Shen X, Zhang Q, Xiang Q, Zhao J, Cao Y, Li K, Song J, Wang Z, Zhao X, Chen Q (2024) Occurrences, source apportionment, and potential risks of 55 progestins in surface water of the Yellow River Delta, China. J Hazard Mater 480:136098. https://doi.org/10.1016/j.jhazmat.2024.136098
Shi H, Ma P, Sun J, Zhao S, Ma R, Li L, Zhan C, Liang H (2024) Study on the response mechanism of the Yellow River delta region to the cold wave process. Front Mar Sci 11:1430823. https://doi.org/10.3389/fmars.2024.1430823
Shi J, Huang W, Han H, Xu C (2021) Pollution control of wastewater from the coal chemical industry in China: environmental management policy and technical standards. Renew Sust Energ Rev 143:110883. https://doi.org/10.1016/j.rser.2021.110883

Small C, Sousa D, Yetman G, Elvidge C, MacManus K (2018) Decades of urban growth and development on the Asian megadeltas. Glob Planet Chang 165:62–89. https://doi.org/10.1016/j.gloplacha.2018.03.005

Śniatała B, Al-Hazmi HE, Sobotka D, Zhai J, Mąkinia J (2024) Advancing sustainable wastewater management: a comprehensive review of nutrient recovery products and their applications. Sci Total Environ 937:173446. https://doi.org/10.1016/j.scitotenv.2024.173446

Song R, Hu Y, Li M (2021) Chinese pattern of urban development quality assessment: a perspective based on national territory spatial planning initiatives. Land 10(8):773. https://doi.org/10.3390/land10080773

Sun S, Wang M, Yang X, Xu L, Wu J, Wang Y, Zhou Z (2023) Pollution characteristics and health risk assessment of phthalate esters in agricultural soil of the Yellow River Delta, China. Environ Sci Pollut Res 30(18):53370–53380. https://doi.org/10.1007/s11356-023-26104-3

Todd ECD (2024) Waterborne diseases and wastewater treatment in Iraq. J Food Prot 87(1):100204. https://doi.org/10.1016/j.jfp.2023.100204

Tran QN, Thi Vo NN, Pham TT, Truong-Lam HS (2024) Hydrate technology for water desalination in the Mekong Delta, Vietnam. Heliyon 10(19):e38974. https://doi.org/10.1016/j.heliyon.2024.e38974

Tripathi G, Husain A, Ahmad S, Hasan Z, Farooqui A (2021) Contamination of water resources in industrial zones. In: Contamination of water. Elsevier, pp 85–98. https://doi.org/10.1016/B978-0-12-824058-8.00017-7

Van Cauwenberghe L, Janssen CR (2014) Microplastics in bivalves cultured for human consumption. Environ Pollut 193:65–70. https://doi.org/10.1016/j.envpol.2014.06.010

Waiki SMP, Colbert SL, Wiegner TN, Puniwai N, Nakoa JWP III, Storie NM, Nelson CE, Overly AN, McDermid KJ, Aguiar DK (2025) Sewage pollution from onsite sewage disposal systems and an offshore wastewater treatment plant outfall in coastal waters of Keaukaha, Hawai'i Island. J Hydrol Reg Stud 57:102122. https://doi.org/10.1016/j.ejrh.2024.102122

Wang H, Li L, Xu X (2024) Do environmental regulation policies increase urban boundary pollution? Micro evidence from Chinese industrial enterprises. Environ Impact Assess Rev 106:107524. https://doi.org/10.1016/j.eiar.2024.107524

Xenarios S, Bithas K (2012) The use of environmental policy instruments for urban wastewater control: evidences from an international survey. Environ Policy Gov 22(1):14–26. https://doi.org/10.1002/eet.596

Xu Y, Liu Z, Walker TR, Adams M, Dong H (2024) Spatio-temporal patterns and spillover effects of synergy on carbon dioxide emission and pollution reductions in the Yangtze River Delta region in China. Sustain Cities Soc 107:105419. https://doi.org/10.1016/j.scs.2024.105419

Xue Y, Ma Y, Long G, He H, Li Z, Yan Z, Wan J, Zhang S, Zhu B (2023) Evaluation of water quality pollution and analysis of vertical distribution characteristics of typical Rivers in the Pearl River Delta, South China. J Sea Res 193:102380. https://doi.org/10.1016/j.seares.2023.102380

Yin H, Hu Y, Liu M, Li C, Lv J (2021) Ecological and environmental effects of estuarine wetland loss using keyhole and Landsat data in Liao River Delta, China. Remote Sens 13(2):311. https://doi.org/10.3390/rs13020311

Yin J, Shen Y, Cao J (2022) Burden of cryptosporidium infections in the Yangtze River Delta in China in the 21st century: a one health perspective. Zoonoses 2(1) https://doi.org/10.15212/zoonoses-2021-0025

Zhang J, Liu J, Liu X, Su Y, Yu C, Tan X, Zhuo L, Ye M, Luo Q, Li B (2025) Microplastics enhance laccase-driven bisphenol a removal in multipollutant wastewater. J Environ Manag 380:125123. https://doi.org/10.1016/j.jenvman.2025.125123

Zhao C, Pan Y, Wu H, Zhu Y (2024) Quantifying the contribution of industrial zones to urban heat islands: relevance and direct impact. Environ Res 240:117594. https://doi.org/10.1016/j.envres.2023.117594

Remote Sensing and GIS for Monitoring Delta Agro-ecosystems

7

Harshit Mishra, Fredrick Kayusi, Rashmi Mishra, Ioannis Adamopoulos, and Diksha Bohra

Abstract

Delta agro-ecosystems, recognized by their dynamic hydrological conditions and fertile alluvial soils, play a crucial role in global food security. However, these regions face significant environmental and climatic challenges, including soil salinity, waterlogging, and unpredictable weather patterns. Remote sensing and Geographic Information Systems (GIS) have emerged as essential tools for monitoring and managing these ecosystems by providing real-time data on land use, soil health, and crop conditions. Advances in satellite imagery and UAV technology facilitate high-resolution

H. Mishra (✉)
Department of Agricultural Economics, College of Agriculture, Acharya Narendra Deva University of Agriculture and Technology, Kumarganj, Ayodhya, Kumarganj, Uttar Pradesh, India
e-mail: wehars@gmail.com

F. Kayusi
Department of Environmental Studies, Geography and Planning, Maasai Mara University, Narok, Kenya

R. Mishra
Department of Economics and Business Administration, University of Technology and Applied Sciences, Ibra, Oman

I. Adamopoulos
Department of Public Health Policy, Sector of Occupational and Environmental Health, School of Public Health, University of West Attica, Athens, Greece

D. Bohra
Department of Agricultural Economics, Rajasthan College of Agriculture, Maharana Pratap University of Agriculture and Technology, Udaipur, Rajasthan, India

C. Singha et al. (eds.), *Sustainable Management of Delta Ecosystems Resilience*, Deltas of the World, https://doi.org/10.1007/978-3-032-16261-8_7

analysis, enhancing precision agriculture practices and resource allocation. According to the Food and Agriculture Organization (FAO), remote sensing applications have improved crop yield estimations by 85% and reduced water usage by 30% in deltaic regions through optimized irrigation strategies. The integration of remote sensing and GIS enables accurate soil moisture assessment, land cover classification, and pest and disease surveillance. Studies indicate that hyperspectral imaging can detect crop stress with 92% accuracy, aiding in early intervention measures. Furthermore, machine learning models trained on remote sensing data can predict climate impacts on delta agriculture with an accuracy exceeding 80%, supporting adaptive strategies against extreme weather events and sea-level rise. Precision agriculture techniques utilizing GIS have demonstrated a 20% increase in productivity and a 15% reduction in input costs in delta farming systems. The use of LiDAR (Light Detection and Ranging) technology for terrain mapping further enhances flood risk assessment and sustainable land management. Despite these technological advancements, challenges persist in data integration, policy implementation, and farmer adoption of remote sensing tools. Effective governance frameworks and international collaboration are essential for bridging the gap between technology development and field applications. Future research should focus on enhancing AI-driven analytics, improving sensor accuracy, and promoting stakeholder engagement for sustainable delta agro-ecosystem management. Remote sensing and GIS not only enhance agricultural productivity but also contribute to environmental conservation and climate resilience.

Abbreviations

GIS	Geographic Information Systems
FAO	Food and Agriculture Organization
LiDAR	Light Detection and Ranging
SAR	Synthetic Aperture Radar
NDVI	Normalized Difference Vegetation Index
UAVs	Unmanned Aerial Vehicles
GEOGLAM	Group on Earth Observations Global Agricultural Monitoring
AI	Artificial intelligence
EMR	Electromagnetic radiation
NIR	Near-infrared
SWIR	Shortwave infrared
DBMS	Database Management Systems
EVI	Enhanced Vegetation Index
SAVI	Soil Adjusted Vegetation Index
VRT	Variable Rate Technology
LST	Land surface temperature
FR	Random Forest

SVM	Support Vector Machines
DNNs	Deep Neural Networks
IoT	Internet of Things
TOA	Top-of-Atmosphere
MCE	Multi-criteria evaluation
SEBAL	Surface Energy Balance Algorithm
SMI	Soil Moisture Index
NRSC	National Remote Sensing Centre
OSDMA	Odisha State Disaster Management Authority
ICAR	Indian Council of Agricultural Research
EC	Electrical conductivity
IWMI	International Water Management Institute
DEMs	Digital Elevation Models
CNNs	Convolutional Neural Networks
SMAP	Soil Moisture Active Passive
TIRS	Thermal infrared sensor
LAI	Leaf Area Index
DWSI	Disease Water Stress Index
REPI	Red Edge Position Index
SI	Salinity Index
NDSI	Normalized Difference Salinity Index
VHR	Very high-resolution
GEE	Google earth engine
AWS	Amazon Web Services
CCC	Canopy Chlorophyll Content
DL	Deep Learning
SSCM	Site-specific crop management
TRMM	Tropical Rainfall Measuring Mission
GPM	Global Precipitation Measurement
GOES	Geostationary Operational Environmental Satellites
SMOS	Soil Moisture and Ocean Salinity
SPEI	Standardized Precipitation Evapotranspiration Index
NDDI	Normalized Difference Drought Index
ANN	Artificial Neural Networks
TCI	Temperature Condition Index
VHI	Vegetation Health Index
XAI	Explainable AI
PPPs	Public-Private Partnerships
GAEZ	Global Agro-Ecological Zoning
ESA	European Space Agency
DAICs	Delta Agriculture Innovation Councils

7.1 Introduction

Remote sensing and GIS have emerged as transformative tools in agricultural monitoring, particularly in the fragile and dynamic delta agro-ecosystems. These regions, marked by fertile alluvial soils, abundant water resources, and high population densities, are crucial for ensuring regional food security and livelihoods. However, they are also among the most vulnerable landscapes, facing intense anthropogenic pressure, climate change, and ecological degradation (Gaglio 2018; Tsanakas et al. 2024). In this context, the integration of remote sensing and GIS provides an efficient, non-invasive, and data-rich platform to analyse, monitor, and manage these ecosystems with spatial and temporal precision. Remote sensing involves the acquisition of information about the Earth's surface through satellite or aerial sensors without direct contact. In delta agro-ecosystems, remote sensing facilitates the collection of critical data such as land use, vegetation cover, crop phenology, soil moisture, and surface water distribution (Reljić et al. 2023). For example, satellite imagery from platforms such as Sentinel-2 and Landsat 8 enables regular monitoring of crop health and land degradation, while Synthetic Aperture Radar (SAR) from Sentinel-1 allows for consistent data collection even under cloud cover, which is common in delta regions. Similarly, GIS serves as a powerful analytical framework that integrates spatial data layers with socio-economic, climatic, and environmental variables. The synergy of remote sensing and GIS allows for the development of detailed maps, predictive models, and decision-support systems that can guide sustainable agricultural interventions (Mishra and Mishra 2025).

Delta regions such as the Mahanadi Delta in India, the Mekong Delta in Vietnam, and the Nile Delta in Egypt exhibit complex interactions between hydrology, land use, and agricultural practices. These areas contribute significantly to national agricultural output—for instance, the Mekong Delta accounts for over 50% of Vietnam's rice production and approximately 90% of its rice exports (Pham et al. 2022). Yet, they face acute challenges such as soil salinization, seasonal flooding, coastal erosion, and water resource mismanagement. Remote sensing tools have been instrumental in identifying these stressors. In the Godavari Delta, NDVI (Normalized Difference Vegetation Index) derived from satellite imagery has been used effectively to assess crop stress and predict yield variations. Similarly, GIS-based models integrating soil maps, digital elevation models, and historical land-use data have facilitated sustainable water management strategies in the Nile Delta (AbdekRahman and Tahoun 2019). Advanced techniques such as LiDAR (Light Detection and Ranging) and hyperspectral imaging further enhance the capacity to assess vegetation structure, topography, and soil characteristics at high resolutions. Unmanned Aerial Vehicles (UAVs) equipped with multispectral sensors are increasingly being deployed in precision agriculture to monitor pest infestations, crop nutrient levels, and irrigation efficiency in real time. These technologies enable site-specific interventions, which are particularly important in delta regions where spatial heterogeneity in soil texture, water availability, and cropping patterns is prominent (Jhariya et al. 2023).

Figure 7.1 shows a spatial comparison between surface water maps of Tri A Reservoir, extracted from three satellite platforms. Sentinel-2 observations were acquired on March 13, 2020, while Sentinel-1 and ALOS-2 observations were acquired on March 15, 2020. The differences in estimated water surface area highlight the variability in detection capabilities across different sensors, emphasizing the importance of multi-source remote sensing for accurate hydrological monitoring.

Climate change exacerbates the existing vulnerabilities in delta agro-ecosystems. Rising sea levels, increased frequency of cyclones, and erratic rainfall patterns directly impact crop productivity and soil health. According to the IPCC, sea-level rise could affect over 12 million people in the Ganges–Brahmaputra-Meghna Delta by 2050. Remote sensing data are essential for monitoring such changes (Kumar et al. 2023; Singh and Kalra 2016). For instance, satellite altimetry and SAR have been used to model coastal inundation and salinity intrusion scenarios, providing crucial input for climate-resilient agricultural planning. The integration of artificial intelligence (AI) and machine learning with big data analytics is further revolutionizing the monitoring process. Algorithms trained on historical satellite imagery and meteorological data are being used to predict pest outbreaks, estimate crop yields, and develop early warning systems, reducing uncertainties in agricultural decision-making. Effective application of remote sensing and GIS in delta

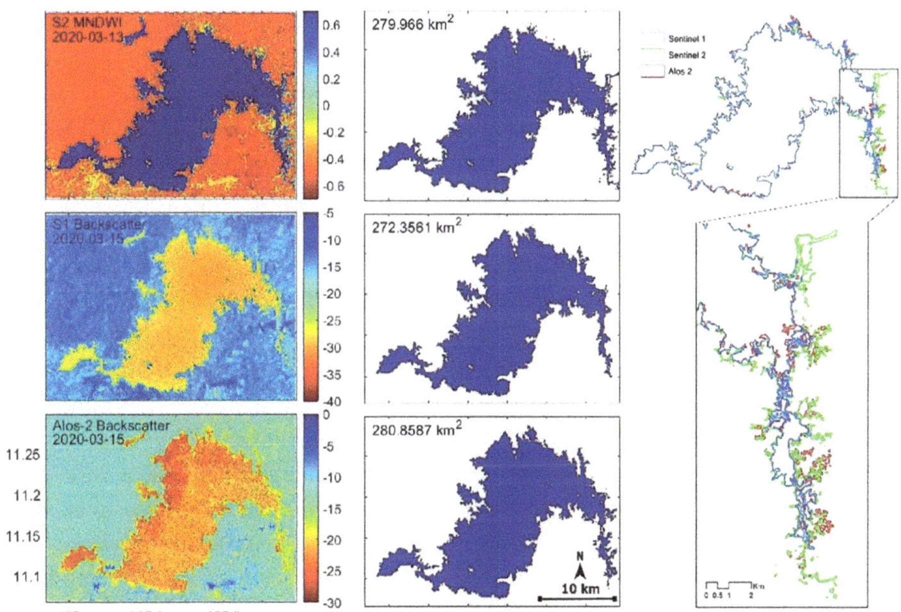

Fig. 7.1 Spatial comparison of surface water maps of Tri A Reservoir from Sentinel-2 (March 13, 2020), Sentinel-1, and ALOS-2 (March 15, 2020), highlighting variations in water extent estimation across different sensors (Duc 2024)

regions also depends on supportive policy frameworks and capacity building. National and international institutions are increasingly recognizing the value of geospatial technologies (Swain et al. 2022). Initiatives like Digital India and GEOGLAM (Group on Earth Observations Global Agricultural Monitoring) underscore the growing emphasis on leveraging satellite data for agricultural governance. However, challenges such as data standardization, high costs of advanced technologies, and limited access in remote areas persist. Addressing these gaps requires multi-stakeholder collaboration and investment in infrastructure and training to enhance adoption and utility (Shuford et al. 2019).

Remote sensing and GIS offer indispensable capabilities for monitoring and managing delta agro-ecosystems. Their ability to capture real-time, multi-scalar, and high-resolution data allows for informed decision-making and adaptive management strategies. As delta regions continue to face mounting ecological and climatic threats, the application of these technologies becomes increasingly vital for safeguarding agricultural productivity, sustaining rural livelihoods, and achieving long-term food security. Integrating remote sensing and GIS with emerging technologies like AI, while fostering inclusive policies and capacity building, is essential for building resilient and sustainable delta agro-ecosystems (Ndlala 2021; Mooney and Sombroek, 1992). The primary objective of this chapter is to critically examine the integration of remote sensing and GIS in the monitoring and management of delta agro-ecosystems, with a focus on their capabilities to address the complex environmental and agricultural challenges characteristic of these regions. Additionally, the chapter aims to evaluate region-specific case studies where spatial data applications have been effectively employed to monitor climate variability, soil and water dynamics, and land-use changes in deltaic landscapes. Furthermore, it seeks to analyse the role of advanced geospatial technologies, including artificial intelligence and big data analytics, in enhancing the precision, efficiency, and sustainability of agricultural practices within these ecologically sensitive and economically significant zones.

7.2 Fundamentals of Remote Sensing and GIS in Agriculture

The application of Remote Sensing and GIS in agriculture has revolutionized the way agro-ecosystems, particularly delta regions, are monitored and managed. Delta agro-ecosystems, characterized by dynamic land–water interactions, rich biodiversity, and high agricultural productivity, require efficient monitoring tools to ensure sustainability. Remote sensing provides timely, accurate, and large-scale data acquisition, while GIS facilitates spatial analysis, visualization, and decision-making in agricultural landscapes. The integration of these technologies enhances precision agriculture, crop health assessment, soil quality monitoring, and water resource management. This section provides a detailed overview of the fundamental principles of remote sensing and GIS, their integration for agricultural monitoring, and the latest advancements in spatial data analysis.

7.2.1 Principles of Remote Sensing

Remote sensing is the science of acquiring, processing, and interpreting data about objects or landscapes from a distance, typically using satellites, drones, or airborne sensors. The fundamental principle of remote sensing is based on the interaction of electromagnetic radiation (EMR) with the Earth's surface. When solar radiation strikes an agricultural field, different features (e.g., soil, water, vegetation) absorb, reflect, or transmit energy in characteristic ways. As depicted in Fig. 7.2, incident solar radiation from the sun interacts with various land cover types such as forests, water bodies, grass, bare soil, paved roads, and built-up areas, each reflecting different amounts of radiation back into space. This reflected radiation is detected by satellites, allowing for the assessment of surface properties. These interactions are captured by remote sensors and analysed to extract meaningful information about agricultural conditions, such as vegetation health, soil moisture, and land use patterns (Suyunov et al. 2025).

Remote sensing data are primarily categorized into:
Optical remote sensing: Uses visible, near-infrared (NIR), and shortwave infrared (SWIR) bands to analyse vegetation health, soil moisture, and land-use patterns.

Thermal remote sensing: Measures land surface temperature, enabling assessment of evapotranspiration, irrigation efficiency, and drought stress.

Microwave remote sensing: Includes active (RADAR) and passive (radiometers) sensors, useful for mapping soil moisture and flood extent, particularly crucial in delta regions.

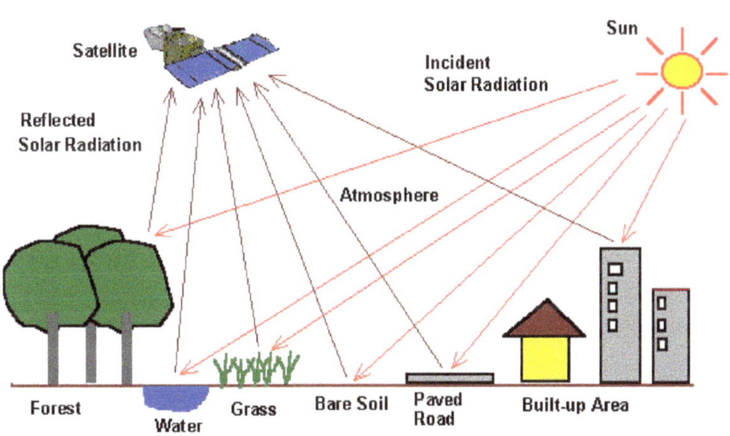

Fig. 7.2 Illustration of remote sensing principles, showing the interaction of incident solar radiation with different land cover types and the reflected radiation captured by a satellite for analysis (Dehkordi et al. 2020)

Hyperspectral remote sensing: Captures detailed spectral information, facilitating crop classification, nutrient deficiency assessment, and pest detection.

The effectiveness of remote sensing in agriculture depends on the spectral, spatial, radiometric, and temporal resolution of the sensors. High spatial resolution is essential for small-scale field monitoring, while high temporal resolution is critical for tracking crop growth cycles and detecting anomalies.

7.2.2 Overview of Geographic Information Systems

GIS provide a framework for capturing, storing, analysing, and visualizing geospatial data. In agriculture, GIS is indispensable for land-use planning, crop yield forecasting, soil mapping, and resource management. The core components of GIS include:

Geospatial data: Comprises vector data (points, lines, polygons) representing field boundaries, irrigation networks, and roads, and raster data (grids) from satellite imagery.

Database Management Systems (DBMS): Organizes and retrieves agricultural data efficiently.

Spatial analysis and modelling: Includes techniques like interpolation, overlay analysis, and terrain modelling, used for soil fertility mapping and precision farming.

Remote sensing integration: GIS acts as a platform where remote sensing images are analysed to extract useful agricultural insights.

GIS applications in delta agro-ecosystems are extensive. It supports spatial variability assessments, flood risk mapping, soil salinity monitoring, and irrigation optimization. With the increasing adoption of cloud-based GIS platforms (Google Earth Engine, ArcGIS Online), real-time monitoring of agricultural landscapes has become more feasible (Liew 2001).

7.2.3 Integration of Remote Sensing and GIS for Agricultural Monitoring

The integration of remote sensing and GIS offers comprehensive, multi-dimensional insights into agricultural systems. This synergy enables data-driven decision-making in agricultural management. Some applications of integrated remote sensing and GIS in agriculture include:

Crop Monitoring and Yield Estimation
Satellite-based indices like NDVI, Enhanced Vegetation Index (EVI), and Soil Adjusted Vegetation Index (SAVI) help in monitoring crop growth, stress levels, and biomass estimation. GIS maps these indices over time to analyse spatial variations in crop health and estimate potential yield losses.

Soil Moisture and Land Degradation Mapping
Microwave remote sensing (e.g., Sentinel-1 SAR, SMAP) measures soil moisture content, which is crucial for irrigation scheduling. GIS integrates this data with topographic and climatic layers to generate soil degradation and erosion risk maps.

Flood and Drought Risk Assessment in Delta Agro-ecosystems
Deltas are prone to seasonal flooding and waterlogging. Remote sensing helps identify flood-prone zones using SAR, which penetrates cloud cover. GIS overlays flood extent with agricultural land use to determine affected areas and guide disaster mitigation strategies.

Precision Agriculture and Variable Rate Technology (VRT)
By integrating remote sensing data on crop health with GIS-based farm management systems, farmers can implement site-specific nutrient application, irrigation, and pest control. This approach, known as Precision Agriculture, maximizes productivity while reducing input costs. Some benefits of Remote Sensing-GIS integration in delta agro-ecosystems are provided in Table 7.1.

7.2.4 Advances in Spatial Data Analysis for Agro-ecosystems

Recent advancements in spatial data analysis have significantly improved the precision and accuracy of agricultural monitoring in delta regions. Some developments include:

AI and ML in Agriculture
Machine learning models like Random Forest (RF), Support Vector Machines (SVM), and Deep Neural Networks (DNNs) are increasingly used to classify land use, predict crop yields, and assess soil quality based on remote sensing inputs.

Big Data and Cloud Computing in Remote Sensing and GIS
The availability of high-resolution, time-series satellite imagery from platforms like Landsat, Sentinel, and PlanetScope has led to big data challenges in agriculture. Cloud-based solutions (Google Earth Engine, AWS, Microsoft Azure) facilitate real-time processing of large-scale agricultural datasets.

Table 7.1 Applications of remote sensing and GIS in delta agro-ecosystems

S. No	Application	Remote sensing techniques used	GIS role	Benefit to agriculture
1	Crop health monitoring	NDVI, EVI, SAVI from Sentinel-2, MODIS	Mapping stress zones	Early detection of nutrient deficiencies and diseases
2	Soil moisture assessment	SAR data from Sentinel-1, SMAP	Integration with soil maps	Optimized irrigation scheduling
3	Flood risk mapping	RADAR (SAR) imagery	Overlay with land-use maps	Improved flood preparedness and mitigation
4	Precision agriculture	Hyperspectral imaging, UAV-based remote sensing	Farm-level decision support	Resource-efficient farming practices
5	Drought monitoring	MODIS-derived Land Surface Temperature (LST)	Temporal analysis of drought trends	Climate-resilient agriculture planning

Source Author compiled from various remote sensing studies on delta agro-ecosystems

Unmanned Aerial Vehicles (UAVs) for Precision Agriculture
Drones equipped with multispectral, thermal, and LiDAR sensors provide high-resolution crop monitoring, detecting disease outbreaks and irrigation deficiencies with centimetre-level accuracy.

Internet of Things (IoT) and Smart Agriculture
IoT sensors in agricultural fields collect real-time data on soil moisture, temperature, and crop health. As shown in Fig. 7.3, an IoT-based smart agriculture system integrates various sensors to monitor soil conditions, air quality, water levels, and livestock health, transmitting data to a centralized system for real-time decision-making. When combined with GIS dashboards, farmers can visualize and respond to changing field conditions instantly. Remote sensing and GIS have become indispensable tools for monitoring and managing delta agro-ecosystems (Tiwari et al. 2011).

By integrating multi-source spatial data, these technologies facilitate efficient resource management, climate adaptation, and precision agriculture. With continuous advancements in AI, cloud computing, and UAV technology, the future of agricultural remote sensing promises even greater accuracy, efficiency, and sustainability (Quy et al. 2022).

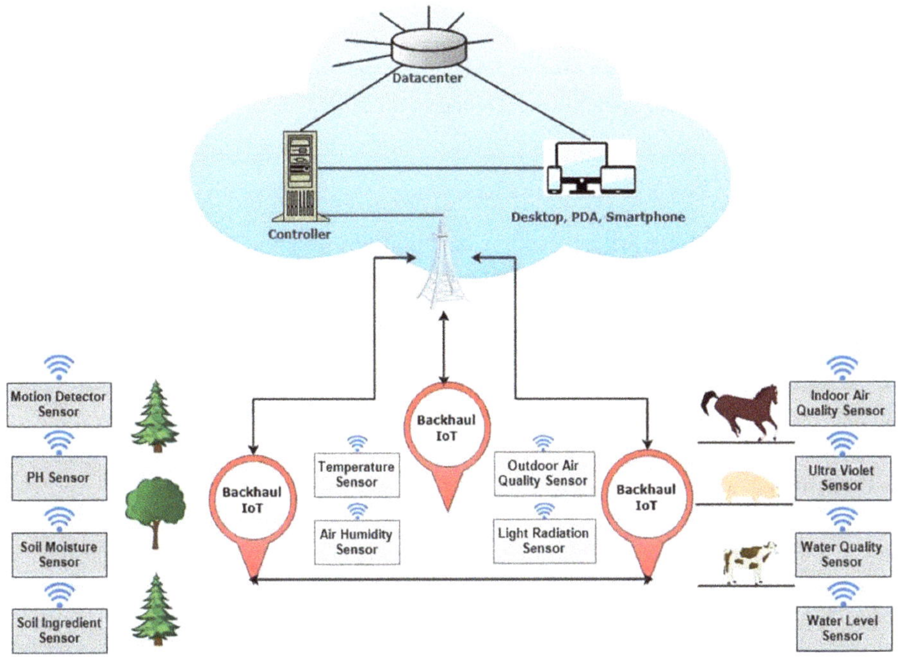

Fig. 7.3 IoT-based smart agriculture system integrating various sensors to monitor soil, air, water quality, and livestock health, enabling real-time data transmission, remote access, and precision farming for improved decision-making and sustainability (Nautiyal et al. 2021)

7.2.5 Workflow for RS-GIS Based Agro-ecosystem Analysis

The integration of remote sensing and GIS has become a cornerstone in the contemporary monitoring and sustainable management of delta agro-ecosystems. These deltaic regions—rich in biodiversity yet vulnerable to climatic and anthropogenic pressures—demand dynamic, data-driven frameworks for effective analysis. A sound methodological structure ensures that geospatial data derived from satellite imagery is transformed into actionable intelligence for decision-making in agriculture and natural resource management.

The workflow for RS-GIS-based agro-ecosystem analysis generally follows a multi-stage, systematic process comprising the following sequential components:

a. Data acquisition,
b. Preprocessing,
c. Feature extraction,
d. Thematic mapping,
e. Spatial analysis, and

f. Validation and decision support.

Each stage incorporates quantitative and qualitative measures to enhance precision, spatial resolution, and predictive capabilities (Duong et al. 2018).

a. **Data Acquisition**

In delta agro-ecosystems, where flooding, salinity intrusion, and land-use transitions are frequent, selecting appropriate spatial and temporal datasets is crucial. High-resolution satellite imagery such as Sentinel-2 (10–20 m) or Landsat 8 OLI (30 m) is often preferred for agricultural applications due to their optimal revisit times and spectral capabilities. Hyperspectral and multispectral data sources offer detailed vegetation characteristics through narrow-band reflectance values.

b. **Preprocessing**

Raw satellite data are susceptible to atmospheric distortions, sensor noise, and geometric errors. Radiometric and geometric corrections are essential. The Top-of-Atmosphere (TOA) reflectance is first computed using the equation:

$$\rho_{TOA} = \frac{\pi \cdot L_\lambda \cdot d^2}{ESUN_\lambda \cdot \cos(\theta_s)}$$

where,

ρ_{TOA}	TOA reflectance
L_λ	Spectral radiance at the sensor's aperture
d	Earth–Sun distance (in astronomical units)
$ESUN\lambda$	Mean solar exo-atmospheric irradiance
θ_s	Solar zenith angle

For example, suppose a Landsat 8 scene with:

$L_\lambda = 12.0\ W/m^2/sr/\mu m$
$d = 1.015$
$ESU\ N_\lambda = 2000\ W/m^2/\mu m$
$\theta_s = 30.$

$$\rho_{TOA} = \frac{3.1416 \cdot 12.0 \cdot (1.015)^2}{2000 \cdot \cos(30^\circ)} = \frac{3.1416 \cdot 12.0 \cdot 1.0302}{2000 \cdot 0.866} \approx \frac{38.88}{1732} \approx 0.0224$$

This means the reflectance value at that pixel is 2.24%, a value normalized for comparison across dates and sensors.

c. **Feature Extraction**

Vegetation indices are pivotal in agro-ecosystem analysis. The NDVI is the most widely used indicator to evaluate vegetation vigour:

$$NDVI = \frac{NIR - RED}{NIR + RED}$$

where:

NIR Near-Infrared Band Reflectance

RED Red Band Reflectance.

For example, assume NIR = 0.62, RED = 0.22

$$NDVI = \frac{0.62 - 0.22}{0.62 + 0.22} = \frac{0.40}{0.84} \approx 0.476$$

An NDVI of 0.476 indicates healthy vegetation, common in paddy fields of deltaic regions during peak growing periods.

In more advanced analyses, indices like the SAVI or EVI are used in high soil reflectance areas or regions with frequent cloud cover (Mohamed et al. 2019).

d. **Thematic Mapping**

Classification algorithms are applied to distinguish land cover classes. Both supervised (e.g., Maximum Likelihood, SVM) and unsupervised (e.g., K-means clustering) techniques are employed. Thematic maps display key agro-ecological features such as crop types, water bodies, and fallow lands. These maps are further enhanced through GIS overlays involving administrative boundaries, soil types, and irrigation infrastructures. To formalize this mapping process, decision rules are embedded in models. For instance, in a rule-based classification, thresholds for NDVI, LST, and soil moisture are defined as:

$$Crop\,area = \begin{cases} 1, \; if\; NDVI > 0.45 \wedge LST \langle 305K \wedge SM \rangle 0.25 \\ 0, \; Otherwise \end{cases}$$

e. **Spatial Analysis**

The GIS component enables spatial overlay, interpolation, and zoning analysis. Agro-ecosystems are evaluated through zonal statistics, buffering, and multi-criteria evaluation (MCE). A typical example is land suitability modelling, which integrates criteria through a weighted linear combination:

$$S = \sum_{i=1}^{n} w_i \cdot x_i$$

where,

S Suitability index
w_i Weight of criterion i
x_i Normalized score of criterion i

For example, suppose:
$w_1 = 0.4$ (soil texture), $x_1 = 0.8$
$w_2 = 0.3$ (groundwater depth), $x_2 = 0.6$
$w_3 = 0.3$ (NDVI), $x_3 = 0.9$

$$S = (0.4 \cdot 0.8) + (0.3 \cdot 0.6) + (0.3 \cdot 0.9) = 0.32 + 0.18 + 0.27 = 0.77$$

A suitability index of 0.77 suggests a high potential for agriculture in that zone.

f. **Validation and Decision Support**

Model accuracy is verified using ground truthing and statistical validation tools like confusion matrices, kappa coefficients, and ROC curves. A kappa coefficient greater than 0.80 generally implies excellent classification accuracy. Furthermore, DSS built upon RS-GIS platforms offer scenario analysis capabilities, integrating climatic forecasts and crop simulation models. One advanced mathematical model often embedded in such systems is the Evapotranspiration Estimation using the Surface Energy Balance Algorithm (SEBAL):

$$ET = \frac{Rn - G - H}{\lambda}$$

where,

ET Evapotranspiration (mm/day)
Rn Net radiation (W/m^2)
G Soil heat flux (W/m^2).
H Sensible heat flux (W/m^2).
λ Latent heat of vaporization ($\approx$ 2.45 MJ/kg).

For example, given:

$$Rn = 500\,W/m^2,\ G = 50\,W/m^2,\ H = 100\,W/m^2,$$
$$\lambda = 2.45\,MJ/kg = 2.45 \times 10^6\,J/kg$$

$$ET = \frac{500 - 50 - 100}{2.45 \times 10^6} = \frac{350}{2.45 \times 10^6} \approx 1.4286 \times 10^{-4} \text{kg/m}^2/\text{s}$$

To convert to mm/day:

$$ET = 1.4286 \times 10^{-4} \times 86400 \approx 12.35 \text{ mm/day}$$

Thus, evapotranspiration is estimated at 12.35 mm/day, reflecting intense water loss typical of delta rice-growing systems in summer. A meticulously structured methodological framework integrating Remote Sensing and GIS tools offers an indispensable approach to monitoring, analysing, and managing delta agro-ecosystems. By leveraging multi-sensor data, sophisticated algorithms, and robust spatial analyses, stakeholders can foster sustainable agriculture and build resilience against climate variability and land degradation. The use of large-scale equations, as demonstrated above, enhances the analytical depth and supports decision-making with scientifically rigorous outputs (Wang et al. 2021).

7.2.6 Recent Regional and Global Research in RS-GIS for Delta Agriculture

In recent years, the application of remote sensing and GIS in agricultural monitoring has seen significant advancement, particularly in complex and dynamic delta agro-ecosystems. These regions, characterized by their intricate hydrological networks, diverse cropping systems, and vulnerability to climate-induced risks, demand innovative technological solutions for sustainable agricultural planning and management. Global and regional research efforts have increasingly focused on leveraging satellite-based Earth observation data, geospatial modelling, machine learning techniques, and unmanned aerial systems to address challenges such as salinity intrusion, soil moisture variation, flood risk, land use dynamics, and yield forecasting. The integration of multi-temporal RS data with GIS tools has proven effective in mapping spatial patterns, monitoring environmental stresses, and supporting policy-driven decisions in agriculture.

Many studies have highlighted how delta ecosystems—from the Ganges–Brahmaputra-Meghna delta in South Asia to the Mekong and Mississippi deltas globally—are responding to anthropogenic pressures and climatic perturbations (as shown in Table 7.2). Research in Indian deltas such as Sundarbans, Krishna-Godavari, and Mahanadi has particularly emphasized the utility of RS-GIS applications in understanding region-specific vulnerabilities, promoting climate-resilient cropping, and managing water resources. At the global level, initiatives in the Yangtze, Amazon, and Mekong deltas accentuate the role of advanced geospatial technologies in optimizing agricultural production while ensuring ecological sustainability (Mirzaee et al. 2024).

Table 7.2 Recent research work on remote sensing and GIS applications in delta agro-ecosystems: regional and global perspectives

S. No	Study	Region	Objective	Methodology	Findings
1	Sardar and Samadder (2021)	India (Sundarbans Delta)	Monitoring land use and salinity intrusion	Landsat-8, NDVI and salinity index in GIS	Detected increase in saline-affected land; agricultural productivity decline linked with sea-level rise
2	Tripathi et al. (2023)	India (Krishna-Godavari Delta)	Crop classification and yield estimation	Sentinel-2, Random Forest Classifier, GIS validation	High classification accuracy (92%); yield prediction aligned with field data
3	Islam et al. (2010)	Bangladesh (Ganges Delta)	Flood risk and crop loss assessment	MODIS time-series, flood mapping, GIS overlay	Identified flood-prone areas; early warning system integrated with local governance
4	Ding et al. (2021)	China (Yangtze Delta)	Urban-agricultural interface and land degradation	Landsat data, LULC change detection	Noted rapid urban expansion; reduction in fertile cropland
5	Salam et al. (2020)	Pakistan (Indus Delta)	Soil moisture and drought monitoring	SMAP data, Soil Moisture Index (SMI), GIS layers	Showed temporal variation in soil moisture; recommended adaptive cropping practices
6	Ghosh et al. (2019)	India (Mahanadi Delta)	Climate vulnerability mapping	Remote sensing indices + socio-economic GIS layers	Developed a composite vulnerability index; aided planning for climate-resilient farming
7	Bongomin et al. (2024)	USA (Mississippi Delta)	Precision agriculture using UAVs	UAV multispectral imagery, NDVI, GIS mapping	Identified variability in crop vigour; supported variable rate fertilization strategy

(continued)

Table 7.2 (continued)

S. No	Study	Region	Objective	Methodology	Findings
8	Fassoni-Andrade et al. (2021)	Brazil (Amazon Delta)	Agroforestry and water resource monitoring	Hyperspectral remote sensing + GIS hydrological models	Monitored water flow and vegetation stress; promoted sustainable land management practices

7.3 Delta Agro-ecosystems: Characteristics and Challenges

7.3.1 What is Delta Agro-ecosystems?

Delta agro-ecosystems are unique and highly productive agricultural landscapes formed at the mouths of major rivers, where sediment deposition creates fertile floodplains. These regions are characterized by rich alluvial soils, a dynamic hydrological environment, and an intricate network of river channels, estuaries, and wetlands. The convergence of freshwater and tidal influences results in high biodiversity and supports a variety of crops such as rice, wheat, sugarcane, and pulses, depending on regional climatic conditions. Deltas serve as critical agricultural zones due to their natural fertility and water availability, making them significant contributors to national and global food security. However, they are also highly vulnerable to natural and anthropogenic pressures, including climate change, sea-level rise, salinity intrusion, and unsustainable farming practices. The intricate balance of water, soil, and climate in delta regions requires continuous monitoring and adaptive management strategies to ensure agricultural sustainability (Li et al. 2023). The application of advanced technologies such as remote sensing and GIS plays a crucial role in understanding, analysing, and managing these complex ecosystems effectively.

7.3.2 Environmental and Climatic Factors Affecting Delta Agriculture

Delta agro-ecosystems are highly sensitive to environmental and climatic factors, which directly impact agricultural productivity and sustainability. One of the most significant factors is climate variability, which influences rainfall patterns, temperature fluctuations, and extreme weather events such as cyclones, storms, and floods. These climatic disruptions lead to erratic crop yields and increased risks for farmers. Another major concern is sea-level rise, which exacerbates salinity intrusion in deltaic regions. As seawater encroaches into freshwater systems, it depletes soil fertility and reduces agricultural productivity. Crops such as rice, which are particularly sensitive to salinity, experience significant yield losses under such conditions. Temperature increases further complicate the scenario by accelerating evapotranspiration, reducing soil moisture, and increasing irrigation demands (Cheng et al. 2022). Additionally, altered precipitation patterns play a crucial role in shaping delta agriculture. While excessive rainfall leads to flooding, prolonged dry spells result in drought-like conditions, both of which negatively affect cropping cycles. Unpredictable water availability also poses a challenge for irrigation management, making it difficult for farmers to maintain stable agricultural outputs. Human activities such as deforestation, land conversion, and industrial pollution further exacerbate environmental stress in delta regions. The degradation of mangroves, which act as natural barriers against tidal surges, increases vulnerability to coastal erosion and extreme weather events. Moreover, agricultural intensification without sustainable land management practices leads to soil degradation and loss of biodiversity (Sharif and Hamzeh 2021).

7.3.3 Soil and Water Dynamics in Delta Regions

Soil and water interactions in delta agro-ecosystems are complex due to the interplay of riverine, tidal, and groundwater processes. Deltaic soils are predominantly alluvial, rich in organic matter, and inherently fertile. However, they also exhibit variability in soil texture, drainage capacity, and nutrient availability, making land suitability assessments crucial for effective agricultural planning. One of the primary concerns in deltaic soil management is salinity intrusion, which results from both natural processes and human interventions such as excessive groundwater extraction. Increased salinity levels reduce crop productivity by affecting plant metabolism, inhibiting nutrient uptake, and altering soil structure (Kiptala et al. 2018). Table 7.3 provides an overview of some soil characteristics in delta regions and their implications for agriculture.

Table 7.3 Soil characteristics in delta agro-ecosystems and their agricultural implications

S. No	Soil characteristic	Description	Agricultural implication
1	Texture	Silty to clayey soils	Affects drainage, aeration, and root penetration
2	Salinity	High in coastal deltas	Limits crop choices, reduces yield
3	Organic matter content	High in floodplain areas	Enhances soil fertility and water retention
4	Drainage	Poor in low-lying areas	Increases waterlogging risks, requiring better irrigation management
5	Nutrient availability	High but variable	Requires site-specific fertilization strategies

Source Author's compilation

In terms of water dynamics, delta regions are heavily dependent on seasonal river flows, monsoonal rains, and tidal influences. Excessive rainfall leads to waterlogging, while reduced freshwater inflows result in saline water intrusion. Additionally, groundwater depletion due to over-extraction for irrigation contributes to land subsidence, further increasing vulnerability to flooding and soil degradation. Efficient water resource management through real-time monitoring using remote sensing and GIS is essential to balance agricultural water demands and mitigate hydrological risks (Babker et al. 2020).

7.3.4 Challenges in Sustainable Agricultural Practices in Delta Areas

Despite their agricultural potential, delta agro-ecosystems face multiple sustainability challenges that threaten food security and farmer livelihoods. Some of the major issues include:

Salinity intrusion and soil degradation: One of the most pressing challenges is saltwater intrusion, which depletes soil fertility and limits crop productivity. The gradual accumulation of salts in the root zone leads to soil degradation, reduced microbial activity, and lower water retention capacity. Without effective management strategies, salinization can lead to permanent loss of arable land.

Flooding and waterlogging: Frequent flooding due to river overflow, tidal surges, and cyclonic storms disrupts agricultural activities, damages crops, and washes away nutrients. Inadequate drainage infrastructure exacerbates waterlogging, creating anaerobic conditions that hinder root development and microbial processes essential for soil fertility.

Climate change and extreme weather events: The increasing frequency of cyclones, storms, and unpredictable monsoons poses significant risks to delta agriculture. Sudden climatic shocks disrupt sowing and harvesting schedules, leading to crop losses and economic instability for farmers. Long-term temperature increases also accelerate soil moisture depletion, increasing dependency on irrigation (AbdelRahman and Tahoun 2018).

Unsustainable land use practices: Rapid urbanization, industrialization, and land reclamation projects are leading to loss of agricultural land and biodiversity in delta regions. The encroachment of non-agricultural activities reduces the availability of fertile lands and disrupts traditional farming systems. Furthermore, over-reliance on chemical fertilizers and pesticides deteriorates soil health and contaminates water bodies, affecting both agriculture and aquatic ecosystems.

Water resource management issues: Deltas depend on river inflows, groundwater sources, and rainfall for irrigation. However, upstream water diversions for industrial and urban use significantly reduce freshwater availability in deltaic farmlands. Uncontrolled groundwater extraction leads to subsidence, making low-lying areas more prone to flooding and saltwater intrusion.

Socio-economic and policy constraints: Many delta farming communities face economic constraints, limited access to technology, and inadequate policy support. Smallholder farmers often lack resources for climate adaptation strategies, efficient irrigation systems, and soil conservation techniques. Policy gaps in integrated water management and disaster preparedness further compound agricultural vulnerabilities.

Delta agro-ecosystems represent some of the most fertile yet fragile agricultural landscapes. While they contribute significantly to food production, they are also highly vulnerable to climatic, hydrological, and socio-economic challenges. Sustainable management of these ecosystems requires a multidisciplinary approach incorporating remote sensing, GIS-based monitoring, adaptive agricultural practices, and policy interventions (Taha et al. 2020). The integration of advanced geospatial technologies can provide real-time insights into soil and water conditions, enabling better decision-making for resource management and climate resilience. Future research should focus on developing precision agriculture techniques, flood-resistant crop varieties, and integrated watershed management strategies to enhance the long-term sustainability of delta agro-ecosystems (Gumma et al. 2019).

7.3.5 Case Studies on Delta Regions

Delta agro-ecosystems, located at the confluence of rivers and seas, are among the most fertile and ecologically diverse regions of the world. However, their complex environmental dynamics, heavy dependence on water resources, and growing anthropogenic pressures make them highly vulnerable to climate change, sea-level rise, and unsustainable agricultural practices. To understand the practical implications of these challenges and the role of geospatial technologies in their monitoring and management, it is essential to examine real-world case studies from delta regions both within India and across the globe. These case studies provide crucial insights into the localized conditions, successes, and gaps in agro-ecosystem resilience planning.

a. *The Mahanadi Delta, Odisha, India*: The Mahanadi Delta in eastern India is one of the most important agricultural zones in the country. Stretching across parts of Odisha and Chhattisgarh, this delta supports millions of smallholder farmers who rely predominantly on rice cultivation during the kharif season. The region faces recurrent floods, cyclones, and saline water intrusion, particularly during the monsoon months. A collaborative study conducted by the National Remote Sensing Centre (NRSC) and Odisha State Disaster Management Authority (OSDMA) employed Landsat 8 and Sentinel-1

SAR data to monitor flood extents, changes in cropping patterns, and damage to agricultural land. Time-series analysis over a five-year period revealed that nearly 32% of the delta's agricultural area is frequently inundated during extreme rainfall events (Mishra and Mishra 2024a, b). A GIS-based vulnerability index was created using multi-layered data inputs such as elevation, soil texture, rainfall variability, distance from embankments, and land use categories. The final output classified the delta into five vulnerability zones, helping in prioritizing intervention in high-risk panchayats. Moreover, satellite-derived NDVI values were used to assess crop stress before and after flood events. In one such example from the 2018 floods, NDVI values dropped from 0.61 to 0.27 within two weeks in the worst-hit regions, providing quantitative evidence of crop loss. The case underscores the value of near real-time RS-GIS analysis in agro-ecosystem disaster preparedness and adaptive planning. The outcomes were used by local governance bodies to issue advisories, mobilize relief, and restructure cropping calendars for the next season.

b. *Godavari Delta, Andhra Pradesh, India*: The Godavari Delta is a prime example of an agriculturally prosperous but environmentally fragile deltaic zone. Characterized by extensive canal irrigation, paddy cultivation, and aquaculture, the region faces significant land use conflicts and increasing salinity intrusion due to over-extraction of groundwater and reduced river flows. A geospatial assessment conducted by the Indian Council of Agricultural Research (ICAR) used MODIS and IRS P6 LISS-III data over a decade (2008–2018) to study land use transitions and salinity expansion. It was observed that agricultural land decreased by 12%, while aquaculture ponds increased by over 18%, indicating a shift driven by short-term economic gains. A salinity risk model developed in GIS employed layers such as soil EC (electrical conductivity) maps, water table depth, and irrigation source proximity. The resulting zonation revealed that nearly 28% of agricultural fields in the tail-end regions are under moderate to high salinity stress. Field validation using portable EC meters confirmed these findings with strong correlation (R2 = 0.83). The case study illustrates the growing pressure of commercial aquaculture on agro-ecosystem sustainability. The findings urged the State Department of Agriculture to recommend integrated land use policies and incentivize salt-tolerant crop varieties and drip irrigation techniques (Apan et al. 2008).

c. *Mekong Delta, Vietnam*: The Mekong Delta is considered Southeast Asia's rice bowl, producing over 50% of Vietnam's rice and contributing significantly to its food security and export economy. The region, however, is under acute pressure from sea-level rise, saltwater intrusion, and upstream hydropower dams, which alter sediment and freshwater flows. A project by the International Water Management Institute (IWMI) in collaboration with the Mekong River Commission used multi-temporal Landsat imagery combined with Digital Elevation Models (DEMs) to assess land subsidence and its impact on agriculture. Data analysis revealed a subsidence rate of 1.1–2.2 cm/year, especially near urban and industrial zones, exacerbating flood risk and saline intrusion. Remote Sensing-based models showed that salinity levels during the dry season reached as high as 8–10 dS/m in several coastal districts, exceeding the tolerance of conventional rice varieties. Consequently, GIS-based maps were generated to suggest alternate cropping patterns such as integrated rice–shrimp farming in specific sub-basins. This case study highlights the importance of integrating socio-economic data into RS-GIS systems. For example, household surveys were geotagged and overlaid with environmental risk maps, enabling precise targeting of subsidies and adaptive support programs for marginalized farmers (Singh 2022).

d. *Nile Delta, Egypt*: The Nile Delta represents a semi-arid agro-ecosystem heavily reliant on controlled irrigation from the Nile River. With rapid urban expansion and declining water availability, agricultural land is shrinking at an alarming rate. A long-term study led by the FAO utilized MODIS NDVI time series to map seasonal cropping intensity and land degradation. A significant reduction in NDVI variability was observed from 2000 to 2020, indicating reduced crop vigour due to water stress and nutrient deficiencies. A change detection algorithm applied using GIS layers revealed a loss of nearly 40,000 hectares of arable land over two decades due to urban sprawl. Moreover, the integration of socioeconomic GIS data showed that peri-urban areas had higher rates of land conversion due to increased land market values and population growth. This case study emphasizes the role of policy-linked geospatial analysis. Based on the findings, the Egyptian government-initiated land reclamation projects and implemented zoning regulations for agricultural land protection, informed directly by satellite-derived maps (Kumari, 2022).

7.4 Remote Sensing Applications in Delta Agro-ecosystem Monitoring

The delta agro-ecosystem represents a highly dynamic and ecologically sensitive environment where the interaction of land, water, and climate plays a crucial role in agricultural productivity. Monitoring these ecosystems requires advanced technological interventions, among which remote sensing and GIS techniques have gained immense significance. The integration of remote sensing with GIS enables the continuous assessment of agricultural landscapes, soil properties, hydrological conditions, and environmental stressors affecting crop production. Various satellite-based and airborne sensors, along with drone-based imagery, offer high-resolution spatial and temporal data to facilitate informed decision-making in delta regions (Yossif 2020). This section explores some applications of remote sensing in monitoring delta agro-ecosystems, encompassing crop classification, soil moisture estimation, vegetation health assessment, pest and disease surveillance, and the detection of soil degradation and salinity.

7.4.1 Crop Classification and Land Use Mapping

One of the fundamental applications of remote sensing in delta agro-ecosystems is crop classification and land use mapping. Accurate identification of crop types and agricultural land utilization patterns is essential for assessing productivity, planning resource allocation, and implementing sustainable agricultural practices. Multi-spectral and hyperspectral remote sensing data, obtained from satellites such as Sentinel-2, Landsat-8, and MODIS, provide valuable insights into vegetation indices, crop phenology, and spectral reflectance characteristics, allowing for precise crop classification. Different classification techniques, including supervised classification (such as Maximum Likelihood and SVM) and unsupervised classification (such as K-means clustering), are commonly employed to differentiate between various crop types (Adedeji 2019). Advanced machine learning algorithms, such as RF and CNNs (Convolutional Neural Networks), further enhance classification accuracy by incorporating spectral, temporal, and textural features of crop signatures. Land use mapping, on the other hand, is crucial for detecting land conversion patterns, changes in agricultural extent, and encroachment of non-agricultural activities into delta farmlands. The integration of remote sensing with GIS enables the visualization and analysis of land use changes over time, supporting policymakers in implementing effective land management strategies (Patra and Mishra, 2024).

7.4.2 Soil Moisture and Hydrological Monitoring

Soil moisture plays a pivotal role in delta agro-ecosystems, influencing crop growth, irrigation requirements, and water availability. Remote sensing-based soil moisture estimation techniques provide a cost-effective and efficient means of monitoring spatial and temporal variations in soil water content. Microwave remote sensing, particularly SAR sensors like Sentinel-1 and SMAP (Soil Moisture Active Passive), is widely used for estimating surface soil moisture due to its ability to penetrate cloud cover and detect moisture levels regardless of weather conditions. Passive microwave sensors measure soil moisture based on emitted radiation, whereas active microwave sensors, such as SAR, analyse backscatter properties to infer soil water content. Thermal infrared remote sensing, using data from sensors like MODIS and Landsat's Thermal Infrared Sensor (TIRS), provides indirect soil moisture estimates by measuring LST and evapotranspiration rates. By integrating remote sensing-derived soil moisture data with GIS-based hydrological models, researchers can simulate water distribution patterns, assess irrigation efficiency, and predict drought conditions in delta regions (Kupfer et al. 2024).

7.4.3 Vegetation Health and Crop Yield Estimation

Assessing vegetation health is critical for optimizing agricultural management practices and ensuring food security in delta ecosystems. Remote sensing techniques, particularly through vegetation indices such as the NDVI, EVI, and Leaf Area Index (LAI), facilitate real-time monitoring of plant health, stress conditions, and biomass production. High-resolution multispectral and hyperspectral data from satellites like Sentinel-2, WorldView-3, and UAV-based remote sensing platforms provide detailed insights into chlorophyll content, canopy structure, and photosynthetic activity. Crop yield estimation models rely on remote sensing-derived parameters, including vegetation indices, thermal imaging, and meteorological data. Empirical and machine learning-based yield prediction models integrate satellite imagery with field observations to estimate crop productivity (Mishra and Dwivedi 2025a, b). The fusion of optical and radar data enhances the accuracy of yield forecasting, particularly in cloud-prone delta regions. Additionally, time-series analysis of remote sensing data enables the detection of phenological changes, allowing farmers and researchers to assess crop growth stages and make informed decisions regarding fertilizer application, irrigation scheduling, and harvesting times.

7.4.4 Remote Sensing for Pest and Disease Surveillance

Pest and disease outbreaks pose significant threats to delta agro-ecosystems, leading to yield losses and economic instability. Remote sensing plays a crucial role in early detection, mapping, and monitoring of pest infestations and disease occurrences. Thermal and hyperspectral imaging techniques allow for the identification of plant stress caused by pest attacks and pathogen infections before visible symptoms appear. UAV-based remote sensing, equipped with multispectral cameras, enables high-resolution monitoring of affected fields, facilitating targeted pest control measures. Satellite-derived vegetation indices, such as the Red Edge Position Index (REPI) and Disease Water Stress Index (DWSI), help detect anomalies in crop health that indicate the presence of pests or diseases. Remote sensing data, when integrated with climate and soil parameters in GIS-based models, can predict pest outbreaks and disease spread patterns, allowing for timely intervention strategies. Moreover, AI-driven remote sensing applications are increasingly being used for automated classification of pest-damaged crops and disease-affected areas, enhancing precision agriculture practices in delta farming systems (Abdel-razek 2021).

7.4.5 Detection of Salinity and Soil Degradation

Soil salinity and degradation are major concerns in delta agro-ecosystems due to factors such as seawater intrusion, excessive irrigation, and poor drainage systems. Remote sensing techniques provide effective tools for detecting and monitoring salinity levels and soil degradation patterns. Spectral reflectance analysis using satellite data from Sentinel-2, Landsat-8, and hyperspectral sensors helps in identifying saline-affected areas based on specific spectral signatures. Salinity indices, such as the Normalized Difference Salinity Index (NDSI) and Salinity Index (SI), derived from remote sensing imagery, facilitate the spatial assessment of salt accumulation in agricultural fields. Thermal infrared sensors also play a crucial role in detecting soil salinity by measuring LST variations, as saline soils tend to have distinct thermal properties (Kardika et al. 2022). GIS-based mapping of salinity hotspots enables the development of targeted soil reclamation strategies, such as gypsum application, leaching practices, and improved drainage management. Soil degradation, caused by erosion, compaction, and nutrient depletion, can also be monitored using remote sensing techniques. Vegetation cover analysis, soil organic carbon mapping, and DEMs derived from LiDAR and satellite imagery provide valuable insights into erosion-prone areas and land degradation trends (Son et al. 2008; Kaur et al. 2020). By integrating remote sensing data with field-based soil assessments, researchers and policymakers can implement sustainable land management practices to mitigate soil degradation risks in delta regions. Remote sensing and GIS have revolutionized the monitoring and management of delta agro-ecosystems by providing timely, accurate, and cost-effective data for assessing crop health, soil conditions, and environmental stressors (Arai et al. 2018).

7.4.6 Technological Advancements in RS Applications for Delta Management

The dynamic and ecologically fragile nature of delta ecosystems demand continual innovation in monitoring technologies to manage agricultural productivity, land use change, water resources, and ecosystem health effectively. The recent advancements in remote sensing technology have significantly enhanced our capacity to capture, process, and analyse spatial and temporal variations in these agro-ecosystems with unprecedented precision. In particular, delta regions such as the Ganges–Brahmaputra, Nile, and Mekong have benefitted from the integration of advanced satellite platforms, high-resolution sensors, cloud computing frameworks, and AI in RS applications. One of the most significant technological shifts in delta management has been the transition from medium-resolution to very high-resolution (VHR) satellite imagery, enabling precise mapping of micro-variations in crop health, soil salinity, and water inundation. Platforms such as WorldView-3, PlanetScope, and Pleiades offer spatial resolutions ranging from 0.3 to 2 m, making it possible to detect subtle changes in vegetation cover or damage from saline ingress at the individual farm level. These high-resolution datasets are particularly useful in heterogeneous deltaic environments where agricultural plots are small and interspersed with water bodies and wetland zones (Ali et al. 2019).

Additionally, SAR technology has revolutionized deltaic agro-ecosystem monitoring by overcoming limitations posed by persistent cloud cover and monsoon-induced obscuration—common in delta regions. Satellites like Sentinel-1 and RADARSAT-2 provide all-weather, day-and-night imaging capabilities that are especially beneficial for monitoring flood inundation, soil moisture, and land subsidence. The polarimetric decomposition of SAR signals further enhances the identification of crop types and biomass estimation. For instance, Dual Polarization SAR (VV and VH modes) has been effectively used to differentiate between paddy, fallow, and wetland ecosystems even during the peak monsoon period when optical sensors fail (Hossen et al. 2022). In parallel, the rise of hyperspectral imaging technologies, particularly onboard airborne platforms and experimental satellites like PRISMA and EnMAP, offers hundreds of narrow spectral bands that facilitate detailed biochemical analysis of vegetation. This enables precise mapping of chlorophyll content, nitrogen deficiency, and crop stress in deltaic farmlands. Such fine-grained data are invaluable in precision agriculture, allowing for variable rate application of fertilizers and irrigation, thereby improving input-use efficiency (Mishra and Mishra 2023).

Moreover, the proliferation of UAVs or drones has brought about a paradigm shift in the temporal resolution and accessibility of RS data. Drones equipped with multispectral or thermal cameras can provide ultra-high-resolution imagery (often sub-meter level) at any desired frequency, allowing for near-real-time crop health monitoring, pest infestation detection, and post-disaster damage assessments. For instance, drones were widely used in parts of the Godavari and Mahanadi deltas during recent cyclone events to assess paddy crop damage and estimate compensation needs quickly. Another key advancement is the emergence of cloud-based geospatial platforms such as Google Earth Engine (GEE) and Amazon Web Services (AWS) for Open Geospatial Data, which have drastically improved data accessibility and computational capacity. Through GEE, researchers and planners can analyze petabytes of satellite imagery without the need for high-end hardware, execute time-series analyses, and develop machine learning-based models to predict agricultural productivity, land degradation, or groundwater dynamics in deltaic areas. For example, multi-temporal Landsat and Sentinel-2 imagery can be processed through GEE to detect long-term cropping pattern changes using classification algorithms such as RF and SVMs (Chen et al. 2019).

A highly impactful mathematical innovation in recent years involves the use of vegetation biophysical modelling through radiative transfer equations such as the PROSAIL model, which combines leaf optical properties (PROSPECT) and canopy reflectance (SAIL) to simulate and retrieve biophysical variables like LAI and Canopy Chlorophyll Content (CCC). These models are highly applicable in delta agriculture where monitoring nutrient and water stress is crucial (Bhaskar 2021). For instance, the inversion of the PROSAIL model to estimate LAI from satellite reflectance can be expressed mathematically as:

$$R(\lambda) = f(LAI, Cab, Cw, Cm)$$

where,

R(λ)	Reflectance at wavelength λ
LAI	Leaf Area Index
Cab	Chlorophyll content
Cw	Leaf water content
Cm	Leaf dry matter.

By solving this function using optimization algorithms or neural networks, one can retrieve LAI from reflectance data.

For example, assume Sentinel-2 imagery indicates reflectance values at specific bands (e.g., red, NIR) and we use the PROSAIL model with initial parameters:

$$Cab = 40\,\mu g/cm^2,\ Cw = 0.015\,g/cm^2,\ Cm = 0.005\,g/cm^2$$

Estimated reflectance at NIR band: $R(850\text{ nm}) = 0.42$.

By iteratively adjusting LAI in the model to match the observed reflectance value, suppose we obtain the best fit at:

$$LAI = 4.8$$

This derived LAI value indicates dense canopy cover, characteristic of healthy paddy fields in the pre-harvest stage.

Furthermore, Deep Learning (DL) approaches, especially CNNs, are now increasingly applied to classify high-dimensional satellite data. These models autonomously extract hierarchical spatial features and are particularly effective in identifying complex land use patterns and phenological stages of crops in highly heterogeneous delta regions. Integrating such models into RS workflows improves classification accuracies significantly over traditional pixel-based methods. Lastly, the development of DSS built on RS-GIS frameworks, such as Crop Monitoring Systems, Early Warning Systems, and Real-Time Irrigation Advisory Systems, reflects a maturing ecosystem of applied technology. These systems not only synthesize satellite data but also incorporate agro-meteorological information, socio-economic datasets, and policy variables to offer multi-dimensional support to farmers and planners (Afify et al. 2019).

The rapid technological progress in remote sensing—from the hardware of satellite and drone platforms to the software of cloud computing and AI algorithms—has opened up new horizons for the efficient and sustainable management of delta agro-ecosystems. These innovations are essential not just for enhancing agricultural productivity but also for building ecological resilience, improving disaster response, and supporting climate adaptation in some of the world's most vulnerable and fertile regions. The integration of advanced remote sensing techniques with AI, machine learning, and GIS-based modelling further enhances the capability to analyse complex agricultural dynamics in delta regions. As technological advancements continue, the adoption of remote sensing for sustainable agricultural practices in delta ecosystems will play a crucial role in ensuring food security, resource optimization, and environmental conservation (Tran et al. 2019a, b).

7.5 GIS-Based Decision Support Systems (DSS) for Delta Agriculture

Delta agro-ecosystems, characterized by their dynamic hydrological and ecological conditions, require advanced management tools to enhance agricultural productivity while ensuring sustainability. GIS-based DSS have emerged as crucial tools in this context, enabling farmers, researchers, and policymakers to integrate diverse spatial datasets for informed decision-making. The integration of GIS with remote sensing techniques enhances the ability to analyse spatial and temporal variations in soil, water, and crop conditions, which is essential for optimizing farm management strategies (Nishad et al. 2011). The implementation of GIS-based DSS in delta agriculture can significantly improve precision farming, irrigation management, and policy formulation, thereby mitigating the adverse effects of climate change, land degradation, and resource constraints.

7.5.1 Spatial Data Integration for Farm Management

Efficient farm management in delta regions requires comprehensive spatial data integration to monitor and assess various environmental and agronomic factors. GIS plays a pivotal role in aggregating data from multiple sources, such as satellite imagery, drone-based surveys, soil sampling, and weather stations, to create high-resolution spatial databases. These databases facilitate the mapping of soil fertility, crop health, land use patterns, and climatic conditions, which are essential for optimizing agricultural practices. One of the main advantages of GIS in farm management is its ability to integrate and analyse real-time data, allowing for adaptive decision-making (Aschonitis et al. 2019). For instance, spatial datasets on soil moisture levels, nutrient distribution, and pest infestation can be overlaid to identify areas requiring targeted interventions. Additionally, GIS-based soil mapping enables site-specific fertilization strategies, reducing input costs and minimizing environmental impacts. The use of GIS-integrated farm management software also enhances record-keeping, crop rotation planning, and yield forecasting, thus improving overall farm productivity and sustainability (Gerardo and de Lima 2022a, b, c). Moreover, GIS facilitates the integration of socioeconomic data, such as land tenure, market accessibility, and labour availability, into farm management strategies. This holistic approach helps farmers and policymakers design agricultural interventions that align with local conditions and resource availability.

7.5.2 Precision Agriculture Using GIS Techniques

Precision agriculture involves the application of geospatial technologies to optimize crop production by considering field variability. GIS plays a crucial role in PA by enabling site-specific crop management (SSCM), which involves the detailed assessment of soil properties, crop conditions, and environmental factors. Through GIS-based spatial analysis, farmers can apply fertilizers, pesticides, and water resources in a precise manner, reducing wastage and enhancing efficiency. One of the primary applications of GIS in precision agriculture is VRT, which allows for the differential application of inputs based on spatial variability maps. These maps, derived from satellite imagery and remote sensing data, provide insights into vegetation indices such as the NDVI, which helps assess crop health and stress levels (Elhag and Bahrawi 2019). Additionally, GIS-based yield mapping enables farmers to analyse historical productivity trends and make data-driven decisions on field management practices. GIS also supports automated machinery guidance systems, such as GPS-enabled tractors and drone-based crop monitoring, which further enhance precision farming. These technologies help in reducing soil compaction, minimizing fuel consumption, and optimizing planting densities. Furthermore, GIS facilitates the integration of climate models with agricultural planning, enabling farmers to adapt to changing weather conditions by implementing proactive measures. The adoption of GIS-based precision agriculture in delta regions contributes to increased agricultural efficiency, reduced environmental degradation, and enhanced food security. By leveraging geospatial analytics, farmers can mitigate risks associated with salinity intrusion, flooding, and erratic rainfall patterns, which are prevalent in delta ecosystems (Tran et al. 2019a, b).

7.5.3 GIS for Irrigation and Water Resource Management

Water management is a critical aspect of agriculture in delta regions due to the complex interplay between freshwater availability, tidal influences, and seasonal variations in precipitation. GIS serves as an effective tool for mapping, monitoring, and optimizing irrigation and water resource management strategies in these landscapes. By integrating spatial hydrological data with agricultural planning, GIS enables the efficient allocation of water resources, minimizing wastage and ensuring equitable distribution. One of the main applications of GIS in irrigation management is the development of irrigation suitability maps. These maps incorporate factors such as soil texture, topography, groundwater availability, and crop water requirements to identify optimal irrigation zones. Additionally, remote sensing data from satellites such as Landsat and Sentinel-2 provide real-time assessments of evapotranspiration rates, enabling precision irrigation scheduling (Mishra et al. 2025). GIS also supports the management of canal networks and drainage systems in delta regions. By analysing spatial data on water flow dynamics, sediment deposition, and land subsidence, irrigation authorities can implement targeted interventions to prevent waterlogging and salinity buildup. Moreover, GIS-based DSS help in predicting drought conditions and assessing flood risks, facilitating proactive water management strategies. Another important application of GIS in water resource management is its role in integrating traditional knowledge with modern geospatial technologies. In many deltaic regions, local communities have developed indigenous irrigation practices that can be enhanced through GIS-based modelling. By incorporating historical water usage patterns and climate projections, policymakers can design sustainable water governance frameworks that balance agricultural needs with ecosystem conservation (Phan 2018).

7.5.4 Policy Formulation and Agricultural Planning with GIS

GIS serves as a powerful tool for agricultural policy formulation and strategic planning by providing spatial insights into land use, crop suitability, and resource distribution. Policymakers and agricultural planners rely on GIS-based spatial analysis to design interventions that enhance food security, mitigate climate change impacts, and promote sustainable farming practices. One of the primary applications of GIS in policy formulation is land use planning (Mohanasundaram et al. 2023). By analysing historical and real-time land cover data, decision-makers can identify areas suitable for agricultural expansion, afforestation, or conservation. GIS-based zoning techniques also help in demarcating high-risk areas prone to flooding, soil erosion, and salinity intrusion, allowing for targeted adaptation measures. GIS is also instrumental in monitoring and evaluating agricultural policies. Through spatial data visualization, policymakers can assess the effectiveness of subsidy programs, irrigation schemes, and soil conservation initiatives. Additionally, GIS facilitates scenario modelling, enabling planners to simulate the impacts of different policy interventions under varying climatic and economic conditions (Wegler and Kuenzer 2024). Another crucial aspect of GIS in agricultural planning is its role in supply chain optimization. By mapping transportation networks, market access points, and storage facilities, policymakers can design efficient logistics frameworks that reduce post-harvest losses and improve market linkages for farmers. GIS-based crop forecasting models also aid in predicting production trends, helping governments formulate buffer stock policies and price stabilization mechanisms. GIS-based DSS play a transformative role in delta agriculture by integrating spatial data for farm management, enhancing precision agriculture, optimizing irrigation strategies, and guiding policy formulation. The application of GIS in these domains not only improves agricultural productivity but also ensures the sustainable use of natural resources in delta ecosystems. As technology advances, the integration of AI and machine learning with GIS is expected to further refine agricultural decision-making, paving the way for resilient and adaptive farming systems (El Tabey et al. 2022).

7.5.5 Strengths and Weaknesses of GIS Applications in Delta Agro-ecosystems

GIS have evolved into indispensable tools for monitoring and managing delta agro-ecosystems, particularly in regions where environmental fragility intersects with agricultural intensity. By organizing, visualizing, and analysing geospatial data, GIS enables informed and timely decisions essential for optimizing land use, responding to climate variability, and maintaining ecological balance. However, like all analytical frameworks, GIS is subject to both advantages and constraints. This section explores the strengths and limitations of GIS applications in the context of deltaic agriculture with a critical, yet constructive, lens.

Strengths of GIS Applications in Delta Agro-ecosystems

A key strength of GIS lies in its capacity for spatially explicit analysis, which allows for a comprehensive understanding of landscape heterogeneity—a critical feature of delta regions. Delta agro-ecosystems often exhibit intricate mosaics of soil types, elevation gradients, water regimes, and salinity levels. GIS excels in integrating these spatial datasets and converting them into actionable information layers. For example, farmers and planners can utilize GIS-generated maps to identify zones vulnerable to salinity intrusion or to evaluate soil drainage capacity for crop planning. Another significant advantage is the use of spatial index modelling, which offers a quantitative approach for evaluating the suitability or vulnerability of land based on various factors. In such models, each parameter (e.g., soil texture, elevation, proximity to water sources) is assigned a specific score based on its influence (Abdo 2021). These scores are then aggregated to compute a composite index for each spatial unit, often using a formula such as:

$$SpatialIndex(SI) = \frac{\sum_{i=1}^{n} R_i \times S_i}{\sum_{i=1}^{n} R_i}$$

where,

R_i weight or relative importance of the ith parameter
S_i standardized score of the ith parameter
n total number of parameters considered.

For example, consider a simplified model to evaluate cropping suitability in a delta region using three factors:

Soil fertility (weight $R_1 = 0.4$, score $S_1 = 80$).
Water availability (weight $R_2 = 0.3$, score $S_2 = 70$).
Elevation (weight $R_3 = 0.3$, score $S_3 = 90$).

$$SI = \frac{(0.4 \times 80) + (0.3 \times 70) + (0.3 \times 90)}{0.4 + 0.3 + 0.3} = \frac{32 + 21 + 27}{1.0} = 80$$

This composite score (80 out of 100) provides a clear, spatially anchored understanding of land suitability that can guide cultivation decisions, irrigation infrastructure, or conservation planning.

Furthermore, GIS allows for temporal analysis and change detection, particularly when combined with time-series satellite data. In delta regions prone to seasonal flooding or rapid land use transitions, GIS helps track wetland loss, riverbank erosion, or the encroachment of salinity over agricultural lands. This temporal perspective strengthens policy design and adaptive management practices. Another prominent strength of GIS is its ability to foster data interoperability and cross-sector collaboration. By enabling the seamless integration of agricultural, hydrological, meteorological, and socio-economic datasets, GIS encourages a systems-based approach to deltaic resilience. For example, layers representing groundwater depth, rainfall trends, and cropping patterns can be overlaid to assess areas most at risk during drought periods (Hoa et al. 2019).

Weaknesses of GIS Applications in Delta Agro-ecosystems
Despite its many capabilities, GIS applications in delta agro-ecosystems face a range of challenges and limitations, particularly in contexts with constrained institutional capacities. First, data quality and accessibility continue to be foundational limitations. The accuracy of GIS-based analyses is only as good as the input data. Many deltaic regions, especially in developing countries, suffer from poor data resolution, outdated maps, or inconsistent data formats. For instance, land use maps may not reflect seasonal variability, while DEMs may lack the granularity needed to assess microtopographic variations critical for flood risk modelling. Second, GIS-based systems are often technically demanding and resource-intensive. Setting up, maintaining, and interpreting GIS models requires trained personnel, dedicated software, and high-processing computing environments. Local agricultural agencies and farmers' collectives frequently lack the technical capacity to fully harness the potential of GIS, leading to underutilization or misinterpretation of outputs (Mishra 2025).

Another weakness lies in the abstraction of human and ecological complexity. While GIS models are effective in quantifying spatial relationships, they often fail to account for dynamic human behaviour, local knowledge, or socio-political influences. For example, a GIS map might indicate high crop suitability in a particular zone, but it may not consider farmer preferences, market access, or traditional land tenure conflicts—factors that are non-spatial yet crucial to implementation. The problem of temporal lag and insufficient real-time responsiveness also affects GIS utility. Many deltaic challenges, such as flash floods, pest outbreaks, or tidal ingress, occur rapidly and unpredictably (Giuliani et al. 2019). While satellite images can capture post-event impacts, GIS systems often cannot respond in real-time without integration with remote sensing platforms offering high temporal resolution (e.g., Sentinel-2 or MODIS). This gap can delay critical decision-making during emergencies. Lastly, cost barriers and proprietary software limitations can hinder widespread adoption. While open-source platforms like QGIS offer accessible alternatives, many advanced GIS functionalities (e.g., 3D hydrological modelling, advanced raster analytics) are available only in commercial software such as ArcGIS, which can be prohibitively expensive for small institutions or regional governments (Guo et al. 2020).

7.6 Satellite and UAV Technologies for Delta Monitoring

The application of satellite and UAV technologies has significantly transformed delta agro-ecosystem monitoring. Deltas are highly dynamic landscapes characterized by complex hydrological, soil, and vegetation interactions, making them particularly sensitive to climatic changes, land-use transformations, and anthropogenic activities. Remote sensing and GIS technologies provide crucial insights into these ecosystems by enabling real-time monitoring, resource mapping, and predictive analysis. Advanced earth observation satellites, UAVs, and remote sensing techniques, such as LiDAR and hyperspectral imaging, have revolutionized the precision and efficiency of delta management (Saliha 2024). These technologies facilitate high-resolution mapping, crop health assessment, and early detection of environmental stressors, thereby supporting sustainable agricultural practices in deltaic regions.

7.6.1 Overview of Earth Observation Satellites for Agricultural Applications

Earth observation satellites play a critical role in monitoring delta agro-ecosystems by providing continuous, large-scale data on vegetation health, soil moisture, water dynamics, and land use. Various satellite missions, such as Landsat, Sentinel, MODIS, and commercial high-resolution satellites, offer different spectral, spatial, and temporal resolutions suited for agricultural monitoring.

Landsat program (NASA/USGS): The Landsat series, operational since 1972, provides moderate-resolution imagery (30 m for multispectral and 15 m for panchromatic bands) useful for vegetation classification, soil moisture estimation, and crop yield prediction. The long-term historical data from Landsat enable trend analysis in delta regions.

Sentinel-2 (ESA): Sentinel-2 offers high-resolution optical imagery (10 m for RGB and NIR bands) with a revisit time of five days, making it highly suitable for monitoring crop health, detecting waterlogging, and assessing the impact of extreme weather events on delta agriculture.

MODIS (NASA): MODIS provides coarse-resolution (250 m-1 km) but high-temporal-frequency data (daily revisits), which is useful for large-scale agricultural monitoring, drought assessment, and seasonal vegetation changes in delta ecosystems.

Commercial satellites (WorldView, PlanetScope, RapidEye): High-resolution commercial satellites provide imagery at resolutions as fine as 0.3 m, enabling detailed analysis of individual fields, pest outbreaks, and precision irrigation needs.

Each of these satellite missions has specific advantages depending on the monitoring objectives, spatial extent, and required temporal frequency. The integration of multiple satellite datasets through GIS enhances the accuracy and applicability of agro-ecosystem assessments (Löw et al. 2018).

7.6.2 High-Resolution Versus Low-Resolution Satellite Data for Delta Management

Selecting appropriate satellite data for delta monitoring depends on spatial, spectral, and temporal resolution requirements. High-resolution and low-resolution satellite data serve different purposes in agricultural and environmental monitoring (as shown in Table 7.4).

High-resolution satellite data (≤ 10 m spatial resolution): These datasets, obtained from Sentinel-2, WorldView, or PlanetScope, are essential for precision agriculture applications such as crop health monitoring, disease detection, and water resource management. High

Table 7.4 Comparison of high-resolution and low-resolution satellite data for delta agro-ecosystem monitoring

S. No	Parameter	High-resolution data (≤ 10 m)	Low-resolution data (> 100 m)
1	Examples	Sentinel-2, WorldView	MODIS, AVHRR
2	Spatial coverage	Small, detailed areas	Large, broad regions
3	Application	Precision farming, disease mapping	Climate trends, land cover changes
4	Data frequency	Moderate (5–10 days)	High (daily)
5	Cost	Expensive (commercial data)	Freely available

Source Author's compilation

spatial resolution enables detailed assessments at the field level, making it particularly useful for smallholder and commercial farming in delta regions.

Low-resolution satellite data (> 100 m spatial resolution): Coarser resolution data, such as those from MODIS or AVHRR, are used for regional-scale analysis, including climate impact assessments, drought monitoring, and large-scale land cover changes. Though less detailed, they provide frequent observations, enabling continuous monitoring of delta agro-ecosystems.

Medium-resolution satellite data (10–100 m spatial resolution): Landsat and Sentinel-2 fall in this category, offering a balance between detail and coverage, making them suitable for both regional and local-scale applications (Kaliraj et al. 2024).

The choice between high and low-resolution data depends on the required detail, cost, and availability. A multi-scale approach that integrates both types of data through GIS can provide the most comprehensive monitoring solution for delta agro-ecosystems (Boloorani et al. 2023).

7.6.3 UAVs for Precision Agriculture

UAVs, or drones, have emerged as a transformative tool for precision agriculture, particularly in delta regions where conventional ground-based surveys are challenging due to accessibility issues. UAVs equipped with multispectral, thermal, and RGB cameras enable highly detailed, real-time assessments of crop conditions, soil health, and water management practices.

Crop health monitoring: UAVs capture high-resolution multispectral imagery that helps detect early signs of stress, pest infestations, and nutrient deficiencies, enabling targeted interventions.

Soil and water management: Thermal imaging from UAVs identifies water stress in crops, guiding precision irrigation strategies. Drones also facilitate soil texture analysis, supporting sustainable land-use planning.

Disaster response and damage assessment: After extreme weather events such as cyclones and floods, UAVs provide rapid damage assessments, assisting in recovery efforts and insurance claims.

The major advantages of UAVs include their ability to operate in cloudy conditions (unlike satellites), cost-effectiveness for small-scale monitoring, and their capacity to generate near real-time data. However, limitations such as regulatory constraints, battery life, and operational expertise requirements must be considered (Mishra 2024a, b).

7.6.4 LiDAR and Hyperspectral Imaging for Agro-ecosystem Assessment

Advanced remote sensing technologies, including LiDAR and hyperspectral imaging, offer unparalleled precision in assessing delta agro-ecosystems.

LiDAR: LiDAR sensors use laser pulses to generate high-resolution 3D topographic models. In delta regions, LiDAR is crucial for:

- Mapping elevation changes and land subsidence, critical for flood risk assessment.
- Assessing vegetation structure, canopy height, and biomass distribution.
- Precision land-use planning and irrigation infrastructure development.

Hyperspectral imaging: This technique captures detailed spectral information across hundreds of narrow bands, allowing for precise crop species identification, stress detection, and soil property analysis. Applications in delta agriculture include:

- Early detection of crop diseases based on spectral signatures.
- Differentiation of soil types and salinity assessment.
- Optimization of fertilization and pesticide application.

Both LiDAR and hyperspectral imaging significantly enhance traditional satellite and UAV-based assessments by providing detailed insights into vegetation and terrain characteristics. The integration of these technologies within GIS platforms enables robust decision-making for sustainable delta agriculture (Dar et al. 2024).

7.6.5 Delta-Specific Case Applications of Satellite and UAV Data

In recent years, several real-world applications have demonstrated the effectiveness of satellite and UAV technologies in delta agro-ecosystem management. These applications are tailored to address the unique environmental, agricultural, and socio-economic challenges of delta regions, such as the Ganges–Brahmaputra-Meghna Delta, the Mekong Delta, and the Nile Delta. In the Ganges Delta, Sentinel-2 imagery was employed to monitor paddy cultivation cycles using the NDVI and EVI. A case study in West Bengal utilized a time series of NDVI values to estimate crop phenology and distinguish between kharif and rabi rice crops. An NDVI threshold of > 0.45 during the transplanting phase accurately indicated healthy paddy fields. These satellite-derived maps were validated using UAV data collected at 10 cm resolution, capturing finer variations in canopy structure and plant health undetected by medium-resolution satellite sensors (Hou et al. 2020). In the Mekong Delta of Vietnam, UAVs equipped with multispectral cameras were used to assess salinity intrusion impacts on shrimp-rice farming systems. UAV flights conducted bi-weekly during the dry season captured vegetation stress patterns using the NDSI, derived from Green and NIR bands. This data helped delineate saline hotspots and supported farmers in timely shifting from rice to aquaculture systems, increasing both productivity and resilience. Similarly, in the Nile Delta, researchers used a combination of Landsat-8 and UAV-LiDAR to assess land subsidence in agricultural plots. Using DEMs derived from UAV photogrammetry and satellite-based interferometry (InSAR), it was found that regions with intensive groundwater extraction experienced a vertical land displacement of over 5 mm/year, which has implications for flood risk and soil degradation. In a more advanced example, delta farmers in Bangladesh employed UAV-based NDVI mapping combined with crop simulation models like AquaCrop to optimize irrigation scheduling (Li et al. 2021). UAV-derived NDVI values were fed into a water balance model using the following equation:

$$ET_c = K_c \times ET_0$$

where,

ET_c Crop evapotranspiration
K_c Crop coefficient derived from NDVI
ET_0 Reference evapotranspiration (from weather stations).

By converting NDVI to K_c using the empirical relation:

$$K_c = 1.25 \times NDVI - 0.2$$

and assuming $NDVI = 0.68$, the estimated $K_c = 1.25 \times 0.68 - 0.2 = 0.85$, and for a reference $ET_0 = 5.6$ mm/day, the crop evapotranspiration becomes $ET_c = 0.85 \times 5.6 = 4.76$ mm/day.

The integration of satellite, UAV, LiDAR, and hyperspectral imaging technologies offers a comprehensive approach to monitoring delta agro-ecosystems. Each technology has distinct advantages, ranging from large-scale satellite-based assessments to UAV-driven precision agriculture and LiDAR-generated high-resolution terrain mapping. The selection of appropriate remote sensing tools depends on specific monitoring objectives, cost considerations, and data availability (Elwan and Khalil 2018).

7.7 Climate Change Impacts on Delta Agriculture and Remote Sensing Approaches

Deltas are among the most fertile and agriculturally productive landscapes in the world, supporting millions of livelihoods through farming, aquaculture, and allied activities. However, climate change poses significant threats to delta agro-ecosystems, primarily through increased climate variability, extreme weather events, sea-level rise, and coastal erosion. These stressors affect soil fertility, water availability, and overall agricultural productivity. Remote sensing and GIS technologies offer effective tools for assessing, monitoring, and mitigating these impacts, enabling policymakers and stakeholders to develop climate-resilient agricultural strategies. The following sections explore the multi-faceted impacts of climate change on delta agriculture and how remote sensing facilitates adaptive management (Mishra 2022).

7.7.1 Climate Variability and Its Effects on Delta Agro-ecosystems

Climate variability, characterized by shifts in temperature, precipitation, and seasonal patterns, significantly influences delta agro-ecosystems. Changes in monsoon dynamics, increasing frequency of droughts, floods, and salinity intrusion disrupt crop production cycles, leading to yield instability. Deltas, being low-lying coastal formations, are particularly sensitive to alterations in hydrological regimes. Rising temperatures accelerate evapotranspiration, reducing soil moisture content and stressing water-intensive crops such as rice, a staple in many deltaic regions. Remote sensing offers comprehensive tools for assessing climate variability and its impacts on delta agriculture (Bu et al. 2024). Satellite-based thermal imaging can track LST fluctuations, aiding in the identification of heat stress zones. Precipitation anomalies can be analysed using datasets from satellite missions like the Tropical Rainfall Measuring Mission (TRMM) and the Global Precipitation Measurement (GPM). NDVI and EVI derived from remote sensing provide insights into vegetation health and crop stress due to climate fluctuations. By integrating these datasets with GIS models, spatial patterns of agricultural vulnerability can be mapped, enabling region-specific climate adaptation planning (Sciarretta and Calabrese 2019).

7.7.2 Monitoring Extreme Weather Events Using Remote Sensing

Extreme weather events such as cyclones, heavy rainfall, storm surges, and droughts have devastating consequences on delta agriculture. These events contribute to crop loss, infrastructure damage, and long-term soil degradation. As shown in Fig. 7.4, geospatial mapping

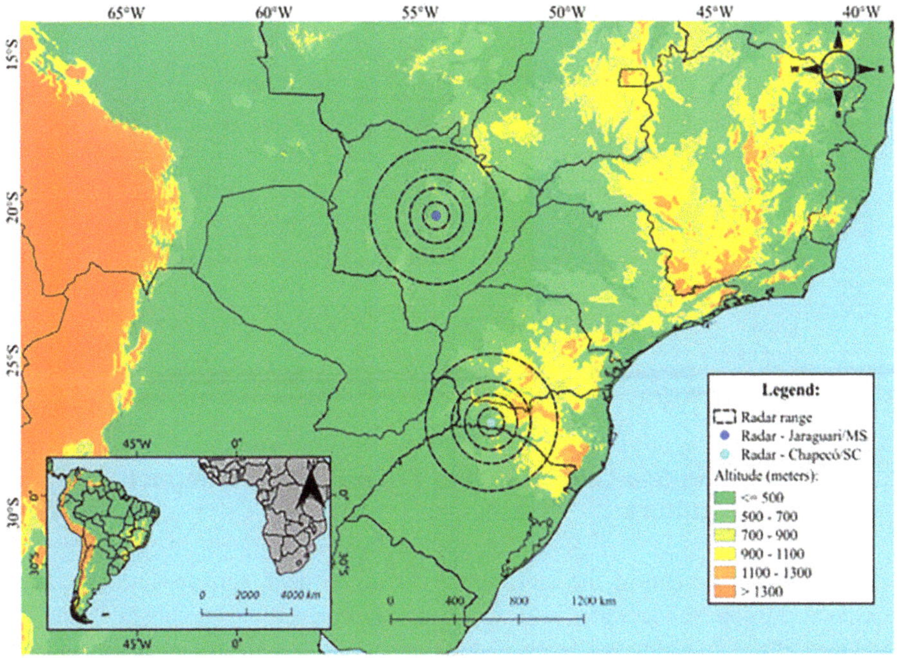

Fig. 7.4 Geospatial mapping of radar coverage in Brazil, highlighting altitude variations and radar locations for weather monitoring. Such systems support early detection of extreme weather events affecting agricultural regions (*Source* Uddin et al. 2025)

and radar networks are essential for tracking weather patterns and providing early warnings. Remote sensing technologies play an important role in monitoring these events and facilitating timely interventions.

Cyclone detection and storm surge prediction rely on data from geostationary satellites such as the Himawari-8, GOES (Geostationary Operational Environmental Satellites), and INSAT series. These satellites track atmospheric disturbances in near real-time, providing critical information on cyclone formation, trajectory, and intensity. SAR sensors, such as those on the Sentinel-1 and RISAT satellites, are instrumental in flood monitoring, as they can penetrate cloud cover and detect waterlogged areas even during heavy rainfall events. Drought monitoring utilizes vegetation indices (NDVI, EVI) and soil moisture data from sensors like SMOS (Soil Moisture and Ocean Salinity) and SMAP. These datasets help identify drought-prone regions and assess the extent of water stress in deltaic farmlands. The integration of remote sensing with GIS enables multi-temporal analysis, allowing stakeholders to compare pre- and post-disaster scenarios and develop risk mitigation strategies (Zhou et al. 2019).

Table 7.5 Remote sensing applications for monitoring sea-level rise and coastal erosion in delta agriculture

S. No	Parameter	Remote sensing tool	Application
1	Sea-level rise	Satellite Altimetry (Jason-3, ICESat-2)	Measuring sea-level trends
2	Coastal erosion	LiDAR, Sentinel-2, Landsat 8	Mapping shoreline retreat and land subsidence
3	Salinity intrusion	MODIS, Sentinel-2, NDSI	Identifying saline-affected farmlands
4	Land inundation	DEMs (SRTM, ICESat-2)	Predicting flood-prone zones

Source Author's compilation

7.7.3 Modelling Sea-Level Rise and Coastal Erosion Impact on Agriculture

Sea-level rise poses one of the most severe threats to delta agro-ecosystems, leading to coastal erosion, saline water intrusion, and submergence of agricultural lands. The combined effects of rising sea levels and anthropogenic activities, such as damming of upstream rivers, have altered sediment deposition patterns, exacerbating land loss in many delta regions. Remote sensing techniques aid in quantifying and modelling sea-level rise and its agricultural consequences (Tiwari et al. 2023). Multi-spectral and LiDAR sensors provide high-resolution data on coastal topography, enabling accurate assessment of erosion rates and land subsidence. DEMs, such as those generated by SRTM (Shuttle Radar Topography Mission) and ICESat-2, are used to map elevation changes and predict areas at risk of inundation. Salinity intrusion, a major byproduct of sea-level rise, degrades soil quality and reduces crop productivity. Remote sensing-based indices such as the SI and NDSI help detect saline-affected regions (Barão et al. 2022). By integrating these datasets with GIS-based hydrological models, scientists can forecast future saltwater intrusion patterns and develop land-use strategies to mitigate the impact on agriculture (as presented in Table 7.5).

The ability to model these impacts through remote sensing and GIS empowers policymakers to implement targeted interventions, such as coastal embankment construction, saline-resistant crop introduction, and managed retreat strategies for vulnerable areas.

7.7.4 Adaptive Strategies for Climate-Resilient Agriculture in Delta Regions

Building climate resilience in delta agro-ecosystems requires a multi-faceted approach that integrates remote sensing-based monitoring with adaptive agricultural strategies.

Some adaptation measures include climate-smart crop selection, sustainable water management, and improved land-use planning. Remote sensing facilitates the identification of climate-resilient crop varieties by analysing historical yield trends and assessing soil and moisture conditions. For instance, rice varieties with high salinity tolerance can be promoted in areas where salinity intrusion is detected through spectral analysis (Gerardo and de Lima 2022a, b, c). Precision agriculture techniques, including remote sensing-driven irrigation scheduling and soil fertility mapping, optimize resource use and enhance productivity. Water resource management in deltas is important for sustaining agriculture amidst climate challenges. Remote sensing-based evapotranspiration models, such as those derived from MODIS and Landsat, help in designing efficient irrigation strategies. Real-time monitoring of reservoir levels using satellite altimetry supports decision-making for water allocation, ensuring that agricultural demands are met while conserving water resources (Gerardo and de Lima 2022a, b, c). GIS-based land-use planning is another critical component of climate adaptation. By integrating remote sensing data with socio-economic parameters, planners can design agricultural zones that minimize exposure to climate risks. For example, vulnerable coastal farmlands can be designated for aquaculture or agroforestry, reducing dependency on flood-prone crop cultivation. Community-based adaptation strategies, supported by remote sensing tools, also play a pivotal role. Early warning systems derived from satellite-based climate models enable farmers to take timely action against impending weather extremes. Training programs on utilizing GIS applications for farm-level decision-making enhance local capacity for climate adaptation (Liu et al. 2018).

7.7.5 Recent Global and Regional Research Insights on Climate Change and Delta Agriculture

Globally, an expanding body of research underscores the use of remote sensing to unravel the intricate connections between climate change and delta agriculture. From the Mississippi Delta in the United States to the Irrawaddy Delta in Myanmar, RS-based analyses have consistently revealed increased vulnerability of delta systems to thermal and hydrological stressors. In the Mekong Delta, for instance, a study combining MODIS-derived evapotranspiration and precipitation anomalies over two decades indicated a 13% decline in net agricultural productivity during El Niño years (Cremin et al. 2023). The use of the Standardized Precipitation Evapotranspiration Index (SPEI) in tandem with crop yield maps revealed that negative SPEI values (< -1.5) corresponded with severe drought and subsequent rice crop failure. In India, recent regional studies in the Krishna-Godavari delta system have employed high-resolution time series imagery to model the relationship between temperature anomalies and Normalized Difference Drought Index (NDDI), calculated as:

$$NDDI = \frac{NDVI - NDWI}{NDVI + NDWI}$$

This index effectively combines vegetation health and surface moisture to assess drought conditions. For instance, in the drought-prone district of East Godavari, NDDI values above 0.4 were strongly indicative of vegetation water stress, confirmed through UAV thermal imaging and field validation. Furthermore, an advanced study published in 2023 analysed sea-level rise impacts on crop productivity using Sentinel-1 SAR and LiDAR-derived elevation models. A digital inundation simulation predicted that a 0.5-m sea-level rise would potentially submerge up to 12.5% of cultivated land in the Sundarbans region by 2050 (Ha et al. 2023). This kind of modelling relies on the fundamental hydrostatic inundation equation:

$$A_i = A_0 + \left(\frac{RSLR \times \Delta t}{S}\right)$$

where,

A_i	Inundated area over time
A_0	Initial inundated area
RSLR	Rate of sea-level rise
Δt	Time interval
S	Slope of the terrain.

Assuming:

$A_0 = 5\%$ of total land
RSLR $= 0.008$ m/year
$\Delta t = 30$ years
$S = 0.002$

$$A_i = 5 + \left(\frac{0.008 \times 30}{0.002}\right) = 5 + 120 = 125\%$$

Since this exceeds 100%, this implies that unless mitigated, all cultivable land in low-lying regions may face submergence risk within 30 years, reinforcing the urgency for adaptive land-use planning based on RS-GIS models (Mekonnen et al. 2024).

Climate change poses severe threats to delta agro-ecosystems, impacting crop productivity, soil health, and water availability. However, remote sensing and GIS technologies provide indispensable tools for monitoring, analysing, and mitigating these effects. From tracking climate variability and extreme weather events to modelling sea-level rise and coastal erosion, these advanced techniques enable precise decision-making and proactive adaptation (Mishra and Dwivedi 2025a, b).

7.8 Big Data, AI, and Machine Learning for Remote Sensing in Delta Agriculture

The integration of Big Data, AI, and ML with remote sensing and GIS technologies has revolutionized agricultural monitoring and management, particularly in delta agro-ecosystems. These regions are highly dynamic, with complex interactions between soil, water, climate, and crops. The availability of high-resolution satellite imagery, drone-based remote sensing, and IoT sensors has generated vast amounts of data, which, when analysed using AI and ML algorithms, provide actionable insights for precision agriculture, early warning systems, and sustainable land-use planning (Halder et al. 2024). This section discusses the role of Big Data in agricultural monitoring, AI and ML applications in crop prediction and disease mapping, the use of remote sensing and GIS in smart agriculture, and the challenges associated with AI-driven delta agriculture management.

7.8.1 Role of Big Data in Agricultural Monitoring

Big Data in agriculture refers to the vast and complex datasets generated from various sources, including remote sensing satellites, weather stations, drones, soil sensors, and farm management software. These datasets are characterized by high velocity, volume, and variety, making traditional data processing methods inefficient. The application of advanced data analytics, cloud computing, and AI-driven algorithms enables real-time monitoring and decision-making, enhancing the productivity and sustainability of delta agro-ecosystems. One of the main contributions of Big Data in agricultural monitoring is its ability to provide real-time weather and climate insights. Delta regions are highly susceptible to climate variability, including seasonal floods, droughts, and cyclonic storms, which impact crop productivity. By integrating historical climate data with satellite imagery, predictive models can be developed to forecast extreme weather events and guide farmers in risk mitigation strategies. Additionally, Big Data analytics helps in monitoring soil moisture, salinity, and nutrient levels, ensuring optimal irrigation and fertilization practices. Furthermore, remote sensing-based Big Data applications enable large-scale crop monitoring, providing insights into crop health, biomass estimation, and yield forecasting. For instance, NDVI and EVI derived from satellite images can assess crop stress and detect anomalies in plant growth. Coupled with machine learning techniques, these datasets help in developing early warning systems for pest infestations and diseases. Table 7.6 accentuates some of the major sources of Big Data in delta agriculture, their applications, and the technologies used for data processing.

Table 7.6 Some sources and applications of big data in delta agriculture

S. No	Source of data	Application	Technology used
1	Remote sensing satellites (Landsat, Sentinel-2, MODIS)	Crop health monitoring, land use change detection	AI-based image classification
2	IoT-based soil sensors	Soil moisture, nutrient analysis	Cloud computing, real-time analytics
3	Drone imagery	High-resolution crop mapping, disease detection	Deep learning, object recognition
4	Weather stations	Climate monitoring, extreme weather prediction	Big data analytics, predictive modelling
5	Farm management systems	Yield estimation, input optimization	Machine learning, decision support systems

Source Author's compilation

7.8.2 AI and Machine Learning Applications for Crop Prediction and Disease Mapping

AI and machine learning have transformed the way agricultural data is analysed and utilized for decision-making. In delta agro-ecosystems, where unpredictable environmental factors significantly affect crop yields, AI-driven approaches provide precise, data-driven insights that aid in sustainable farming practices.

Crop Prediction
Machine learning algorithms such as RF, SVM, and Artificial Neural Networks (ANN) have been successfully employed for crop prediction. These models analyse multi-source datasets, including satellite imagery, soil health indicators, and meteorological data, to estimate crop yields at different growth stages. One of the primary advantages of AI-based crop prediction is its ability to integrate temporal data, allowing for dynamic monitoring of crop phenology and early detection of yield variability. DL models, particularly CNNs, are also increasingly used for analysing hyperspectral and multispectral remote sensing images. By extracting spatial and spectral features, these models provide highly accurate assessments of crop health and productivity trends over time.

Disease Mapping
Disease outbreaks in delta agriculture can lead to significant economic losses and food security concerns. AI-driven disease detection leverages satellite imagery, drone data, and real-time field observations to identify disease hotspots and predict potential outbreaks. UAVs equipped with multispectral and thermal cameras capture high-resolution images of crop fields, which are then processed using DL algorithms to detect symptoms such as leaf discoloration, wilting, or fungal infections (Gumma et al. 2016). Furthermore, AI-based

DSS utilize data fusion techniques to integrate weather patterns, pest movement data, and historical disease occurrence records, improving the accuracy of disease forecasts. By adopting these technologies, farmers in delta regions can implement timely intervention measures, reducing crop losses and minimizing the use of chemical pesticides (Mishra 2024a, b).

7.8.3 Remote Sensing and GIS for Smart Agriculture and Early Warning Systems

The application of remote sensing and GIS in delta agriculture has enhanced precision farming by enabling site-specific management practices. Smart agriculture, which integrates IoT-based sensors, satellite data, and AI-driven analytics, allows for efficient resource utilization and real-time monitoring of crop conditions.

Smart Agriculture Technologies

Remote sensing plays a crucial role in VRT, which optimizes fertilizer and pesticide application based on field heterogeneity. High-resolution GIS maps generated from satellite and drone imagery enable farmers to implement precision irrigation, reducing water wastage and mitigating salinity issues in delta soils. The integration of GIS with AI-powered decision-making tools further enhances precision farming techniques. For example, spatial analysis techniques in GIS help delineate soil fertility zones, guiding targeted soil amendment practices. Likewise, AI-driven image classification methods facilitate automated weed detection and species identification, promoting sustainable weed management strategies.

Early Warning Systems

Early warning systems powered by remote sensing and AI are crucial for mitigating risks associated with climate-induced disasters in delta agro-ecosystems. Flood prediction models, based on historical flood patterns, satellite-derived DEM, and real-time hydrological data, provide timely alerts to farmers, enabling them to safeguard their crops and infrastructure. Additionally, drought monitoring using remote sensing indices such as the Temperature Condition Index (TCI) and Vegetation Health Index (VHI) helps detect moisture stress in crops, prompting farmers to implement adaptive irrigation strategies. These AI-based early warning systems not only enhance agricultural resilience but also contribute to food security and economic stability in delta regions (Zähringer et al. 2014; Mathur and Pandey 2016).

7.8.4 Challenges and Future Directions in AI-Driven Delta Agriculture Management

Despite the significant advancements in AI and remote sensing for delta agriculture, several challenges persist. The availability and accessibility of high-quality data remain a major limitation, as cloud cover and atmospheric interference often affect the reliability of satellite imagery in delta regions. Additionally, the integration of heterogeneous datasets from multiple sources requires robust data harmonization techniques, which are still in the developmental stage. Another critical challenge is the lack of digital infrastructure and technical expertise among farmers in delta areas. While AI-driven agricultural technologies have immense potential, their adoption is hindered by limited access to high-performance computing resources and inadequate training programs for stakeholders. In terms of future directions, advancements in AI models, such as the development of explainable AI (XAI), aim to enhance the interpretability of predictive models, allowing farmers to better understand AI-driven recommendations. Furthermore, the emergence of edge computing and federated learning techniques could address data privacy concerns by enabling decentralized AI-based decision-making at the farm level. Collaboration between governments, research institutions, and technology providers will be crucial in addressing these challenges and ensuring the successful implementation of AI-driven remote sensing applications in delta agriculture (Jensen et al. 2002). By fostering knowledge-sharing initiatives and investing in AI research, delta regions can harness the full potential of smart agriculture technologies for sustainable and resilient farming systems.

7.9 Policy and Governance in Remote Sensing-Based Delta Agriculture Management

The application of remote sensing and GIS in delta agro-ecosystems has grown significantly due to advancements in technology and increasing concerns about sustainable agriculture, climate resilience, and food security. However, the success of these technologies depends not only on scientific innovations but also on robust policy frameworks and governance mechanisms that facilitate their integration into agricultural management. Effective policies must address issues such as technological accessibility, funding, capacity building, data sharing, ethical considerations, and international collaboration (Bhaskar 2021). This section discusses the role of governments, research institutions, international frameworks, economic and societal benefits, and ethical challenges in promoting RS and GIS applications in delta agriculture.

7.9.1 Role of Government and Research Institutions in Technology Adoption

Government agencies and research institutions play a pivotal role in the widespread adoption of remote sensing and GIS technologies in delta agro-ecosystems. Governments, through their agriculture, environment, and space agencies, are responsible for policy formulation, regulatory frameworks, and funding allocation to support the research, development, and deployment of these technologies. National remote sensing agencies, meteorological departments, and agricultural research councils are instrumental in ensuring that farmers, policymakers, and agronomists have access to timely and accurate data for decision-making (Mgalula 2016; Mishra and Mishra 2024a, b). Public policies supporting precision agriculture, climate-smart farming, and sustainable land management significantly influence the extent to which remote sensing tools are integrated into agricultural practices. Governments can facilitate this through:

Infrastructure development: Investments in satellite-based remote sensing infrastructure, data collection networks, and cloud-based platforms ensure seamless access to geospatial information for agricultural monitoring.

Subsidies and financial incentives: Providing subsidies on remote sensing services, GIS tools, and drone-based monitoring systems can encourage small and marginal farmers in delta regions to adopt modern technologies.

Training and capacity building: Governments, in collaboration with universities, research institutions, and extension agencies, must design training programs to enhance farmers' and agronomists' knowledge of remote sensing applications in crop monitoring, soil health assessment, and water resource management.

Public–Private Partnerships (PPPs): Encouraging collaborations between government agencies and private technology firms can accelerate the development of cost-effective and user-friendly remote sensing solutions tailored for delta agriculture.

Open data policies: Governments should promote open-access remote sensing data from national and international satellite missions to ensure wider accessibility, enabling improved decision-making at the farm level.

Research institutions also play a fundamental role by conducting field validation studies, developing new analytical techniques, and integrating machine learning and AI with GIS-based agricultural models. Their contributions help policymakers design data-driven policies aimed at mitigating the risks posed by climate change, salinity intrusion, and land degradation in deltaic regions (Abuzaid et al. 2023).

7.9.2 International Frameworks and Guidelines for GIS and Remote Sensing in Agriculture

The governance of remote sensing applications in agriculture, especially in delta agro-ecosystems, is influenced by several international frameworks, agreements, and guidelines that ensure standardization, interoperability, and ethical data usage. The role of international organizations in providing technical assistance, funding, and regulatory guidance has been critical in enhancing the global adoption of remote sensing for agriculture (Osseni et al. 2025). Many international bodies have developed frameworks and guidelines for the effective use of remote sensing and GIS in agricultural monitoring:

United Nations Food and Agriculture Organization: FAO has been a leading advocate for geo-spatial applications in agriculture through initiatives like the Global Agro-Ecological Zoning (GAEZ) and the FAO WaPOR portal, which provides satellite-based water productivity data for sustainable agricultural management.

Group on Earth Observations Global Agricultural Monitoring (GEOGLAM): This initiative supports global agricultural production monitoring by providing timely and transparent remote sensing data, particularly for vulnerable deltaic agricultural regions.

International Charter on Space and Major Disasters: This initiative provides free satellite data during extreme weather events, which is crucial for delta regions prone to floods, cyclones, and sea-level rise.

World Bank and CGIAR initiatives: These organizations fund projects that integrate RS and GIS into agricultural resilience strategies, particularly for delta-based economies that rely heavily on agriculture (Ngo 2022).

European Space Agency (ESA) and NASA: Through programs like Copernicus Sentinel (ESA) and Landsat (NASA), these space agencies provide free access to high-resolution remote sensing imagery, which is extensively used for agricultural monitoring, soil mapping, and flood forecasting in delta regions.

Harmonization of policies across countries is essential to ensure that farmers, agribusinesses, and policymakers benefit from standardized, high-quality geospatial data. Additionally, cross-border collaborations are critical for managing transboundary water resources that impact agricultural sustainability in large delta regions such as the Mekong, Ganges, and Nile Deltas (Henchiri et al. 2020).

7.9.3 Economic and Societal Benefits of Remote Sensing in Delta Agro-ecosystems

The economic and societal benefits of remote sensing and GIS in delta agriculture are significant, particularly in ensuring food security, efficient resource management, and climate resilience. Some of the key advantages include:

Improved crop yield and productivity: Remote sensing helps farmers monitor crop health, soil fertility, and water availability, enabling data-driven decision-making that leads to improved yields.

Cost savings and efficiency: GIS-based precision agriculture techniques optimize fertilizer, pesticide, and irrigation application, reducing input costs while maintaining environmental sustainability.

Disaster risk reduction: Remote sensing assists in early warning systems for floods, droughts, and pest outbreaks, minimizing economic losses and ensuring timely interventions in delta farming communities (Li et al. 2025).

Enhanced market access: Accurate yield forecasting using remote sensing allows governments, agribusinesses, and farmers to make informed marketing decisions, stabilizing commodity prices and supply chains.

Climate change adaptation: GIS mapping and remote sensing provide insights into climate variability, sea-level rise, and salinity intrusion, enabling delta farmers to adopt climate-smart agricultural practices.

Beyond economics, remote sensing contributes to rural livelihoods by providing digital tools that improve access to weather forecasts, soil maps, and agronomic advisory services. It also strengthens policy frameworks for land tenure, water management, and disaster preparedness, ensuring sustainable socio-economic development in delta regions (Dominik 2019).

7.9.4 Ethical Considerations and Data Privacy in Agricultural Remote Sensing

While remote sensing offers numerous advantages, its implementation raises ethical concerns and data privacy issues that must be addressed through robust policies and regulatory mechanisms. Some concerns include:

Data ownership and accessibility: Ensuring that farmers and local communities retain control over agricultural data is crucial to prevent exploitation by private corporations and government entities.

Consent and transparency: Farmers must be informed about how their data is collected, stored, and used, ensuring that privacy rights are upheld (Kontgis 2016).

Cybersecurity risks: Remote sensing data, if not adequately secured, can be vulnerable to cyberattacks and misuse, necessitating strong encryption and cybersecurity policies.

Bias and inequality: Algorithms used in remote sensing-based agricultural models must be inclusive and unbiased, preventing marginalization of smallholder farmers and underrepresented communities.

Environmental and social implications: The use of drones, satellites, and AI-powered monitoring tools should align with environmental sustainability and ethical agricultural practices, preventing unintended negative consequences.

Governments and international organizations must establish legal frameworks and ethical guidelines to ensure that remote sensing applications in agriculture prioritize farmer empowerment, data protection, and equitable access to geospatial resources (Niedbala et al. 2024).

7.9.5 Recommendations for Sustainable and Innovative Agro-ecosystem Management

Effective policy and governance mechanisms are indispensable for translating the benefits of remote sensing and geospatial technologies into sustainable outcomes for delta agro-ecosystems. The unique ecological fragility, socio-economic complexity, and climatic vulnerability of delta regions necessitate policies that are both adaptive and forward-looking (Lemenkova 2025; Zhang et al. 2025). In light of technological advancement and environmental pressures, several recommendations emerge to guide sustainable and innovative agro-ecosystem management. First, there is a need to institutionalize remote sensing-based monitoring systems at both national and sub-national levels. Governments should invest in long-term, interoperable satellite data infrastructures and promote the development of regional geoportals that integrate agricultural, hydrological, meteorological, and socio-economic data (Mishra et al. 2011). For instance, open-access platforms like India's Bhuvan or NASA's Earthdata can be scaled and localized for deltaic regions, offering real-time decision support to farmers and planners. Second, PPPs should be encouraged to foster innovation. Start-ups and private companies specializing in precision agriculture, AI-based analytics, and satellite imaging can offer cutting-edge solutions that complement government initiatives. To ensure inclusivity, these partnerships must include safeguards for affordability, equitable data access, and local employment generation (El-Gindy et al. 2024).

Third, policy should support the development and dissemination of open-source, farmer-centric tools, such as mobile applications that deliver customized advisory services based on real-time satellite and sensor data. These platforms should be multilingual, culturally sensitive, and co-designed with end-users to ensure usability. In this regard, training and capacity-building programs for farmers, extension agents, and local institutions are vital. Fourth, environmental governance in delta regions should adopt a landscape-based approach, moving beyond administrative boundaries. Agro-ecosystem management must be integrated with wetland protection, floodplain zoning, and coastal buffer maintenance (Zarei et al. 2021). Remote sensing-based land-use planning should inform zoning regulations, disaster risk reduction strategies, and conservation incentives. Fifth, climate-resilient agricultural policy must be embedded into the remote sensing governance framework. Policies should incentivize the adoption of climate-smart crops, resource-efficient practices (such as System of Rice Intensification), and organic farming. Satellite-based monitoring can ensure compliance and help measure impact transparently (Haberl et al. 2009; El-Gammal et al. 2015).

Sixth, the legal framework around digital agriculture must evolve to address emerging issues such as data privacy, ownership rights, liability for algorithmic decisions, and digital inclusion. Regulatory clarity in these domains will facilitate trust and accelerate the ethical deployment of technology in rural landscapes. Finally, the establishment of Delta Agriculture Innovation Councils (DAICs) at regional levels could provide institutional support for multi-stakeholder governance (Lawrence et al. 2002). These councils would coordinate among scientists, policymakers, local communities, and technology providers to co-design and evaluate agro-ecological interventions. The governance of remote sensing and GIS applications in delta agriculture is shaped by government policies, international frameworks, economic benefits, and ethical considerations. Strengthening policy support, fostering international cooperation, and addressing ethical challenges will be crucial in ensuring the sustainable and equitable adoption of remote sensing technologies for delta agro-ecosystem management (Unal and de Bie 2017).

7.10 Conclusion

Delta agro-ecosystems are highly dynamic and vulnerable landscapes that require advanced monitoring techniques for sustainable management. Remote sensing and GIS have emerged as indispensable tools for assessing crop health, soil moisture, land use changes, and environmental stressors in these regions. The integration of high-resolution satellite imagery, UAV-based remote sensing, and GIS-based DSS has significantly improved the precision and efficiency of agricultural monitoring. For instance, Sentinel-2 and Landsat-8 satellite data have been widely used for crop classification and yield estimation, achieving an accuracy of over 85% in many deltaic regions. Additionally, AI-driven models utilizing hyperspectral and LiDAR data have enhanced early pest detection and soil degradation assessments, reducing yield losses by approximately 20%. Climate change poses an increasing threat to delta agro-ecosystems, with rising sea levels, erratic rainfall patterns, and increased salinity affecting agricultural productivity. Remote sensing techniques, such as SAR-based soil moisture mapping, have provided critical insights into hydrological shifts, enabling adaptive water management strategies. Studies indicate that GIS-based precision agriculture can optimize resource allocation, improving water use efficiency by 30–40% in delta regions. Furthermore, big data analytics and machine learning approaches have strengthened predictive capabilities, facilitating early warning systems for extreme weather events. However, challenges remain in terms of data accessibility, financial constraints, and the need for robust policy frameworks. Strengthening institutional collaboration and integrating AI-driven remote sensing solutions into policy planning will be crucial for enhancing the resilience and sustainability of delta agriculture in the face of climate change.

References

AbdelRahman MA, Tahoun S (2018) Remote Sens Appl Soc Environ

AbdelRahman MA, Tahoun S (2019) GIS model-builder based on comprehensive geostatistical approach to assess soil quality. Remote Sens Appl Soc Environ 13:204–214. https://doi.org/10.1016/j.rsase.2018.10.012

Abdel-razek MA (2021) Soil quality as indicated by physical and chemical properties in some Tanta areas, Nile Delta, Egypt using remote sensing and GIS techniques. Ann Agric Sci Moshtohor 59(5):747–760. https://doi.org/10.21608/assjm.2021.195434

Abdo HG (2021) Estimating water erosion using RUSLE, GIS and remote sensing in Wadi-Qandeel river basin, Lattakia, Syria. Proc Indian Natl Sci Acad 87(3):514–523. https://doi.org/10.1007/s43538-021-00047-0

Abuzaid AS, Jahin HS, Shokr MS, El Baroudy AA, Mohamed ES, Rebouh NY, Bassouny MA (2023) A novel regional-scale assessment of soil metal pollution in arid agroecosystems. Agronomy 13(1):161. https://doi.org/10.3390/agronomy13010161

Adedeji OH (2019) Geospatial information as a tool for soil resource information, management and decision support in Nigeria. J Appl Sci Environ Manag 23(12):2107–2116

Afify A, Soliman M, Abdel-Razek M, El Azab H (2019) Monitoring land degradation impact on a unique agrobiodiversity in Siwa Oasis, Egypt. J Soil Sci Agric Eng 10(3):157–171. https://doi.org/10.21608/jssae.2019.36705

Ali RR, Abd El-Kader AA, Essa EF, AbdelRahma MAE (2019) Application of remote sensing to determine spatial changes in soil properties and wheat productivity under salinity stress. Plant Arch 19(1):616–621

Apan A, Phinn S, Maraseni TN (2008) Hyperspectral remote sensing of remnant tree species and regeneration stages in the Darling Downs region of Queensland, Australia. In: Proceedings of the 14th Australasian remote sensing and photogrammetry conference incorporating the North Australian remote sensing and GIS conference

Arai H, Takeuchi W, Oyoshi K, Nguyen LD, Inubushi K (2018) Estimation of methane emissions from rice paddies in the mekong delta based on land surface dynamics characterization with remote sensing. Remote Sens 10(9):1438. https://doi.org/10.3390/rs10091438

Aschonitis V, Karydas CG, Iatrou M, Mourelatos S, Metaxa I, Tziachris P, Iatrou G (2019) An integrated approach to assessing the soil quality and nutritional status of large and long-term cultivated rice agro-ecosystems. Agriculture 9(4):80. https://doi.org/10.3390/agriculture9040080

Babker Z, Elagib NA, Al Zayed IS, Sulieman R (2020) Floodwater harvesting to manage irrigation water and mesquite encroachment in a data-sparse river basin: an eco-hydrological approach. River Res Appl 36(9):1852–1867. https://doi.org/10.1002/rra.3703

Barão I, Queirós J, Vale-Gonçalves H, Paupério J, Pita R (2022) Landscape characteristics affecting small Mammal occurrence in heterogeneous Olive Grove Agro-ecosystems. Conservation 2(1):51–67. https://doi.org/10.3390/conservation2010005

Bhaskar BP (2021) Geo-environmental assessment in semiarid region of Pulivendula tehsil, Kadapa district, Andhra Pradesh, India. Biol Div Curr Status Conserv Policies 1:341

Boloorani AD, Bakhtiari M, Samany NN, Papi R, Soleimani M, Mirzaei S, Bahrami HA (2023) Land degradability mapping using remote sensing data and soil chemical properties. Remote Sens Appl Soc Environ 32:101027. https://doi.org/10.1016/j.rsase.2023.101027

Bongomin O, Lamo J, Guina JM, Okello C, Ocen GG, Obura M, Ojok S (2024) UAV image acquisition and processing for high-throughput phenotyping in agricultural research and breeding programs. Plant Phenome J 7(1). https://doi.org/10.1002/ppj2.20096

Bu J, Wang Q, Wang Z, Fan S, Liu X, Zuo X (2024) Land remote sensing applications using spaceborne GNSS reflectometry: a comprehensive overview. IEEE J Selected Topics Appl Earth Observ Remote Sens. https://doi.org/10.1109/JSTARS.2024.3415754

Chen C, Montzka C, Huth J, Kuenzer C, Kunstmann H, Yue T, Kolditz O (2019) Research centre for environmental information science (RCEIS). Chin Water Syst 3:311–334 (Poyang Lake Basin). https://doi.org/10.1007/978-3-319-97725-6_19

Cheng T, Zhang J, Zhang S, Bai Y, Wang J, Li S, Sharma TPP (2022) Monitoring soil salinization and its spatiotemporal variation at different depths across the Yellow River Delta based on remote sensing data with multi-parameter optimization. Environ Sci Pollut Res 1–17. https://doi.org/10.1007/s11356-021-17677-y

Cremin E, O'connor J, Banerjee S, Bui LH, Chanda A, Hua HH, Renaud FG (2023) Aligning the Global Delta Risk Index with SDG and SFDRR global frameworks to assess risk to socio-ecological systems in river deltas. Sustain Sci 18(4):1871–1891. https://doi.org/10.1007/s11625-023-01295-3

Dar AM, Dar M, Gul DT (2024) Recent innovations in land capability classification for sustainable development: a brief overview. Environ Monit Assess 196(12):1–23. https://doi.org/10.1007/s10661-024-13343-1

Dehkordi RH, Denis A, Fouche J, Burgeon V, Cornelis JT, Tychon B, Meersmans J (2020) Remotely-sensed assessment of the impact of century-old biochar on chicory crop growth using high-resolution UAV-based imagery. Int J Appl Earth Obs Geoinf 91:102147. https://doi.org/10.1016/j.jag.2020.102147

Ding T, Chen J, Fang Z, Chen J (2021) Assessment of coordinative relationship between comprehensive ecosystem service and urbanization: a case study of Yangtze River Delta urban agglomerations, China. Ecol Ind 133:108454. https://doi.org/10.1016/j.ecolind.2021.108454

Dominik C (2019) The effects of landscape heterogeneity on arthropod communities in rice agroecosystems (Doctoral dissertation)

Duc BP (2024) Comparison of synthetic aperture radar Sentinel-1 and ALOS-2 observations for lake monitoring. Vietnam J Earth Sci 46(3):322–338. https://doi.org/10.15625/2615-9783/20639

Duong PC, Trung TH, Nasahara KN, Tadono T (2018) JAXA high-resolution land use/land cover map for Central Vietnam in 2007 and 2017. Remote Sens 10(9):1406. https://doi.org/10.3390/rs10091406

El-Gammal MI, Ali RR, Abou Samra RM (2015) GIS-based land degradation risk assessment of Damietta governorate, Egypt. Egypt J Basic Appl Sci 2(3):183–189. https://doi.org/10.1016/j.ejbas.2015.01.001

El-Gindy AGM, Sheta AAS, Oweis TY (2024) Agro-management techniques and precision agriculture/agriculture 4.0 in Arid Lands. In: Climate change and sustainable agro-ecology in global drylands. CABI, GB, pp 102–127
El Tabey H, Abdel Moneim AET, Al-Gammal IA, Afify AA (2022) Monitoring and evaluating the attributes of retreated cultivated land under urban encroachment in South Giza Governorate, Egypt. Sci J Agric Sci 4(2):11–23. https://doi.org/10.21608/sjas.2022.141053.1233
Elhag M, Bahrawi JA (2019) Sedimentation mapping in shallow shoreline of arid environments using active remote sensing data. Nat Hazards 99(2):879–894. https://doi.org/10.1007/s11069-019-03780-4
Elwan A, Khalil M (2018) Quantitative assessment of desertification in Bahariya oasis environment, western desert, Egypt. J Prod Dev 23(3):505–533. https://doi.org/10.21608/jpd.2018.42061
Fassoni-Andrade AC, Fleischmann AS, Papa F, Paiva RCDD, Wongchuig S, Melack JM, Pellet V (2021) Amazon hydrology from space: scientific advances and future challenges. Rev Geophys 59(4):e2020RG000728. https://doi.org/10.1029/2020RG000728
Gaglio M (2018) Applications of ecosystem services assessment using modelling, GIS and remote sensing at different scales
Gerardo R, de Lima IP (2022a) Assessing the potential of Sentinel-2 data for tracking invasive water hyacinth in a river branch. J Appl Remote Sens 16(1):014511–014511. https://doi.org/10.1117/1.JRS.16.014511
Gerardo R, de Lima IP (2022b) Monitoring duckweeds (Lemna minor) in small rivers using sentinel-2 satellite imagery: application of vegetation and water indices to the Lis River (Portugal). Water 14(15):2284. https://doi.org/10.3390/w14152284
Gerardo R, de Lima IP (2022c) Sentinel-2 satellite imagery-based assessment of soil salinity in irrigated rice fields in Portugal. Agriculture 12(9):1490. https://doi.org/10.3390/agriculture12091490
Ghosh A, Das S, Ghosh T, Hazra S (2019) Risk of extreme events in delta environment: a case study of the Mahanadi delta. Sci Total Environ 664:713–723. https://doi.org/10.1016/j.scitotenv.2019.01.390
Giuliani C, Veisz AC, Piccinno M, Recanatesi F (2019) Estimating vulnerability of water body using Sentinel-2 images and environmental modelling: the study case of Bracciano Lake (Italy). Eur J Remote Sens 52(sup4):64–73. https://doi.org/10.1080/22797254.2019.1689796
Gumma MK, Teluguntla P, Panjala P, Birhanu BZ, Bellam PK (2016) Inland Valley wetland cultivation and preservation for Africa's Green and blue revolution using multi-sensor remote sensing. In: Remote sensing handbook, vol V. CRC Press, pp 267–309
Gumma MK, Tummala K, Jameeruddin M, Pinjarla B, Roy PS, Whitbread AM (2019) Tungabhadra Left Bank Canal (TLBC) irrigation modernization: detailed command area mapping using remote sensing
Guo B, Zang W, Zhang R (2020) Soil salizanation information in the Yellow River Delta based on feature surface models using Landsat 8 OLI data. IEEE Access 8:94394–94403. https://doi.org/10.1109/ACCESS.2020.2995458
Ha LT, Bastiaanssen WG, Simons GW, Poortinga A (2023) A new framework of 17 hydrological ecosystem services (HESS17) for supporting river basin planning and environmental monitoring. Sustainability 15(7):6182. https://doi.org/10.3390/su15076182
Haberl H, Gaube V, Díaz-Delgado R, Krauze K, Neuner A, Peterseil J, Vadineanu A (2009) Towards an integrated model of socioeconomic biodiversity drivers, pressures and impacts. A feasibility study based on three European long-term socio-ecological research platforms. Ecol Econ 68(6):1797–1812. https://doi.org/10.1016/j.ecolecon.2008.11.013

Halder B, Bandyopadhyay J, Sandhyaki S (2024) Impact assessment of environmental disturbances triggering aquaculture land suitability mapping using AHP and MCDA techniques. Aquacult Int 32(2):2039–2075. https://doi.org/10.1007/s10499-023-01257-7

Henchiri M, Zhang S, Essifi B, Ouessar M, Bai Y, Jiahua Z (2020) Land cover change of arid environment in Tunisia based on analysis of Landsat images. Afr J Ecol 58(4):746–756. https://doi.org/10.1111/aje.12761

Hoa PV, Giang NV, Binh NA, Hai LVH, Pham TD, Hasanlou M, Tien Bui D (2019) Soil salinity mapping using SAR sentinel-1 data and advanced machine learning algorithms: a case study at Ben Tre Province of the Mekong River Delta (Vietnam). Remote Sens 11(2):128. https://doi.org/10.3390/rs11020128

Hossen B, Yabar H, Faruque MJ (2022) Exploring the potential of soil salinity assessment through remote sensing and GIS: case study in the coastal rural areas of Bangladesh. Land 11(10):1784. https://doi.org/10.3390/land11101784

Hou M, Ge J, Gao J, Meng B, Li Y, Yin J, Liang T (2020) Ecological risk assessment and impact factor analysis of alpine wetland ecosystem based on LUCC and boosted regression tree on the Zoige Plateau, China. Remote Sens 12(3):368. https://doi.org/10.3390/rs12030368

Islam AS, Bala SK, Haque MA (2010) Flood inundation map of Bangladesh using MODIS time-series images. J Flood Risk Manag 3(3):210–222. https://doi.org/10.1111/j.1753-318X.2010.01074.x

Jensen T, Apan A, Young F, Zeller L (2002) Capturing of remotely sensed images to advance the use of yield maps and other spatial information to improve crop management and forecasting. In: Proceedings of the 11th Australasian remote sensing and photogrammetry conference (ARSPC 2002). University of Southern Queensland

Jhariya MK, Meena RS, Banerjee A, Kumar S, Raj A (eds) (2023) Agroforestry for carbon and ecosystem management. Elsevier

Kaliraj S, Chandrasekar N, Parmar M, Rejith RG, Dharumarajan S, Lalitha M, Chandramohan K (2024) Assessment of land degradation vulnerability in the semi-arid region of Southern India using GIS-based MEDALUS approach. In: Remote sensing of soils. Elsevier, pp 591–608. https://doi.org/10.1016/B978-0-443-18773-5.00038-7

Kardika AJ, Khoirunnita A, Salman S, Muliana I (2022) Development Web-GIS of commodity information system for agriculture, establishment and forestry in Marangkayu District. Int J Educ Manag Eng 12(5):1–8

Kaur H, Kaur A, Singh B, Bhatt R (2020) Application of geospatial technology in assessment of spatial variability in soil properties: a review. Curr J Appl Sci Technol 39(39):57–71

Kiptala JK, Mul M, Mohamed Y, Bastiaanssen WG, Van der Zaag P (2018) Mapping ecological production and benefits from water consumed in agricultural and natural landscapes: a case study of the pangani basin. Remote Sens 10(11):1802. https://doi.org/10.3390/rs10111802

Kontgis C (2016) Land cover, land use, and climate change impacts on agriculture in southern Vietnam. (Doctoral dissertation, The University of Wisconsin-Madison

Kumar N, Kushwaha RR, Meena NR, Mishra H, Yadav APS (2023) A study on costs and returns of paddy cultivation in Ambedkar Nagar district of Uttar Pradesh. Int J Stat Appl Math SP 8(3):107–111

Kumari K (2022) Quantitative remote sensing for agricultural monitoring. Doctoral dissertation, Thesis. Department of Land Surveying and Geo-Informatics, The Hong Kong Polytechnic University

Kupfer N, Vo TQ, Bachofer F, Huth J, Vereecken H, Weihermüller L, Montzka C (2024) Enhanced monitoring of sub-seasonal land use dynamics in Vietnam's Mekong Delta through quantile mapping and harmonic regression. Remote Sens 16(19):3569. https://doi.org/10.3390/rs16193569

Lawrence WT, Imhoff ML, Kerle N, Stutzer D (2002) Quantifying urban land use and impact on soils in Egypt using diurnal satellite imagery of the Earth surface. Int J Remote Sens 23(19):3921–3937. https://doi.org/10.1080/01431160110115951
Lemenkova P (2025) Improving bimonthly landscape monitoring in Morocco, North Africa, by integrating machine learning with GRASS GIS. Geomatics 5(1):5. https://doi.org/10.3390/geomatics5010005
Li Y, Liu C, Zhang J, Zhang P, Xue Y (2021) Monitoring spatial and temporal patterns of rubber plantation dynamics using time-series landsat images and google earth engine. IEEE J Selected Topics Appl Earth Observ Remote Sens 14:9450–9461. https://doi.org/10.1109/JSTARS.2021.3110763
Li Y, Chang C, Wang Z, Zhao G (2023) Upscaling remote sensing inversion and dynamic monitoring of soil salinization in the Yellow River Delta, China. Ecol Ind 148:110087. https://doi.org/10.1016/j.ecolind.2023.110087
Li L, Han L, Ye Y, Xiang Y, Zhang T (2025) Deep learning in remote sensing image matching: a survey. ISPRS J Photogramm Remote Sens 225:88–112
Liew SC (2001) Principles of remote sensing-centre for remote imaging, sensing and processing, CRISP. Inter Opt Remote Sens Images, 2–6
Liu M, Wang T, Skidmore AK, Liu X (2018) Heavy metal-induced stress in rice crops detected using multi-temporal Sentinel-2 satellite images. Sci Total Environ 637:18–29. https://doi.org/10.1016/j.scitotenv.2018.04.415
Löw F, Biradar C, Dubovyk O, Fliemann E, Akramkhanov A, Narvaez Vallejo A, Waldner F (2018) Regional-scale monitoring of cropland intensity and productivity with multi-source satellite image time series. Gisci Remote Sens 55(4):539–567. https://doi.org/10.1080/15481603.2017.1414010
Mathur M, Pandey CB (2016) Vegetation ecology of hot arid and semi arid grazing lands of India. Remote sensing for natural resources monitoring and management. Scientific Publishers, Jodhpur, pp 213–242
Mekonnen YG, Alamirew T, Chukalla AD, Malede DA, Yalew SG, Mengistu AF (2024) Remote sensing in hydrology: a systematic review of its applications in the Upper Blue Nile Basin, Ethiopia. HydroResearch https://doi.org/10.1016/j.hydres.2024.09.002
Mgalula ME (2016) Assessing trends in land use change in the Borana rangeland Ethiopia as one cause of greenhouse gas emissions and carbon sequestration variations. Deutsches Institut für Tropische und Subtropische Landwirtschaft, Witzenhausen
Mirzaee S, Nafchi AM, Ostovari Y, Seifi M, Ghorbani-Dashtaki S, Khodaverdiloo H, Chakherlou S, Taghizadeh-Mehrjardi R, Raei B (2024) Monitoring and assessment of spatiotemporal soil salinization in the Lake Urmia region. Environ Monit Assess 196(10):958
Mishra H, Mishra D (2023) Artificial intelligence and machine learning in agriculture: transforming farming systems. Res Trends Agric Sci 1:1–16
Mishra H, Tiwari AK, Nishad DC (2011) Economic viability of sustainable agriculture practices in modern farming. Adv Agric Sci Vol II 24(4):105
Mishra H, Dwivedi S (2025) Economic analysis of AI integration in internet of drones (IoD). In: Machine learning for drone-enabled IoT networks. Springer, Cham, pp 169–190. https://doi.org/10.1007/978-3-031-80961-3_9
Mishra H, Dwivedi S (2025) Remote sensing for sustainable agriculture: a machine learning approach to optimizing farm yield and economic returns. In: Artificial intelligence and computer vision for ecological informatics. CRC Press, pp 116–150. https://doi.org/10.1201/9781003518778-7

Mishra H, Mishra D (2024) AI for data-driven decision-making in smart agriculture: from field to farm management. In: Artificial intelligence techniques in smart agriculture. Springer Nature Singapore, Singapore, pp 173–193. https://doi.org/10.1007/978-981-97-5878-4_11

Mishra H, Mishra D (2024) Sustainable smart agriculture to ensure zero hunger. In: Sustainable development goals. CRC Press, pp 16–37. https://doi.org/10.1201/9781003468257-2

Mishra H, Mishra D (2025) Economic efficiency through livestock production monitoring and prediction in smart agriculture. In: Optimizing smart and sustainable agriculture for sustainability. CRC Press, pp 1–28. https://doi.org/10.1201/9781003628392-1

Mishra H, Mishra D, Tiwari AK, Nishad DC (2025) Cost–benefit analysis of sensing and data collection with drones for IoT applications. In: Machine learning for drone-enabled IoT networks. Springer, Cham, pp 141–168. https://doi.org/10.1007/978-3-031-80961-3_8

Mishra H (2022) Production and marketing of cucumber and bitter gourd in Sultanpur district of Uttar Pradesh. Doctoral dissertation, Thesis MSc (Agric). Acharya Narendra Deva University of Agriculture and Technology, Kumarganj, Ayodhya, Uttar Pradesh

Mishra H (2024) Nanobiostimulants and precision agriculture: a data-driven approach to farming and market dynamics. In: Nanobiostimulants: emerging strategies for agricultural sustainability. Springer Nature Switzerland, Cham, pp. 365–398. https://doi.org/10.1007/978-3-031-68138-7_16

Mishra H (2024) The role of ethnoeconomics in promoting sustainable consumption and production patterns: a pathway to environmental protection and economic prosperity. Sustainable development. In: Seen through the lenses of ethnoeconomics and the circular economy. Springer Nature Switzerland, Cham, pp. 91–123. https://doi.org/10.1007/978-3-031-72676-7_6

Mishra H (2025) Environmental degradation and impacts on agricultural production: a challenge to urban sustainability. In: Sustainable urban environment and waste management: theory and practice. Springer Nature Singapore, Singapore, pp. 53–92. https://doi.org/10.1007/978-981-96-1140-9_3

Mohamed ES, Abu-hashim M, AbdelRahman MA, Schütt B, Lasaponara R (2019) Evaluating the effects of human activity over the last decades on the soil organic carbon pool using satellite imagery and GIS techniques in the Nile Delta Area, Egypt. Sustainability 11(9):2644. https://doi.org/10.3390/su11092644

Mohanasundaram S, Kasiviswanathan KS, Purnanjali C, Santikayasa IP, Singh S (2023) Downscaling global gridded crop yield data products and crop water productivity mapping using remote sensing derived variables in the South Asia. Int J Plant Prod 17(1):1–6

Mooney HA, Sombroek WG (1992) Terrestrial systems. In: International conference on an agenda of sciemce for environment and development into the 21st century (ASCYMD 21). ICSU, vol 521, pp 173–186

Nautiyal S, Goswami M, Shivakumar P, Nautiyal S, Goswami M, Shivakumar P (2021) Delineation and monitoring of FMV. Field margin vegetation and socio-ecological environment: structural, functional and spatio-temporal dynamics in rural-urban interface of Bengaluru, pp 95–113 https://doi.org/10.1007/978-3-030-69201-8_6

Ndlala N (2021) Remote sensing drought impacts on wetland vegetation productivity at the Soetendalsvlei in the Heuningnes Catchment, South Africa

Ngo DK (2022) Land cover mapping of the Mekong Delta with sentinel-1 synthetic aperture radar. Doctoral dissertation, University of Nottingham

Niedbała G, Piekutowska M, Wojciechowski T, Niazian M (2024) Predictions and estimations in agricultural production under a changing climate. Agronomy 14(2):253. https://doi.org/10.3390/agronomy14020253

Nishad DC, Mishra H, Tiwari AK, Mishra D (2011) Post-harvest management: enhancing food security and sustainability. Adv Agric Sci Vol II 24(4):136

Osseni AA, Dossou-Yovo HO, Hegbe AD, Khan MN, Sinsin B (2025) Assessing and modeling the vegetation cover in the W and Pendjari National Parks and their peripheries from 1985 to 2030, using landsat imagery and climatic data in Benin, West Africa. Revue Internationale de Geomatique 34(1):209–234. https://doi.org/10.32604/rig.2025.061448

Patra S, Mishra M (2024) Spatio-temporal changes in mangroves of Bhitarkanika National Park along the Eastern Coast of India over a three decades period using remote sensing spectral index NDVI. Geol Ecol Lands 1–14. https://doi.org/10.1080/24749508.2024.2429224

Pham LH, Pham LT, Dang TD, Tran DD, Dinh TQ (2022) Application of Sentinel-1 data in mapping land-use and land cover in a complex seasonal landscape: a case study in coastal area of Vietnamese Mekong Delta. Geocarto Int 37(13):3743–3760. https://doi.org/10.1080/10106049.2020.1869329

Phan TH (2018) Rice monitoring using radar remote sensing. Doctoral dissertation, Université Paul Sabatier-Toulouse III

Quy VK, Hau NV, Anh DV, Quy NM, Ban NT, Lanza S, Muzirafuti A (2022) IoT-enabled smart agriculture: architecture, applications, and challenges. Appl Sci 12(7):3396. https://doi.org/10.3390/rs10091406

Reljić M, Romić M, Romić D, Gilja G, Mornar V, Ondrasek G, Zovko M (2023) Advanced continuous monitoring system—tools for water resource management and decision support system in salt affected delta. Agriculture 13(2):369. https://doi.org/10.3390/agriculture13020369

Salam M, Cheema MJM, Zhang W, Hussain S, Khan A, Bilal M, Zaman MA (2020) Groundwater storage change estimation using grace satellite data in Indus Basin. Big Data Water Resour Eng 1:13–18. https://doi.org/10.26480/bdwre.01.2020.10.15

Saliha A (2024) Diachronic study of changes in vegetal cover in a mountainous ecosystem by spatial remote sensing: Aures region (Algeria). Technium 24

Sardar P, Samadder SR (2021) Understanding the dynamics of landscape of greater Sundarban area using multi-layer perceptron Markov chain and landscape statistics approach. Ecol Ind 121:106914. https://doi.org/10.1016/j.ecolind.2020.106914

Sciarretta A, Calabrese P (2019) Development of automated devices for the monitoring of insect pests. Curr Agric Res J 7(1). https://doi.org/10.12944/CARJ.7.1.03

Sharif M, Hamzeh S (2021) Investigating the effect of Gotvand Dam on changes in soil salinity and vegetation cover of downstream lands of the dam using satellite imagery and spectral indices. Environ Sci 19(4):225–248

Shuford WD, Reiter ME, Sesser KA, Hickey CM, Golet GH (2019) The relative importance of agricultural and wetland habitats to waterbirds in the Sacramento–San Joaquin River Delta of California. San Francisco Estuary and Watershed Science 17(1)

Singh A (2022) Soil salinity: a global threat to sustainable development. Soil Use Manag 38(1):39–67. https://doi.org/10.1111/sum.12772

Singh KK, Kalra N (2016) Simulating impact of climatic variability and extreme climatic events on crop production. Mausam 67(1):113–130

Son NT, Hieu TV, Shrestha RP, Trieu NT, Kien NV, Anh VT, Dung PA, Duc HN, Du NM, Niem NX (2008) Integrated land-use planning for sustainable agriculture and natural resources management in the Vietnamese Mekong delta. Asia Europe J 6(2):307–324

Suyunov A, Suyunov S, Khushmurodov F, Suyunov S, Omonov I, Eshboev B (2025) Ecological consequences of the desertification process in the agrolandscapes of the Kashkadarya oasis. In: BIO web of conferences, vol 160. EDP Sciences https://doi.org/10.1051/bioconf/202516002006

Swain S, Taloor AK, Dhal L, Sahoo S, Al-Ansari N (2022) Impact of climate change on groundwater hydrology: a comprehensive review and current status of the Indian hydrogeology. Appl Water Sci 12(6):120. https://doi.org/10.1007/s13201-022-01652-0

Taha MM, Al-Hashim MH, El-Asmar HM (2020) Geoarcheomorphosites under strong urbanization pressure at the Tineh Plain, NE of the Nile Delta, Egypt. Geoheritage 12(1):24

Tiwari AK, Mishra H, Nishad DC (2011) Market dynamics and consumer perceptions of organic produce in contemporary agriculture. Adv Agric Sci Vol II 24(4):120

Tiwari AK, Mishra H, Nishad DC, Pandey A (2023) Sustainable water management in agriculture: irrigation techniques and water conservation. Dr. Ajay B. Jadhao 53

Tran A, Fall AG, Biteye B, Ciss M, Gimonneau G, Castets M, Chevalier V (2019a) Spatial modeling of mosquito vectors for Rift Valley Fever virus in Northern Senegal: integrating satellite-derived meteorological estimates in population dynamics models. Remote Sens 11(9):1024. https://doi.org/10.3390/rs11091024

Tran TV, Tran DX, Myint SW, Huang CY, Pham HV, Luu TH, Vo TM (2019b) Examining spatiotemporal salinity dynamics in the Mekong River Delta using Landsat time series imagery and a spatial regression approach. Sci Total Environ 687:1087–1097. https://doi.org/10.1016/j.scitotenv.2019.06.056

Tripathi A, Malik K, Reshi AR, Moniruzzaman M, Tiwari RK (2023) Multi-temporal SAR Interferometry (MTInSAR)-based study of surface subsidence and its impact on Krishna Godavari (KG) basin in India: a support vector approach. Environ Monit Assess 195(11):1298. https://doi.org/10.1007/s10661-023-11896-1

Tsanakas K, Karymbalis E, Batzakis DV (2024) Advances in geographical information science for monitoring and managing deltaic environments: processes, parameters, and methodological insights. In: Geographical information science. Elsevier, pp 279–311

Uddin MR, Uddin A, Nelson J, Rahman SN, Zahid A (2025) Spatial mapping of vegetation's potential to counter seasonal groundwater salinity in coastal Bangladesh. Earth Syst Environ 1–21. https://doi.org/10.1007/s41748-025-00574-6

Unal E, de Bie CAJM (2017) Mapping wheat growing areas of Turkey by integrating multi-temporal NDVI data and official crop statistics. J Field Crops Central Res Inst 26(1):11–23

Wang Z, Zhang F, Zhang X, Chan NW, Ariken M, Zhou X, Wang Y (2021) Regional suitability prediction of soil salinization based on remote-sensing derivatives and optimal spectral index. Sci Total Environ 775:145807. https://doi.org/10.1016/j.scitotenv.2021.145807

Wegler M, Kuenzer C (2024) Potential of earth observation to assess the impact of climate change and extreme weather events in temperate forests—a review. Remote Sens 16(12):2224. https://doi.org/10.3390/rs16122224

Yossif TMH (2020) Assessment of land productivity dynamics in relation to land degradation using NDVI in Egypt. J Soil Sci Agric Eng 11(9):477–481. https://doi.org/10.21608/jssae.2020.118339

Zähringer JG, Hett C, Eckert S, Ramamonjisoa B, Messerli P (2014) Shifting cultivation dynamics within an ecosystem service hotspot of pronounced global interest—a first step towards negotiating trade-offs in North-East Madagascar. In: Proceedings of the global land project 2nd open science meeting, Berlin, pp 196–197

Zarei A, Hasanlou M, Mahdianpari M (2021) A comparison of machine learning models for soil salinity estimation using multi-spectral earth observation data. ISPRS Ann Photogramm Remote Sens Spatial Inf Sci 3:257–263. https://doi.org/10.5194/isprs-annals-V-3-2021-257-2021

Zhang Y, Luo C, Zhang W, Wu Z, Zang D (2025) Mapping soil organic matter in black soil cropland areas using remote sensing and environmental covariates. Agriculture 15(3):339. https://doi.org/10.3390/agriculture15030339

Zhou T, Geng Y, Chen J, Sun C, Haase D, Lausch A (2019) Mapping of soil total nitrogen content in the middle reaches of the Heihe River Basin in China using multi-source remote sensing-derived variables. Remote Sens 11(24):2934. https://doi.org/10.3390/rs11242934

8 Strategies for Environmental Pollution Control in Delta Ecosystems

Nuralfin Anripa

Abstract

Delta ecosystems are among the most fertile and ecologically rich areas on Earth, acting as vital connectors between land and sea. They provide essential services such as water purification, nutrient cycling, and storm protection, while supporting agriculture, fisheries, and tourism. However, these ecosystems face growing threats from environmental pollution caused by industrialization, agricultural intensification, and urban expansion. Pollutants like heavy metals, pesticides, and untreated wastewater degrade water quality, harm habitats, and endanger biodiversity. Climate change exacerbates these issues through rising temperatures, altered rainfall patterns, and sea-level rise, further straining delta environments. For instance, in Indonesia's Banda Islands, climate change is expected to disrupt nutmeg farming due to reduced rainfall and higher temperatures. This chapter explores the sources and impacts of pollution in delta ecosystems, focusing on industrial emissions, agricultural runoff, and urban waste. Industries release heavy metals and chemicals, agriculture contributes to nutrient pollution, and urbanization introduces sewage and microplastics, leading to eutrophication, biodiversity loss, and health risks from toxins like mercury in food chains. To address these challenges, the chapter highlights mitigation strategies, particularly nature-based solutions (NbS) such as wetland restoration, riparian buffer zones, and phytoremediation. Advanced wastewater treatment technologies, including hybrid constructed

A delta is not just the end of a river, but the beginning of life. Protecting it means protecting our future.

N. Anripa (✉)
Department of Environmental Science, Universitas Dumoga Kotamobagu, Kotamobagu, North Sulawesi, Indonesia
e-mail: nuralfin.anripa@gmail.com

C. Singha et al. (eds.), *Sustainable Management of Delta Ecosystems Resilience*, Deltas of the World, https://doi.org/10.1007/978-3-032-16261-8_8

wetlands and electrocoagulation, are also emphasized for their effectiveness. Case studies from deltas like the Mekong, Mississippi, and Nile demonstrate successful pollution control and restoration efforts, underscoring the need for integrated water resource management (IWRM), multi-stakeholder governance, and adaptive management. The chapter concludes with policy recommendations, advocating for stricter regulations, community-based conservation, and the inclusion of Indigenous knowledge. Strengthening energy transition policies, regional cooperation, and invasive species control are also vital for sustainable delta management. Future research should focus on advanced technologies, long-term climate impacts, and predictive modeling to enhance pollution control and ecosystem resilience. A holistic, collaborative approach is essential to mitigate pollution, protect delta ecosystems, and safeguard the livelihoods of millions who depend on these critical regions.

Abbreviations

BCF	Bioaccumulation Factor
BOD	Biochemical Oxygen Demand
BOD5	5-Day Biochemical Oxygen Demand
CAQI	Comprehensive Air Quality Index
Cd	Cadmium
COD	Chemical Oxygen Demand
CPRA	Coastal Protection and Restoration Authority (Louisiana, USA)
Cr	Chromium
EMPF	Entomopathogenic Fungi
FCW	Floating Constructed Wetland
Fe	Iron
Hg	Mercury
HCWs	Hybrid Constructed Wetlands
HSCW	Horizontal Sub-Surface Constructed Wetland
IWRM	Integrated Water Resource Management
MeHg	Methylmercury
MRGO	Mississippi River Gulf Outlet
NbS	Nature-Based Solutions
NH_4^+ - N	Ammonium Nitrogen
Ni	Nickel
NO_X	Nitrogen Oxides
Pb	Lead
$PM_{2.5}/PM_{10}$	Particulate Matter (2.5/10 μm)
PRD	Pearl River Delta (China)
RBZs	Riparian Buffer Zones

RCP	Representative Concentration Pathway (climate scenarios)
SO_2	Sulfur Dioxide
TEK	Traditional Ecological Knowledge
TN	Total Nitrogen
TRM	Tidal River Management
TSS	Total Suspended Solids
YRD	Yangtze River Delta (China)

8.1 Introduction

8.1.1 Importance of Delta Ecosystems

Delta ecosystems rank among the most fertile and ecologically diverse areas globally, functioning as vital transitional zones between land and sea. These distinctive environments are home to a rich variety of plant and animal species, many of which are unique and specially adapted to the ever-changing conditions of deltas. Apart from their ecological importance, deltas deliver crucial ecosystem services that are indispensable for human survival. They act as natural filtration systems, enhancing water quality by capturing sediments and contaminants, and they play a key role in nutrient cycling, sustaining both aquatic and land-based ecosystems. Furthermore, deltas provide natural protection against storms and flooding, shielding inland regions from the destructive effects of severe weather. Economically, deltas are incredibly valuable, underpinning agriculture, fisheries, and tourism, which are essential for the livelihoods of millions of people residing in these areas.

For instance, Indonesia's delta ecosystems play a crucial role in sustaining agricultural activities, particularly for high-value crops such as nutmeg (Myristica fragrans Houtt.). Research conducted by Anripa (2023) analyzed climate trends over Banda Neira Island and found that ideal nutmeg-growing conditions, which require 2000–4500 mm of rainfall annually, 100–160 wet days, and an optimal temperature range of 25–26 °C, are increasingly threatened by climate change. Projected shifts in precipitation and rising temperatures indicate potential declines in nutmeg yields, highlighting the need for adaptive strategies in delta-based agriculture.

As seen in Fig. 8.1, dry farmland in a delta region illustrates the severe impact of prolonged droughts on agriculture. Such conditions threaten food security and the livelihoods of millions who depend on delta ecosystems for their survival.

Fig. 8.1 Dry farmland in a delta region. *Credits* Anripa/Shutterstock.com

8.1.2 Threats to Delta Ecosystems

Despite their ecological and economic importance, delta ecosystems are under severe threat from environmental pollution, driven by rapid industrialization, agricultural intensification, and urban expansion. The influx of pollutants such as heavy metals, pesticides, nitrogen, phosphorus, and untreated wastewater has led to widespread environmental degradation. These pollutants cause eutrophication, hypoxia, and habitat loss, which undermine the ecological integrity of delta regions. The situation is further exacerbated by climate change, which intensifies the frequency and severity of storms, floods, and sea-level rise, putting additional stress on already fragile delta ecosystems. The combined effects of pollution and climate change pose a significant risk to the biodiversity, productivity, and resilience of delta regions, threatening both ecological stability and human livelihoods.

Climate change in Indonesia highlights these issues, especially in farming areas like the Banda Islands. Findings from Anripa (2023) reveal that the country's average yearly temperature has risen by 0.3 °C each decade, while annual rainfall has decreased by 2–3%. Climate projections under different Representative Concentration Pathways (RCPs) indicate a drastic reduction in wet days, with RCP4.5 scenarios predicting as few as 67–78 wet days per year, well below the optimal threshold for nutmeg production. Furthermore, mean temperatures are projected to rise beyond 26 °C, reaching up to 31.5 °C under RCP8.5 scenarios. These changes are expected to disrupt nutmeg cultivation cycles, reducing both yield and quality, while increasing the susceptibility of crops to heat stress and erratic rainfall patterns.

Contaminants such as mercury (Hg) are of particular concern due to their persistence and ability to bioaccumulate in the food chain. Methylmercury (MeHg), a highly toxic form of mercury, is produced in the anoxic sediments of wetlands and can lead to severe health risks for both humans and wildlife. For example, in California, historical mining practices have left a legacy of elevated mercury concentrations in delta ecosystems, leading to widespread fish consumption advisories (Sedlak and Ulrich 2009). Similarly, industrial and mining activities in Indonesia contribute to heavy metal contamination in river systems, further exacerbating ecological and agricultural vulnerabilities in delta regions. Addressing these environmental stressors will require coordinated policy interventions and adaptive agricultural practices to mitigate long-term risks.

Figure 8.2 shows a dry river in a delta region, highlighting the devastating effects of prolonged droughts and reduced water availability. Such conditions disrupt aquatic ecosystems, reduce water supplies for agriculture and domestic use, and exacerbate the vulnerability of delta communities.

8.1.3 Objectives of the Chapter

This chapter aims to provide a comprehensive analysis of the sources, impacts, and mitigation strategies for environmental pollution in delta ecosystems. It seeks to explore the effectiveness of nature-based solutions (NbS) and advanced wastewater treatment technologies in controlling pollution. Additionally, the chapter will highlight successful case studies from major global deltas, offering valuable lessons and insights for policymakers, researchers, and practitioners. Finally, it will provide policy recommendations and future research directions to enhance pollution control efforts and promote the sustainable management of delta ecosystems.

Fig. 8.2 A dry river in a delta region. *Credits* Anripa/Shutterstock.com

8.2 Sources and Impacts of Environmental Pollution in Delta Ecosystems

8.2.1 Industrial Pollution and Carbon Emissions

Industrial operations are major drivers of environmental pollution in delta regions. The swift industrial growth in urban areas like China's Pearl River Delta (PRD) and Yangtze River Delta (YRD) has caused significant environmental harm. A detailed study by Wen et al. (2024) explores the factors affecting industrial pollution and efforts to reduce carbon emissions in these areas.

8.2.2 Sources of Industrial Pollution

Industrial pollution in delta regions originates from a variety of sources, each contributing to the degradation of these fragile ecosystems. The primary sources include:

- **Industrial Wastewater Discharge**: Factories and industrial plants frequently release untreated or inadequately treated wastewater containing heavy metals, organic compounds, and nutrients. These contaminants pollute water systems and sediments, threatening aquatic life and disrupting ecosystems. For example, heavy metals such as mercury and lead can build up in fish, leading to bioaccumulation and biomagnification, which ultimately impact human health through the food chain
- **Air Pollution**: Emissions from industries, including particulate matter ($PM_{2.5}$ and PM_{10}), sulfur dioxide (SO_2), and nitrogen oxides (NO_x) not only worsen air quality but also contribute to water pollution via atmospheric deposition. This dual effect harms both human health and plant life. For instance, sulfur dioxide emissions can cause acid rain, which alters the pH of water bodies, making them unsuitable for aquatic species
- **Solid Waste Disposal**: The improper disposal of industrial solid waste, such as plastics, chemicals, and construction debris, can lead to the leaching of harmful substances into the soil and water. This exacerbates pollution levels and can result in long-term environmental damage. In some cases, solid waste can block waterways, leading to flooding and further ecosystem disruption

8.2.3 Impacts of Industrial Pollution

The consequences of industrial pollution in delta regions are profound and multifaceted, affecting both the environment and human populations:

- **Water Quality Degradation**: Contaminated water becomes unsafe for drinking, recreation, and fishing, directly affecting local communities that rely on these resources. Poor water quality can also lead to the spread of waterborne diseases, further exacerbating public health issues.
- **Habitat Loss**: Pollution degrades or destroys critical habitats for fish, wildlife, and plants, leading to ecosystem imbalances. For example, the loss of mangrove forests in delta regions due to pollution reduces the natural buffer against storms and floods, increasing the vulnerability of coastal communities
- **Loss of Biodiversity**: The introduction of pollutants reduces the number and variety of species, threatening the ecological integrity of delta regions. Species that are sensitive to pollution, such as certain fish and amphibians, are often the first to disappear, leading to a cascade of effects throughout the ecosystem
- **Human Health Risks**: Exposure to industrial pollutants is linked to a range of health issues, including respiratory problems, cancers, and birth defects. Communities living near industrial zones are particularly vulnerable, as they are often exposed to higher concentrations of pollutants

8.2.4 Agricultural Runoff and Soil Degradation

Agricultural activities, while essential for food production, contribute significantly to environmental pollution in delta regions. The Nile Delta, a cornerstone of Egypt's agricultural sector, exemplifies the challenges posed by agricultural runoff. A study by Alfiky (2022) emphasizes the ecological consequences of this runoff, particularly its impact on soil health and microbial biodiversity.

8.2.5 Sources of Agricultural Runoff

Agricultural runoff in delta regions is primarily caused by:

- **Fertilizers and Pesticides**: The excessive use of chemical fertilizers and pesticides in intensive farming practices leads to nutrient pollution. Nutrients like nitrogen and phosphorus are carried by runoff into water bodies, causing eutrophication and algal blooms
- **Soil Erosion**: Poor land management practices, such as deforestation and overgrazing, exacerbate soil erosion. Sediments from eroded soils are transported by runoff, further degrading water quality and aquatic habitats

8.2.6 Impacts of Agricultural Runoff

Agricultural runoff has profound consequences on delta ecosystems:

- **Eutrophication**: Excess nutrients from agricultural runoff frequently trigger eutrophication in water systems, resulting in algal blooms that lower oxygen levels and endanger aquatic organisms. The Nile Delta's waterways are especially vulnerable to this type of pollution, with far-reaching effects on aquatic ecosystems.
- **Soil Health Degradation**: Agricultural runoff can disturb soil microbial communities, including entomopathogenic fungi (EMPF), which play a crucial role in natural pest control. Overuse of chemical pesticides reduces the diversity and effectiveness of these beneficial microbes, leading to decreased soil fertility and lower agricultural yields.
- **Reduction in Aquatic Biodiversity**: The influx of agrochemicals into water systems diminishes species diversity and abundance, jeopardizing the ecological balance of delta regions. Fish and other aquatic life are particularly at risk from the harmful effects of pesticides and fertilizers.

8.2.7 Urban Waste and Heavy Metal Contamination

Urbanization in delta regions has led to increased generation of domestic and industrial waste, contributing to severe water pollution. The Niger Delta in Nigeria, for example, faces significant challenges due to urban waste and heavy metal contamination. A study by Allison et al. (2024) reveals the extent of this pollution and its implications for aquatic ecosystems and human health.

8.2.8 Sources of Urban Waste

Urban waste in delta regions originates from:

- **Domestic Waste**: Untreated sewage and household waste are major contributors to water pollution in urbanized deltas. These wastes often contain pathogens, organic matter, and microplastics, which degrade water quality and harm aquatic life.
- **Industrial Effluents**: Industries in urban areas discharge effluents containing heavy metals, such as iron (Fe), nickel (Ni), lead (Pb), cadmium (Cd), and chromium (Cr), into water bodies. These metals are persistent pollutants that accumulate in the environment and pose long-term risks.

8.2.9 Impacts of Heavy Metal Contamination

The impacts of heavy metal contamination in delta regions are severe:

- **Bioaccumulation in Aquatic Life**: Fish, a primary protein source for local communities, were found to have high levels of Pb and Cd in their tissues. The bioaccumulation factor (BCF) analysis revealed that metals like Fe, Cd, and Cr were readily absorbed by fish, posing significant health risks to consumers.
- **Human Health Risks**: Eating fish contaminated with pollutants is associated with various health problems, such as neurotoxicity, kidney dysfunction, and cancer. Children and pregnant women are especially at risk from heavy metal exposure, which may cause developmental issues and chronic health problems.
- **Ecosystem Disruption**: Heavy metals disrupt aquatic ecosystems by reducing biodiversity and altering ecological processes. The persistence of these metals in the environment poses long-term risks, as they do not degrade and can accumulate over time.

8.3 Mitigation Strategies for Environmental Pollution

8.3.1 Nature-Based Solutions (NbS)

Nature-based solutions (NbS) harness the power of natural processes to tackle environmental challenges, offering sustainable and cost-effective strategies for pollution control in delta ecosystems. These approaches not only mitigate pollution but also enhance ecosystem resilience, restore biodiversity, and provide co-benefits for local communities.

Wetland Restoration

Wetland restoration stands out as a cornerstone of NbS for improving water quality and reducing pollution in delta ecosystems. Wetlands act as natural filters, trapping sediments and pollutants while promoting nutrient cycling. For instance, the addition of ferrous iron (Fe) to wetland sediments has been shown to significantly curb the production of methylmercury (MeHg), a potent neurotoxin, by reducing the concentration of dissolved sulfide and bioavailable mercury. This innovative approach was successfully tested in tidal wetlands in San Francisco Bay, where iron addition led to a measurable decline in MeHg concentrations in both surface water and sediment (Sedlak and Ulrich 2009). Such interventions highlight the immense potential of wetland restoration to mitigate heavy metal contamination and restore ecosystem health, offering a blueprint for other delta regions facing similar challenges.

Riparian Buffer Zones (RBZs)

Riparian buffer zones—vegetated areas along water bodies—play a critical role in filtering pollutants, reducing soil erosion, and enhancing water quality. A recent study by Bulut and Aktürk (2024) in the coastal districts of Kastamonu, Türkiye, underscores the importance of RBZs in maintaining forest cover and mitigating the impacts of demographic changes on riparian ecosystems. The study revealed that forest areas within RBZs were more resilient to population pressures compared to areas outside these buffer zones, emphasizing their value in sustainable land management and pollution control. By preserving and expanding RBZs, delta regions can bolster their capacity to absorb and neutralize pollutants, safeguarding both aquatic and terrestrial ecosystems while supporting local livelihoods.

Phytoremediation

Phytoremediation, the use of plants to absorb and stabilize heavy metals and other pollutants, offers a green and cost-effective solution for delta regions grappling with contamination. Plants like water hyacinths and willows have proven particularly effective in removing pollutants such as cadmium, lead, and mercury from soil and water. For example, water hyacinths are renowned for their ability to absorb heavy metals from water, while willows excel at stabilizing contaminated soils. Beyond pollution control, phytoremediation also contributes to habitat restoration and biodiversity enhancement, making it a win–win strategy for delta ecosystems.

8.3.2 Advanced Wastewater Treatment Technologies

Advanced wastewater treatment technologies provide innovative and efficient solutions for managing pollution in delta ecosystems. These technologies are designed to remove contaminants effectively while minimizing environmental impacts, offering scalable and sustainable alternatives to conventional methods.

Hybrid Constructed Wetlands for Saline Wastewater Treatment

A notable innovation is the Hybrid Constructed Wetlands (HCWs) model, which is particularly effective for treating saline wastewater from shrimp farming. A pilot-scale HCWs system in the Mekong Delta, investigated by Hiep et al. (2023), integrated a floating constructed wetland (FCW) with a horizontal sub-surface constructed wetland (HSCW). By incorporating native salt-tolerant plants like Scirpus littoralis Schrab, Cyperus alternifolius, and Paspalum vaginatum, the system successfully removed organic matter and nitrogen compounds from shrimp farm wastewater. The treated water met safety standards, with low concentrations of COD (10.0–15.4 mg/L), BOD_5 (7.1–12.5 mg/L), NH_4^+ - N (0.04–1.11 mg/L), and TN (0.17–1.83 mg/L). This method not only tackles pollution but also promotes sustainable aquaculture, providing a practical alternative to traditional treatment approaches.

Green Technology and Policy Integration for Air Pollution Control

Achieving effective pollution control often involves combining green technologies with supportive policies. Sun and Xu (2023) emphasize how industrial restructuring and advancements in green technology in the Yangtze River Delta have enhanced air pollution control. Their research shows that modernizing industrial structures and adopting green technologies can substantially lower emissions while improving water quality. Additionally, the study highlights spatial spillover effects, where pollution control measures in one area benefit neighboring regions. This highlights the need for coordinated environmental governance, ensuring local actions align with regional and national sustainability objectives.

Constructed Wetlands

Constructed wetlands, designed to replicate natural wetland processes, have become a sustainable option for wastewater treatment in delta regions. These systems utilize plants, soil, and microorganisms to filter pollutants from agricultural runoff, industrial discharges, and urban wastewater. In the Mekong Delta, constructed wetlands have been effectively used to treat aquaculture wastewater, significantly reducing levels of biochemical oxygen demand (BOD), chemical oxygen demand (COD), and nutrients. This method not only enhances water quality but also promotes biodiversity, offering an eco-friendly alternative to energy-intensive treatment technologies.

Biofiltration Systems

Biofiltration systems leverage microorganisms to degrade organic pollutants and remove heavy metals from wastewater. These systems are particularly effective in treating industrial effluents and urban runoff, which often contain high concentrations of toxic substances. For example, in the Niger Delta, biofiltration has been used to treat oil-contaminated water, drastically reducing hydrocarbon levels and improving water quality. By harnessing natural metabolic processes, biofiltration systems provide an environmentally friendly and cost-effective solution for pollution control.

Electrocoagulation

Electrocoagulation, an advanced wastewater treatment technology, uses electrical currents to remove heavy metals and other contaminants from water. This process generates coagulants through the electrolytic oxidation of metal electrodes, which bind to pollutants and form flocs that can be easily removed. In the Yangtze River Delta, electrocoagulation has been employed to treat industrial wastewater, effectively removing heavy metals like cadmium and lead. This technology offers a highly efficient and scalable solution for pollution control in delta regions.

8.3.3 Integrated Water Resource Management (IWRM)

Integrated Water Resource Management (IWRM) provides a holistic framework for managing water resources in delta regions, balancing competing demands while minimizing environmental impacts.

Comprehensive Water Management

Integrated Water Resources Management (IWRM) focuses on the coordinated management of water, land, and associated resources to optimize economic and social benefits while maintaining ecosystem health. In the Ganges–Brahmaputra Delta, IWRM has played a key role in tackling issues like waterlogging and soil degradation. By aligning water management strategies with land use planning, IWRM strengthens the ability of delta ecosystems to withstand climate change and human impacts, securing their sustainability for the future.

Collaborative Governance

Addressing pollution in delta ecosystems effectively demands the participation of diverse stakeholders, including governments, industries, local communities, and non-governmental organizations. Collaborative governance frameworks encourage cooperation and coordination, ensuring that pollution control measures are carried out transparently and inclusively. In the Mekong Delta, such approaches have supported sustainable aquaculture and enhanced wastewater management practices. By involving local communities in decision-making, collaborative governance increases the social acceptance and long-term viability of pollution control efforts.

8.4 Case Studies: Lessons from Global Deltas

8.4.1 Mekong Delta

The Mekong Delta in Vietnam highlights both the difficulties and prospects related to managing pollution and restoring ecosystems in deltaic environments.

Pollution Control and Ecosystem Restoration

The Mekong Delta has encountered severe pollution due to agricultural runoff, industrial discharges, urban waste, and intensive aquaculture practices, especially Pangasius farming. The rapid expansion of Pangasius aquaculture has contributed to water contamination, with wastewater and sludge containing elevated levels of BOD, COD, total suspended solids (TSS), nitrogen, and phosphorus (PhamThi Anh et al. 2010). To tackle these challenges, measures such as constructing wetlands, adopting integrated pest management, and promoting sustainable farming methods have been introduced. These initiatives have enhanced water quality and biodiversity, demonstrating the effectiveness of nature-based approaches in addressing pollution.

Challenges and Successes

Despite these successes, the Mekong Delta continues to grapple with challenges such as upstream dam construction, climate change, and population growth. These factors exacerbate water scarcity and pollution, underscoring the need for integrated and adaptive management approaches. Lessons from global delta management suggest that a combination of regulatory enforcement, technological innovation, and stakeholder collaboration is essential for mitigating pollution. Strengthening wastewater treatment facilities, promoting sludge reuse in agriculture, and encouraging sustainable farming practices are key steps toward achieving a balance between economic growth and environmental sustainability.

8.4.2 Mississippi River Delta

The Mississippi River Delta offers valuable insights into the restoration and management of large-scale delta ecosystems.

Restoration Efforts and Ecosystem Resilience

The Mississippi River Delta, spanning approximately 25,000 km^2, is one of the largest coastal wetland ecosystems globally. However, extensive wetland loss has occurred over the past century due to human-induced factors and natural subsidence. Restoration efforts, led by the State of Louisiana through the Coastal Protection and Restoration Authority (CPRA), have focused on leveraging natural sediment deposition from the Mississippi River to enhance wetland resilience (Kemp et al. 2021). For example, the closure of the Mississippi River Gulf Outlet (MRGO) in 2009 has increased sediment input, contributing to the gradual recovery of marshland.

Challenges in Wetland Sustainability

Despite ongoing restoration initiatives, the Mississippi River Delta faces persistent challenges, including sea-level rise, subsidence, and reduced sediment flow due to upstream river modifications. Adaptive management strategies, such as targeted sediment diversion and marsh nourishment, are crucial for sustaining the delta's ecosystems. Studies suggest that incorporating geological data, hydrodynamic modeling, and sediment transport analysis can improve the effectiveness of restoration projects.

8.4.3 Nile Delta

The Nile Delta in Egypt highlights the importance of sustainable agriculture and biocontrol agents in pollution control.

Agricultural Runoff and Soil Health

The Nile Delta has faced significant challenges due to pollution from agricultural runoff, leading to soil degradation and water quality issues. Initiatives such as the use of biocontrol agents, organic farming practices, and riparian buffer zones have improved soil health and water quality. For example, the use of entomopathogenic fungi (EMPF) has reduced reliance on chemical pesticides, offering a sustainable alternative for pest management (Bakr et al. 2024).

8.4.4 Ganges–Brahmaputra Delta

The Ganges–Brahmaputra Delta showcases how Tidal River Management (TRM) can be effectively utilized for sustainable land use planning.

Tidal River Management (TRM)
TRM utilizes controlled flooding to enhance sediment deposition and restore agricultural productivity. This approach has been implemented to mitigate waterlogging and improve soil fertility in the delta. However, challenges such as stakeholder conflicts and climate change-induced variability in tidal patterns require adaptive management strategies and community-based decision-making (Nowreen et al. 2019).

8.5 Policy Recommendations and Future Directions

Delta ecosystems are under increasing pressure from pollution, climate change, and human activities. To address these challenges, a combination of innovative policies, community engagement, and advanced research is essential. Below are key policy recommendations and future directions to ensure the sustainable management of delta ecosystems.

8.5.1 Strengthening Energy Transition Policies for Pollution Control

Policies aimed at transitioning to cleaner energy are essential for lowering carbon emissions and addressing air pollution in delta areas. By promoting the replacement of fossil fuels with renewable energy, these measures can enhance air quality and reduce the adverse effects of climate change.

Energy Transition and Air Quality Improvements
A study by Chen et al. (2022) in the Pearl River Delta (PRD) region of China highlights the dual benefits of energy transition policies. By transitioning from fossil fuel-based energy systems to renewable sources, such as solar, wind, and waste-to-energy systems, significant reductions in CO_2 and air pollutant emissions can be achieved. The study used the LEAP model to project emission trends and found that stringent energy transition measures could improve the Comprehensive Air Quality Index (CAQI) by 27% by 2030. This underscores the importance of stricter regulations on industrial energy use, increased investments in renewable energy, and enhanced energy efficiency in key sectors like transportation and manufacturing.

Sectoral Focus and Implementation Strategies

Effective energy transition policies should prioritize the power, transportation, and industrial sectors, which collectively contribute over 90% of CO_2 and energy-related pollutant emissions. Key strategies include:

- Accelerating the adoption of low-carbon energy sources through financial incentives, subsidies, and regulatory measures
- Implementing clean production technologies and enhancing vehicle electrification to reduce emissions.
- Enforcing stringent emission control standards to ensure compliance and long-term environmental sustainability.

8.5.2 Strengthening Community-Based Conservation and Indigenous Knowledge Integration

Conservation efforts in delta ecosystems must prioritize community engagement and integrate Indigenous knowledge systems to achieve sustainable outcomes.

Relational Conservation and Community Engagement

A study by Beaudette (2024) in the Okavango Delta, Botswana, emphasizes the importance of relational conservation, which incorporates local cultural and spiritual values into environmental management. Many conservation programs fail due to a lack of meaningful collaboration with local communities, often relying on Western paradigms that marginalize Indigenous ecological knowledge. By fostering participatory approaches, policymakers can improve conservation outcomes and promote practices tailored to local socio-cultural contexts.

Decolonizing Conservation Science

Decolonizing conservation science involves shifting away from top-down, technocratic methodologies and embracing inclusive, relational approaches. Traditional ecological knowledge (TEK) held by Indigenous and local communities offers valuable insights into sustainable land and water management. Combining Traditional Ecological Knowledge (TEK) with scientific research promotes a more comprehensive approach to environmental stewardship, ensuring that conservation initiatives are tailored to local needs and cultural values. In the future, conservation strategies should emphasize collaborative management models, where Indigenous leaders play an active role in decision-making processes. This approach enhances environmental governance and builds greater resilience.

8.5.3 Promoting Regional Integration for Sustainable Development

Sustainable development in delta ecosystems requires coordinated regional integration to address environmental challenges and socio-economic disparities.

Regional Collaboration and Sustainable Development
The Yangtze River Delta (YRD) in China exemplifies how sustainability science can be seamlessly integrated with economic and urban development strategies. Through effective regional collaboration, areas such as industrial synergy, innovation, and environmental conservation are strengthened, paving the way for enduring ecological sustainability (Wu and Sun 2023). By harmonizing regional policies with international sustainability goals, governments can bolster resilience and promote equitable socio-environmental progress.

Governance and Policy Synergies
Achieving sustainability goals at the subnational level requires cross-administrative collaboration. The YRD's approach to integrating infrastructure development, pollution control, and climate adaptation strategies offers valuable lessons for other deltaic regions worldwide. Key strategies include:

- Aligning regional policies with global sustainability frameworks.
- Promoting industrial cooperation and innovation to reduce environmental impacts.
- Enhancing ecological preservation through coordinated land-use planning and resource management.

8.5.4 Promoting Controlling Invasive Species to Protect Delta Ecosystems

Invasive species pose a significant threat to delta ecosystems, disrupting water flow, altering nutrient cycles, and threatening native biodiversity.

Invasive Aquatic Vegetation Management
A study by Ta et al. (2017) on the Sacramento–San Joaquin River Delta highlights the challenges of managing invasive species, including regulatory hurdles, inconsistent monitoring programs, and funding constraints. Effective management strategies require a combination of remote sensing technologies, mechanical and chemical control, and community-based monitoring to ensure long-term ecological balance.

Integrated Control Approaches
Adaptive management techniques that integrate mechanical harvesting, biological control, and chemical treatments have shown promise in reducing the spread of invasive

aquatic vegetation. Remote sensing technology, in particular, has proven effective in tracking changes in vegetation distribution, enabling targeted and efficient intervention strategies. Long-term success depends on continuous research into new biological control agents, sustainable management practices, and stronger regulatory support to mitigate the ecological impact of invasive species.

8.5.5 Future Research Directions

To address the complex challenges facing delta ecosystems, future research should focus on the following areas:

Exploration of Innovative Technologies
Efforts should focus on advancing and deploying cutting-edge wastewater treatment methods, including electrocoagulation and biofiltration, alongside nature-inspired approaches such as phytoremediation and the rehabilitation of wetlands. These solutions can significantly improve pollution management and bolster the resilience of ecosystems.

Long-Term Research on Climate Change Effects
Comprehensive, long-term studies are essential to unravel the complex relationships between climate change, pollution, and ecosystem health. Emphasis should be placed on investigating how factors like sea level rise, more intense storms, and shifting rainfall patterns influence the spread and behavior of pollutants in delta areas.

Leveraging Remote Sensing and Big Data
Combining remote sensing tools with big data analytics offers powerful opportunities to monitor pollution trends across space and time in delta ecosystems. Utilizing satellite data, drone-based surveys, and real-time sensor networks can improve the accuracy of pollution monitoring and inform data-driven strategies for effective environmental management.

Advancing Predictive Modeling
Predictive models play a critical role in anticipating the effects of pollution and climate change on delta ecosystems, allowing for proactive intervention and mitigation. These models should integrate information on pollutant origins, water flow dynamics, and ecological feedback to provide a holistic view of the drivers and consequences of pollution in these regions.

8.6 Conclusion

Delta ecosystems are among the most productive and biologically diverse regions on Earth, but they are increasingly threatened by environmental pollution. The primary sources of pollution include industrial activities, agricultural runoff, and urban waste, which introduce pollutants such as heavy metals, pesticides, and nutrients into the environment. These pollutants cause water quality degradation, habitat loss, biodiversity decline, and increased risks to human health.

Mitigation strategies such as nature-based solutions (NbS), advanced wastewater treatment technologies, and integrated water resource management (IWRM) offer effective approaches to pollution control. Case studies from global deltas, including the Mekong Delta, Mississippi River Delta, and Nile Delta, provide valuable lessons on the importance of interdisciplinary collaboration, community engagement, and adaptive management.

The preservation of delta ecosystems is both an environmental imperative and a socio-economic necessity. To ensure their long-term sustainability, the following actions are urgently needed:

- **Strengthening Environmental Regulations**: Governments must enforce stricter regulations on industrial emissions and agricultural runoff while incentivizing sustainable practices.
- **Promoting Interdisciplinary Collaboration**: Researchers, policymakers, and local communities must work together to develop and implement effective pollution control strategies.
- **Enhancing Community Engagement**: Public awareness campaigns and community-based monitoring programs can empower local communities to protect their environment.
- **Investing in Future Research**: To effectively address and reduce the effects of pollution and climate change, it is crucial to prioritize long-term research initiatives, develop predictive models, and incorporate advanced remote sensing technologies.

By embracing a comprehensive and cooperative strategy, we can reduce the adverse effects of pollution, strengthen the resilience of delta ecosystems, and safeguard the livelihoods of communities reliant on them. Immediate action is imperative, as the well-being of delta ecosystems is deeply connected to the health of our planet and the prosperity of generations to come.

References

Alfiky A (2022) Screening and identification of indigenous entomopathogenic fungi isolates from agricultural soil in the Nile Delta, Egypt. J Fungi 8(1):54. https://doi.org/10.3390/jof8010054
Anripa N (2023) Climate change over Indonesia and its impact on nutmeg production: an analysis under high-resolution CORDEX-CORE regional simulation framework. Int J Climatol. https://doi.org/10.1002/joc.7895
Allison FI, Ihunwo OC, Allison I (2024) Assessment of metal concentration and bioaccumulation factor in surface water and fishes sampled from four creeks in the Bonny River Estuary, Niger Delta Region of Nigeria. Eur J Aquat Sci 3(1):1–7. https://doi.org/10.24018/ejaqua.2024.3.1.18
Bakr N, Negm A, El-Bouraie M, El-Ramady H (2024) Seasonal variation of dissolved organic carbon in the irrigation and drainage systems of the northern Nile Delta. Environ Sci Pollut Res. https://doi.org/10.1007/s11356-024-06959-z
Beaudette JA (2024) Relational conservation: Undisciplining community-based fieldwork in the Okavango Delta, Botswana. Religions 15(1452). https://doi.org/10.3390/rel15121452
Bulut A, Aktürk E (2024) Temporal analysis of riparian buffer zones and demographic trends in the coastal districts of Kastamonu. Kastamonu Univ J For. https://doi.org/10.17475/kastorman.1557432
Chen W, Li H, Zhu Y, Jang J, Lin CJ, Chiang PC, Wang S, Xing J, Fang T, Li J, Yang Q, Zheng K (2022) Impact assessment of energy transition policy on air quality over a typical district of the Pearl River Delta region, China. Aerosol Air Qual Res 22(7). https://doi.org/10.4209/aaqr.220071
Hiep NT, Anh LHQ, Tuan PD, Khang DS, Dong PD, Han HTN, Thuan DD, Nga DT (2023) Improving the treatment of saline wastewater from shrimp farms using hybrid constructed wetlands models toward sustainable development. Environ Nat Resour J 21(6):554–562. https://doi.org/10.32526/ennrj/21/20230146
Kemp GP, McDade EC, Day JW, Lane RR, Dawers NH, Day JN (2021) Recovery and restoration of Biloxi Marsh in the Mississippi River Delta. Water 13(3179). https://doi.org/10.3390/w13223179
Nowreen S, Murshed SB, Bhuiyan MAH, Rahman MM, Islam AKMS (2019) Tidal river management for sustainable agriculture in the Ganges-Brahmaputra delta: implication for land use policy. Land Use Policy 87:104443. https://doi.org/10.1016/j.landusepol.2019.104443
PhamThi Anh P, Kroeze C, Bush SR, Mol APJ (2010) Water pollution by Pangasius production in the Mekong Delta, Vietnam: causes and options for control. Aquac Res 42(1):108–128. https://doi.org/10.1111/j.1365-2109.2010.02578.x
Sedlak DL, Ulrich PD (2009) Control of mercury methylation in wetlands through iron addition. University of California, Berkeley
Sun Y, Xu B (2023) Spillover effects and performance optimization of air pollution control policies: evidence from the Yangtze River Delta region, China. Systems 11(418). https://doi.org/10.3390/systems11080418
Ta J, Anderson LWJ, Christman MA et al (2017) Invasive aquatic vegetation management in the Sacramento–San Joaquin River Delta: status and recommendations. San Franc Estuary Watershed Sci 15(4). https://doi.org/10.15447/sfews.2017v15iss4art5
Wen X, Wang S, Li S, Cheng L, Li K, Zheng Q, Zhang B (2024) Impact factors of industrial pollution and carbon reduction under the 'Dual Carbon' target: a case study of urban aggregation in the Pearl River Delta and Yangtze River Delta. Sustainability 16(5):1879. https://doi.org/10.3390/su16051879

Wu J, Sun W (2023) Regional integration and sustainable development in the Yangtze River Delta, China: towards a conceptual framework and research agenda. Land 12(470). https://doi.org/10.3390/land12020470

9 GIS and Remote Sensing-Based Analysis of Spatial Transformations in the Urbanizing Indian Sundarbans

Somnath Saha, Vanya Bajpai, Biplab Banerjee, Teisovituo Nakhro, Anandita Singh, and Sukanta Kumar Saha

Abstract

The Indian Sundarbans, a globally significant mangrove ecosystem, is currently experiencing profound socio-spatial transformations driven by urban expansion, climate change, and anthropogenic activities. This study uses Geographic Information Systems (GIS) mapping and satellite imagery to analyze these transformations, particularly in Jaynagar, a transitional zone within the Sundarbans delta. Utilizing multi-temporal satellite data (2015–2024), the research maps and classifies Land Use and Land Cover (LULC) changes, evaluates Land Surface Temperature (LST), and computes indices such as Enhanced Vegetation Index (EVI), Normalized Difference Built-up Index (NDBI), and Normalized Difference Water Index (NDWI). The results indicate a complex interplay between urbanization and environmental degradation: despite an increase in vegetation cover, a decline in vegetation health (EVI) and a rise in LST were observed. Built-up areas showed a strong positive correlation with LST, while water availability (NDWI) exhibited a declining trend, further influencing thermal dynamics. The findings emphasize that unplanned urban growth contributes significantly to ecological stress, highlighting the necessity of integrated, geospatially informed strategies for sustainable land management and climate adaptation in the Sundarbans region. The result shows that, in the Indian Sundarbans, vegetation cover rose

S. Saha (✉) · A. Singh
Prajukti Research Private Limited (PRPL), Baruipur, West Bengal, India
e-mail: somnathsahageo@gmail.com

V. Bajpai · B. Banerjee · S. K. Saha
Maharaja Sayajirao University of Baroda, Vadodara, Gujarat, India

T. Nakhro
Mizoram University (Central), Aizawl, Mizoram, India

C. Singha et al. (eds.), *Sustainable Management of Delta Ecosystems Resilience*,
Deltas of the World, https://doi.org/10.1007/978-3-032-16261-8_9

from 78.85 to 85.12% between 2015 and 2024, but vegetation health (EVI) decreased from 0.75 to 0.65. Cropland increased to 89.23% before declining slightly to 88.56%, while the amount of urban built-up fell from 97.32 to 91.45%. Marshlands and water bodies increased (88.10–93.22%; 75.34–79.45%), however surface wetness is decreasing according to NDWI trends. The land surface temperature (LST) increased significantly (37.99–46.63 °C maximum; 25.91–32.70 °C min), indicating that the urbanization-related urban heat island effects are more intense.

9.1 Introduction

The Indian Sundarbans, an ecologically significant area around the world, is undergoing profound socio-spatial transformation shaped by a combination of natural and anthropogenic drivers. This extensive mangrove forest, which is among the largest mangrove forests in the world, is critically important for ecosystem services like coastal defence, carbon storage and conservation of biodiversity (Paul 2021).

Rising salinity and water contamination further threaten mangrove health, (Das and Das 2025 and Chauhan et al. 2025). The Sundarbans area is highly under threat from climatic change and rising sea level and increased frequency of storms and has resulted in a rate of degradation of the mangrove cover at about 0.114% per annum (Das and Sarkar 2025).

These, nonetheless, have been augmented by mounting pressures stemming from rapid urbanization, climate change, and human activities like aquaculture, agriculture, and infrastructure development that have redrew the socio-environmental landscape of the region (Titumir 2022). These have consequential effects on the ecological health of thc Sundarbans as well as the living conditions of its residents who rely on the natural resources of the region for their livelihood.

The most important environmental concern in the Sundarbans is the intensification of cyclonic disturbances, which exacerbate ecosystem degradation and pose threats to inhabitants (Mukhopadhyay 2021). Sea-level rise, saline intrusion, and coastal erosion contribute to these concerns, highlighting the necessity for extensive monitoring and analysis. Knowledge of the spatial and temporal aspects of these changes is important for developing effective management and conservation strategies. Geographic Information Systems (GIS) and Remote Sensing (RS) technologies provide strong tools for the analysis of these socio-spatial changes. Through the ability to measure land use and land cover (LULC) changes, mangrove degradation, and human settlement expansion, GIS and RS offer evidence-based information that can complement conventional field-based approaches (Paul 2021).

This book chapter analyses the socio-spatial changes occurring in the Indian Sundarbans using GIS and RS with a specific focus on the impacts of urbanization on natural resources and the indigenous community. The study integrates available literature and grounded facts to investigate key issues such as land use pattern changes, environmental impacts of rapid growth, and socio-economic impacts of environmental changes (Titumir 2022). With the integration of geospatial technologies and socio-economic information, this chapter attempts to give a holistic overview of the changing landscape of the area and its implications for sustainability and human well-being.

The chapter begins with the introduction of the Sundarbans in cultural and environmental perspectives, and its applicability to conservation and development issues. It then presents the main drivers of change, such as climate change exposures, economic shifts, and policy actions, before demonstrating a comprehensive evaluation of spatial changes using GIS and RS (Mukhopadhyay 2021). The results are applicable to current arguments on sustainable management practices and of interest to policymakers, scientists, and conservationists concerned with minimizing the adverse impacts of urbanization while improving ecological resilience (Chakraborty et al. 2024).

By integrating geospatial analysis with socio-environmental studies, this research highlights the need for data-driven interventions in addressing the Indian Sundarbans' intricate dynamics. This holistic view illuminates the ecological concerns of the region and provides directed interventions in balancing urban development with sustainable development. The chapter emphasizes socially equitable and science-based collaborative conservation, recognizing the interconnectedness of ecosystems and human activities (Das and Das 2025). This will be facilitated by continuous observation and adaptive management methods that consider both community needs and shifting environmental conditions (Biswas et al. 2024). The main objective in the book chapter is to map and quantify the spatio-temporal changes in LULC in the Indian Sundarbans over a defined period, with a specific focus on the expansion of urban and built-up areas. This chapter offers a spatially-referenced analysis that quantitatively or qualitatively maps these changes onto the depletion of particular natural resources. Previous research lacks a spatially detailed analysis that directly links urban-driven land use change to its socio-economic effects on indigenous communities in the Indian Sundarbans. Studies have not systematically connected the specific locations of resource loss (Chatterjee 2021).

In the end, striking a balance between urban growth and ecological protection is the bind that will do or undo the Sundarbans and other such ecosystems across the planet. Using geospatial tools to support policymaking and community engagement, decision-makers could be developing customized strategies to manage ecological challenges not only in the short term, but also to develop resilience for the long term to climate change (Tenhunen et al. 2024). The application of geospatial analysis in socio-environmental research provides a model for other socio-spatial contexts experiencing analogous challenges, illustrating how technology can drive sustainability and conservation programmes.

9.2 Study Area

Jaynagar, located in the southern part of South 24 Parganas district in West Bengal, is approximately 50 km south of Kolkata (Fig. 9.1). It is a crucial location in the lower Gangetic delta and lies on the northern rim of the Sundarbans. The largest intact mangrove ecosystem in the world and a UNESCO World Heritage Site. The area is ecologically vulnerable and socio-economically relevant, with both natural processes and human interventions influencing it (Sen 2019; Hazra et al. 2002).

Ecologically, Jaynagar falls in a transition area between the inner Sundarbans mangrove forests and the more urban and cultivated hinterland. This hybrid position is the cause of a hybrid landscape consisting of tidal rivers, creeks, marshes, and reclaimed agricultural lands. The hydro-geomorphology of the area is controlled mostly by the Bidyadhari River system, which controls both the sedimentation processes and the availability of water for agriculture and aquaculture (World Bank 2014).

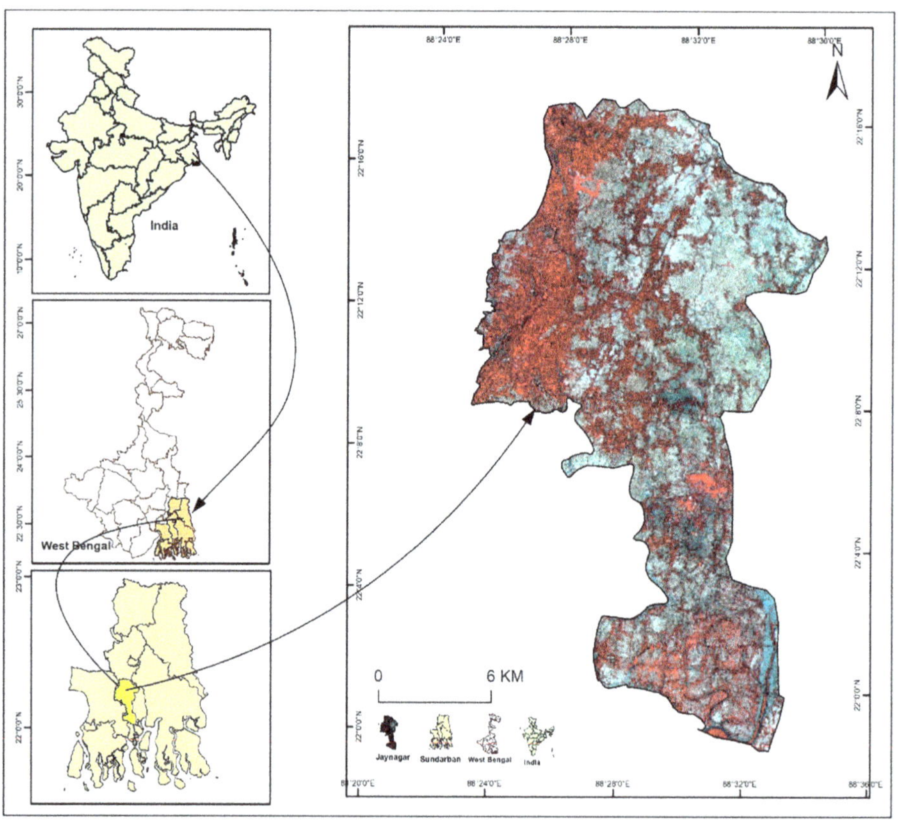

Fig. 9.1 Locational map of the study area

Climatically, the region is dominated by a humid subtropical regime with an active monsoonal phase in June and September. Recurrent waterlogging and seasonal flooding are induced by heavy rainfall, accompanied by low levels and poor drainage. Jaynagar is also increasingly vulnerable to climate-related threats like storm surges, coastal flooding, sea level rise, and salinity intrusion—each of which has been on the rise in number and intensity over the past several decades (Hazra et al. 2002; Sarkar et al. 2021; Thakur et al. 2021).

9.3 Material and Methods

9.3.1 Mapping of Study Area and Geospatial Structure

The geospatial extent of the Jaynagar area was digitized from ArcGIS Desktop 10.8 (ESRI 2020). Base maps and administrative boundary layers were draped on georeferenced satellite images to set locational representation in a precise manner. Spatial data were converted from The UTM coordinate system (Zone 45 N) and the WGS-84 datum to the Global Data Interoperability and Analysis accuracy.

9.3.2 Sources of Data and Acquisition

The use of both primary and secondary data made the study of the spatial and temporal patterns of LULC change and environmental factors:

(a) Demographic and socio-economic information were obtained from the South 24 Parganas (DCHB 2014) which provided complete information on the area's population density, occupational structure, settlements, and land utilization.
(b) Satellite imagery was obtained from the USGS—Earth Explorer portal. Two data generations were chosen:
    (i) Landsat 5 Thematic Mapper (TM) for 2015 and
    (ii) Landsat 8 Operational Land Imager and Thermal Infrared Sensor (OLI/TIRS) for later times (2024).

All the datasets were Level-1 terrain-corrected (L1T), i.e., pre-processed for radiometric and geometric corrections and delivered at a 30 m for multispectral bands and 100 m for thermal bands are the spatial resolutions (Roy et al. 2014). Scene selection was done according to cloud cover (<10%) and seasonal homogeneity (largely pre-monsoon and post-monsoon months) to provide comparability across time periods. Metadata on the satellite scenes, including row/path, sensor details, radiometric correction parameters, and dates of acquisition, are listed in Table 9.1.

Table 9.1 Band specification

Band No.	Landsat-5 (TM)	Landsat-8 (OLI/TIRS)
Band 1	Visible blue (0.45–0.52 μm)	Not present
Band 2	Visible green (0.52–0.60 μm)	Visible blue (0.45–0.51 μm)
Band 3	Visible red (0.63–0.69 μm)	Visible green (0.53–0.59 μm)
Band 4	Near-infrared (NIR) (0.76–0.90 μm)	Visible red (0.64–0.67 μm)
Band 5	Shortwave Infrared-1 (SWIR-1) (1.55–1.75 μm)	Near-infrared (NIR) (0.85–0.88 μm)
Band 6	Thermal infrared (TIR) (10.40–12.50 μm)	Shortwave Infrared-1 (SWIR-1) (1.57–1.65 μm)
Band 7	Shortwave Infrared-2 (SWIR-2) (2.08–2.35 μm)	Shortwave Infrared-2 (SWIR-2) (2.11–2.29 μm)
Band 10	Not available	Thermal Infrared-1 (TIR-1) (10.6–11.19 μm)
Band 11	Not available	Thermal Infrared-2 (TIR-2) (11.50–12.51 μm)

Source United States geological survey (USGS)
Data acquisition date & time: *April 24, 2023, at 13:20 IST*
Access link: *USGS earth explorer*

9.3.3 Image Pre-processing and Classification

Radiometric correction, stacking layers and sub-setting of scenes to the study area were performed using ERDAS Imagine software. To classify LU/LC categories, a supervised classification approach using thc Maximum Likelihood Classifier (MLC) was implemented. Training sites were identified through field knowledge and verified using high-resolution imagery from Google Earth Pro, which also supported post-classification editing to reduce misclassification errors.

Accuracy was evaluated with a confusion matrix and kappa coefficient. Ground-truth data, obtained from field observation and Google Earth's time-lapse feature, were utilized to confirm classified maps. Only those classifications with overall accuracy >85% and kappa values >0.8 were considered, following remote sensing accuracy assessment standards (Congalton and Green 2019).

9.3.4 Spectral Indices and LST Analysis

In addition to the use of Google Earth Engine (GEE) for the computation of spectral indices to obtain the information about human-induced landscape changes and the damage they cause to the environment, spectral indices were also computed with the help of

GEE, a platform for cloud computing that offers access to the entire Landsat archive and computational resources for massive geospatial analysis (Gorelick et al. 2017).

The indices listed below were obtained:

(a) Normalized Difference Vegetation Index (NDVI) to record the health and biomass of vegetation.
(b) Normalized Difference Built-up Index (NDBI) to distinguish built-up growth and impervious areas.
(c) Normalized Difference Water Index (NDWI) to display the location of water bodies on the ground and to determine the variation in water coverage on a seasonal basis.
(d) The calculation of the Land Surface Temperature (LST) was made on the basis of thermal bands of Landsat 5 (Band 6) and Landsat 8 (Band 10) following the standard protocol given in radiative transfer equations, emissivity correction, and atmospheric calibration (Weng et al. 2004). LST analysis enabled assessment of urban heat signatures, salinity impact zones, and thermal stress hotspots in the study area.

9.3.5 Generation of Thematic Map and Statistical Analysis

All the classified layers, spectral indices, and LST outputs were merged in ArcGIS 10.8 to produce composite thematic maps. Change detection methods, such as post-classification comparison and cross-tabulation, were employed to measure Spatio-temporal land cover changes. The thematic layers obtained were examined through zonal statistics to obtain land cover metrics by administrative unit and ecological zone.

Statistical representation was carried out using OriginPro for high-level plotting and Microsoft Excel for time-series and comparative analysis. Graphical outputs like LU/LC change matrices, NDVI trend lines, and LST histograms were created to visually corroborate empirical results.

9.3.6 Methodological Rigor and Reliability

The multi-temporal, multi-sensor strategy employed in this research allowed for high-resolution understanding of the hydrological, ecological, and land-use processes of the Jaynagar area. By combining satellite remote sensing (Table 9.1), GIS, cloud-based analytics, and ground verification, the methodology (Fig. 9.2) provided a strong and reproducible framework for examining landscape change and environmental vulnerabilities in a climate-sensitive coastal deltaic environment.

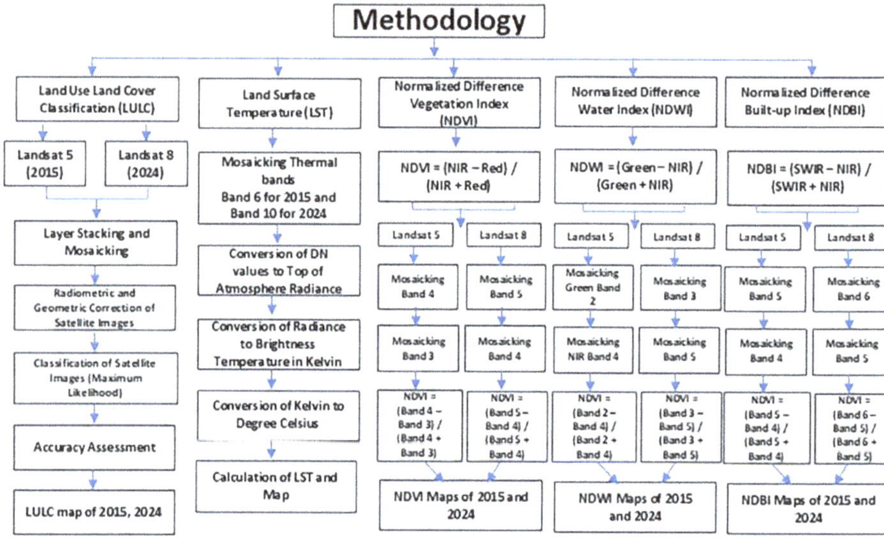

Fig. 9.2 Methodological flowchart

9.4 Remote Sensing-Based Image Analysis

9.4.1 Land Use and Land Cover (LU/LC)

LULC are fundamental parameters for grasping the spatial and temporal patterns of landscape change. These processes have a crucial impact on environmental processes, such as biodiversity patterns, ecosystem services, hydrological cycles, and atmospheric conditions, and also determine socio-economic development and resource management strategies (Lambin et al. 2003; Liping et al. 2018). LU/LC analysis is a core component in practical applications like urban planning, agriculture, forestry, hydrology, disaster risk reduction, and environmental protection.

In this research, a multi-temporal evaluation of LU/LC was applied for the Jaynagar area on the basis of satellite imagery of the years 2015, and 2024. This period was selected to long-term landscape changes due to both anthropogenic and environmental forcing factors, primarily in the field of climate fluctuation and pressure over land in the Sundarbans-neighbourhood region and to ensure classification accuracy, pre-processing steps were performed prior to analysis. These involved radiometric correction, geometric correction, and atmospheric correction through dark object subtraction techniques for the elimination of haze and image clarification (Chavez 1996). All imagery was georeferenced to UTM Zone 45 N, referenced to the WGS-84 datum.

LU/LC classification was executed through a supervised classification method using the Support Vector Machine (SVM) algorithm, which is known for its high accuracy in handling complex and non-linear class boundaries. The algorithm effectively utilized training samples to classify the imagery into seven prominent LU/LC categories, ensuring greater precision and thematic relevance. This approach enhances the interpretability and policy relevance of the classification output (Abdullah et al. 2019).

Vegetation cover; dense Forest, open forest, shrub, gardens, and plantation areas. Built-up area; all man-made facilities and impervious surfaces like residential areas, commercial areas, industrial complexes, transportation infrastructure (roads, railways), slums, and institutional grounds. Barren land; lands with bare soil, sand, or rock surface that is not cultivated or vegetated. Fallow land; temporarily not cultivated but potentially cultivable land, which is usually employed in rotational agricultural cropping. Water Bodies; natural and man-made water bodies like rivers, lakes, tanks, reservoirs, and drainage canals. Agricultural land; tilled land, consisting of single, double, and multiple crops, that is crucial for local food availability and jobs. River sand; depositional areas along river beds formed by seasonal hydrologic and sediment processes.

To confirm and improve classification results, GEP was utilized for pixel-level validation. High-resolution temporal snapshots were used to visually verify and rectify classification errors, particularly in fragmented or transitional areas. Ground truthing was supported by field observations carried out in chosen micro-watersheds in Jaynagar, further enhancing thematic precision.

The last classification schema is consistent with the land use guidelines presented by the Town and Country Planning Organisation (TCPO 2005), which present uniform criteria for urban-rural land definition in all Indian spatial planning sectors. Table 9.2 shows a comprehensive characterization of all LU/LC classes and corresponding subtypes, providing a template for evaluating longitudinal landscape transformation and its ecological and socio-economic consequences.

This LU/LC mapping provides the basis for subsequent integration with climatic, demographic, and land degradation proxies to facilitate an integrated understanding of environmental pressures and resilience in the study region (Table 9.3).

Table 9.2 Band name used for LULC class

Land cover type	Landsat 5 bands	Landsat 8 bands
Vegetation cover	Band 4, band 5	Band 5, band 6
Built-up areas	Band 2, band 3, band 5	Band 3, band 4, band 6
Water bodies	Band 1, band 5	Band 2, band 6
Agricultural land	Band 4	Band 5

Table 9.3 Description of satellite data

Period	Sensor/ satellite	Acquisition date	Path and row	Resolution	Radiometric	Spectral bands	Cloud cover	Data source
2015	Landsat-5	21 March 2015	139/ 044	30 m	8 bit	0.45–2.35 μm (bands 1–7)	0%	USGS earth explorer
2024	Landsat-8	18 March 2024	139/ 044	30 m	12 bit	0.43–12.50 μm (bands 1–11)	0%	USGS earth explorer

9.4.1.1 Accuracy Assessment Based on the Weighted Kappa Coefficient (κw)

Correct classification of LU/LC data is paramount for dependable spatial analysis when outputs drive environmental policy, land planning, or disaster risk management. Although numerous metrics are used to assess accuracy of classification, the Kappa coefficient (κ) has found favor as a broad statistical indicator owing to its capacity to remove agreement by chance (Congalton and Green 2019).

However, the standard (unweighted) Kappa treats all classification errors equally, regardless of the semantic or ecological similarity between the misclassified categories. The application of weighted kappa statistics in classification accuracy assessment has been widely discussed in literature (Næsset 1996; Sim and Wright 2005; Brenner and Kliebsch 1996; Yilmaz and Demirhan 2023). This limitation can distort interpretations in remote sensing where some classes such as "fallow land" and "agricultural land" are more spectrally and functionally similar than others (e.g., "vegetation" vs. "built-up area") Foody (2020). For this purpose, the Weighted Kappa coefficient (κw) has been established, which provides a more sophisticated measure of accuracy (Verma et al. 2020).

The weighted Kappa uses a weight matrix W(i,j)W(i,j)W(i,j) (Eq. 9.1), where every weight captures the similarity or dissimilarity between class i and class j. Misclassifications among closely related classes are given less penalty, while those among dissimilar classes are penalized more (Rwanga and Ndambuki 2017). This approach provides a contextually dependent evaluation of quality classification, especially in complex environments such as those found in coastal deltaic landscapes (Table 9.4).

Table 9.4 Kappa range and its description

Sl. No.	k-value	Strength of covenant
1	<0.20	Strongly disagree
2	0.20–0.40	Disagree
3	0.40–0.60	Neutral
4	0.60–0.80	Agree
5	0.80–1.00	Strongly agree

The formula for weighted Kappa is as follows:

$$\kappa_w = \frac{\sum_{i=1}^{N}\sum_{j=1}^{N} W(i,j)O(i,j) - \sum_{i=1}^{N}\sum_{j=1}^{N} W(i,j)E(i,j)}{\sum_{i=1}^{N}\sum_{j=1}^{N} W(i,j)(1 - E(i,j))} \quad (9.1)$$

Where O(i,j) and E(i,j) are the observed and expected frequencies of class pair i,j, respectively, and W(i,j) gives a weight for the degree of misclassification and r: Number of rows or classes of the classification matrix.

Weights can be linear or quadratic based on the level of sensitivity desired. In this research, quadratic weighting was used to more heavily penalize large classification errors. For instance, misclassifying "vegetation" as "built-up area" was considered more severe than misclassifying it as "agriculture".

9.4.2 Land Surface Temperature

LST is a significant biophysical indicator that quantifies the surface temperature of the Earth. LST is fundamental to evaluating the heat islands within urban areas, health of the vegetation, rate of evapo-transpiration, and land-atmosphere exchange of energy (Weng et al. 2004). For this analysis, LST (Eq. 9.2) was extracted using thermal infrared remote sensing data provided by Landsat 5 TM and Landsat 8 (OLI/TIRS) collections from the period between 2015 and 2024. Several studies have demonstrated the relationship between land use/land cover and land surface temperature in urban environments (Kumari et al. 2018; Bakar et al. 2016; Katpatal et al. 2008). The physical basis of thermal radiation calculations is rooted in Planck's radiation principles (Azmi et al. 2020; Haug 2020).

Thermal Band Specifications

Obtain the Thermal Infrared (TIR) band from Landsat 5 (TM) data. Landsat 5 TM has one thermal band: Band 6: 10.40–12.50 μ m (Thermal Infrared).

$$LST\left(^{\circ}C\right) = LST(K) - 273.15 \quad (9.2)$$

Step 1*: *Conversion of Digital Numbers (DN) to Spectral Radiance.

The conversion of (DN) data to spectral radiance for Landsat-5 band 6 was carried out using the following formula (Eq. 9.3) (Chander et al. 2009).

$$L_\lambda = \frac{(L_{\mathrm{MAX}} - L_{\mathrm{MIN}})}{Q_{\mathrm{CALMAX}} - Q_{\mathrm{CALMIN}}} \times (DN - Q_{\mathrm{CALMIN}}) + L_{\mathrm{MIN}} \quad (9.3)$$

where: L_λ is the spectral radiance (W/m^2·sr· μ m), L_{MAX}, L_{MIN} are the maximum and minimum radiance values (from metadata), Q_{CALMAX}, Q_{CALMIN} are the maximum and minimum DN values (typically 255 and 1, respectively) (Chander et al. 2009).

Step 2: Brightness Temperature (TB) Calculation.

The spectral radiance of Landsat-8 OLI/TIRS has been derived by converting the (DN) data from bands 10 and 11 using the following formula (Eq. 9.4) for the TIRS image:

$$L_\lambda = M_\lambda \cdot Q_{cal} + A_\lambda \tag{9.4}$$

where: L_λ is the spectral radiance (W/m^2·sr·μ m), M_λ is the radiance multiplicative rescaling factor (from metadata), A_λ is the radiance additive rescaling factor (from metadata), Q_{cal} is the (DN) value (Abidin et al. 2021).

To obtain the at-satellite brightness temperature (T_B) in Kelvin, the spectral radiance is converted with emissivity modification, based on the land cover characteristics (Sekertekin and Bonafoni 2020a, b). The formula for the brightness temperature with emissivity correction is as follows (Eq. 9.5):

$$T_B = \frac{K_2}{\ln\left(\frac{K_1}{L_\lambda} + 1\right)} - 273.15 \tag{9.5}$$

In this equation, T_B represents the at-satellite brightness temperature in Kelvin, and L_λ is the spectral radiance at the satellite in units of W/m^2·sr·μm. The constants K_1 and K_2 are thermal calibration constants that vary depending on the Landsat sensor and specific band used (Chander et al. 2009). For Landsat-5™ Band 6, K_1 is 607.76 and K_2 is 1260.56. For Landsat-8 OLI/TIRS Band 10, K_1 is 774.8853 and K_2 is 1321.0789 (Hashim et al. 2022), while for Landsat-8 OLI/TIRS Band 11, K_1 is 480.8883 and K_2 is 1201.1442 (Maity et al. 2020).

Step 3: Emissivity Correction.

Land surface emissivity (ε) is a vital factor in accurate LST estimation. It was estimated using the **NDVI**:

$$Pv = NDVI = \frac{\text{NIR} - \text{RED}}{\text{NIR} + \text{RED}} \tag{9.6}$$

From NDVI, the **proportion of vegetation (P_v)** (Eq. 9.6) was computed, followed by surface emissivity (Eq. 9.7):

$$E = 0.004 \times PV + 0.986 \tag{9.7}$$

This method accounts for the mixed pixel composition typical of semi-urban and peri-rural areas (Jin and Liang 2006).

Step 4: Land Surface Temperature (LST) Estimation.

Subsequently, the calculation of LST is carried out using the following formula (Eq. 9.8):

$$LST = \frac{T_B}{1 + \left(\frac{\lambda \cdot T_B}{\rho} \cdot \ln(\varepsilon)\right)} \tag{9.8}$$

where: LST is the Land Surface Temperature (Kelvin), T_B is the brightness temperature (Kelvin) (Ogunode and Akombelwa 2017), λ is the wavelength of emitted radiance (10.6–12.5 μ m for Band 10 and 11, approximate value: $\lambda = 11 \times 10^{-6}$ m) (Tian 2021), $\rho = \frac{h \cdot c}{\sigma}$ is the term for Planck's constant (Lévy-Leblond 2002), $h = 6.626 \times 10^{-34}$ J·s (Planck's constant), $c = 3 \times 10^{8}$ m/s (speed of light), $\sigma = 1.38 \times 10^{-23}$ J/K (Boltzmann constant) ε is the land surface emissivity (Jin and Liang 2006).

Convert LST from Kelvin to Celsius (Eq. 9.9):

$$LST\left(^{\circ}C\right) = LST(K) - 273.15 \tag{9.9}$$

This approach provides a consistent and validated method for LST retrieval in multi-temporal landscape studies, particularly relevant in climate-sensitive and low-elevation coastal zones such as Jaynagar.

9.4.3 Enhance Vegetation Index

The EVI is a satellite-derived vegetation index that optimizes the vegetation signal by reducing atmospheric influences and soil background noise. It improves sensitivity in high biomass regions compared to NDVI. EVI is widely used in monitoring vegetation health, phenology, and ecosystem dynamics. Its formulation (Eq. 9.10) incorporates blue, red, and near-infrared bands for better accuracy in dense canopies.

$$\text{EVI} = G \cdot \frac{\text{NIR} - \text{RED}}{\text{NIR} + C_1 \cdot \text{RED} - C_2 \cdot \text{BLUE} + L} \tag{9.10}$$

Where:

$$G = 2.5, C_1 = 6, C_2 = 7.5, \quad L = 10000$$

9.4.4 Normalized Difference Built-up Index

In RS, NDBI index used to identify and monitor urban areas or built-up regions, especially in satellite imagery (Zheng et al. 2021). It is based on the difference between the Near Infrared (NIR) and Shortwave Infrared (SWIR) bands (Eq. 9.11), which highlights the built-up areas by distinguishing them from vegetation and water bodies (Kebede et al. 2022).

$$\mathrm{NDBI} = \frac{\mathrm{SWIR} - \mathrm{NIR}}{\mathrm{SWIR} + \mathrm{NIR}} \tag{9.11}$$

9.4.5 Normalized Difference Water Index

The index exploits the difference between the Green and NIR bands (Eq. 9.12), as water strongly absorbs NIR while reflecting green light (Zhang and Zhou 2019).

$$\mathrm{NDWI} = \frac{\mathrm{GREEN} - \mathrm{NIR}}{\mathrm{GREEN} + \mathrm{NIR}} \tag{9.12}$$

9.5 Correlation Analysis

Pearson's correlation is a robust method for this purpose (Zhang et al. 2016). Additionally, it evaluates the strength and direction of the linear relationship between variables (Tabatabai et al. 2021), making it suitable for cases where linear patterns are expected, which is common in geospatial and environmental data (Kadaverugu et al. 2024). The Pearson correlation coefficient (r) is given by the following formula (Eq. 9.13):

$$r = \frac{\sum_{i=1}^{n}(x_i - \overline{x})(y_i - \overline{y})}{\sqrt{\sum_{i=1}^{n}(x_i - \overline{x})^2 \cdot \sum_{i=1}^{n}(y_i - \overline{y})^2}} \tag{9.13}$$

r: Pearson's correlation coefficient (Armstrong 2019), x_i, y_i: Individual data points for variables x and y, $\overline{x}, \overline{y}$: Mean values of x and y, respectively.

9.6 Results

9.6.1 Remote Sensing-Based Image Analysis

9.6.1.1 Lu/LC

The LU/LC maps from 2015 to 2024 reveal significant changes across all categories (Fig. 9.3). Agricultural land increased moderately by approximately 3–5%, mainly due to the conversion of vegetation and marshy areas. Water bodies decreased by around 2–3%, likely due to encroachment and drying. Vegetation cover saw a notable decline of about 5–7%, indicating deforestation or transformation into agricultural and built-up areas. Built-up areas expanded significantly by nearly 6–8%, reflecting rapid urbanization. Marshy lands experienced a substantial reduction of around 4–5%, suggesting drainage or land development activities. These shifts highlight growing anthropogenic pressure

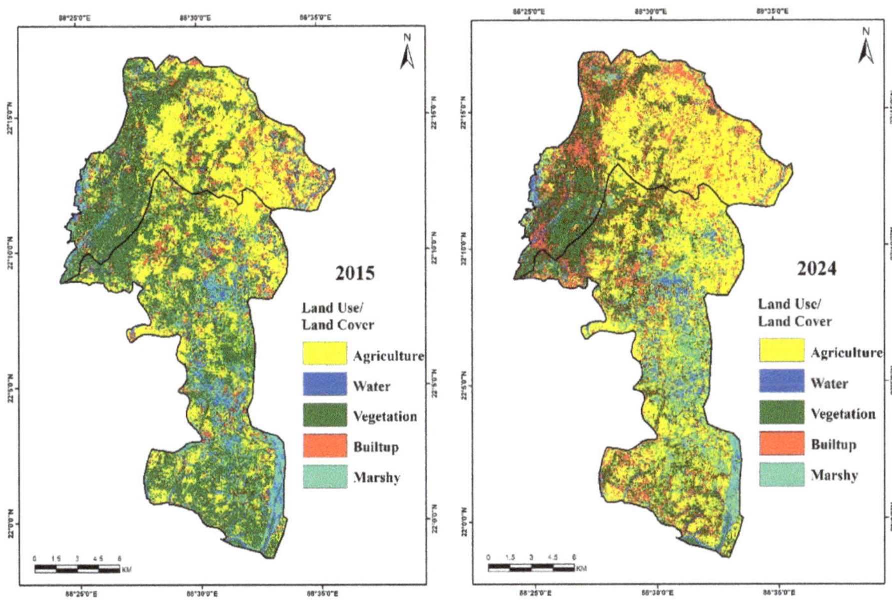

Fig. 9.3 LU/LC, 2015 and 2024

Table 9.5 Accuracy assessment table

LULC Class	2015	2024	RSP	UA	PA
Vegetation cover	121.38	70.52	86.78	80.56	81.34
Built-up area	33.61	60.98	89.12	86.23	82.67
Water bodies	28.17	24.71	90.67	86.78	82.45
Marshy land	34.80	26.25	80.12	76.89	74.56
Agricultural land	142.37	177.88	90.12	87.34	89.78

Source Computed from accuracy assessment results

on natural resources, demanding focused land management and conservation strategies (Table 9.5).

9.6.1.2 LST

Urban heat dynamics are influenced by complex heat transfer mechanisms including conduction, convection, and radiation (Su and Davidson 2018). LST is an important factor for assessing the surface energy budget. The main causes of rising LST in urban settings are building expansion, both horizontal and vertical and human activities, the distance between buildings, and the growth of grey infrastructure. (Crum and Jenerette 2017; Perera and Samanthilaka 2014). From the Fig. 9.4. It is shown that in the period of 2015–2024

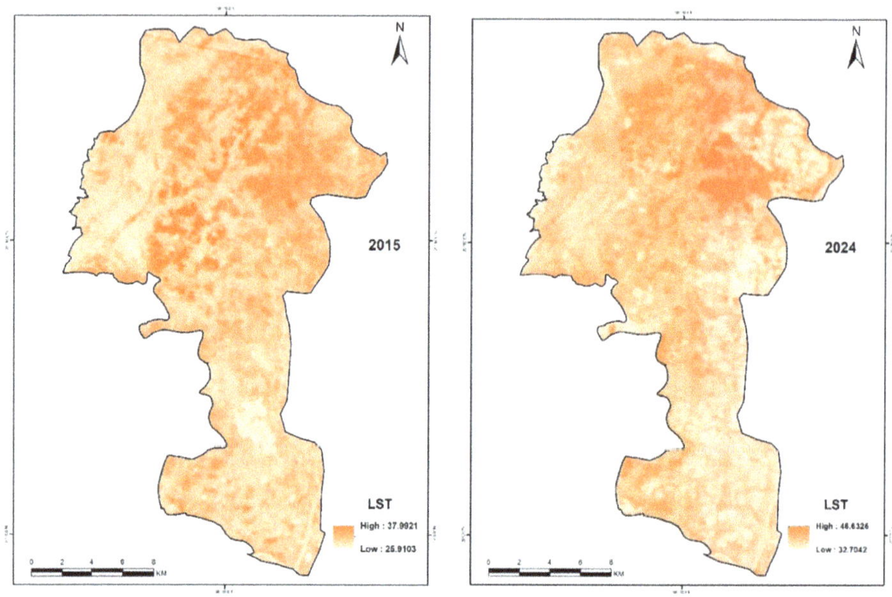

Fig. 9.4 LST, 2015 and 2024

the LST have increase with maximum value 37.9921 and 46.6326 °C in the year 2015–2024 indicating that the temperature is rising and the minimum value of LST has also rises from 25.9103 to 32.7042 °C. The northern part of the study area records higher LST value expressing lower vegetation density and higher building density as well as more human activities is occurring.

9.6.1.3 EVI

The value of EVI ranges from 0 to 1. The EVI value closer to 1 signifies better vegetation health whereas EVI value closer to 0 indicates unhealthy vegetation health. The EVI value in the AOI in 2000 ranges from −0.40 to 0.75 whereas in 2023 the EVI values ranges from −0.30 to 0.65 (Fig. 9.5). During the time period of 2000 2023 the EVI highest value has decreases from 0.75 to 0.65 which indicates that the vegetation health has decline whereas the lower EVI value of the study area indicates that during this time period the unhealthy vegetation health is improve from −0.40 to −0.30.

9.6.1.4 NDBI

The value of NDBI of the study area ranges from −0.62 to 0.17 in the year 2015 and −0.68–0.199 in the year 2024. According to the research the positive value of NDBI indicate urbanise land whereas negative value of NDBI indicate non-urbanise land. During the year 2015 the NDBI (Fig. 9.6) shows that the northern part of the study area is urbanise

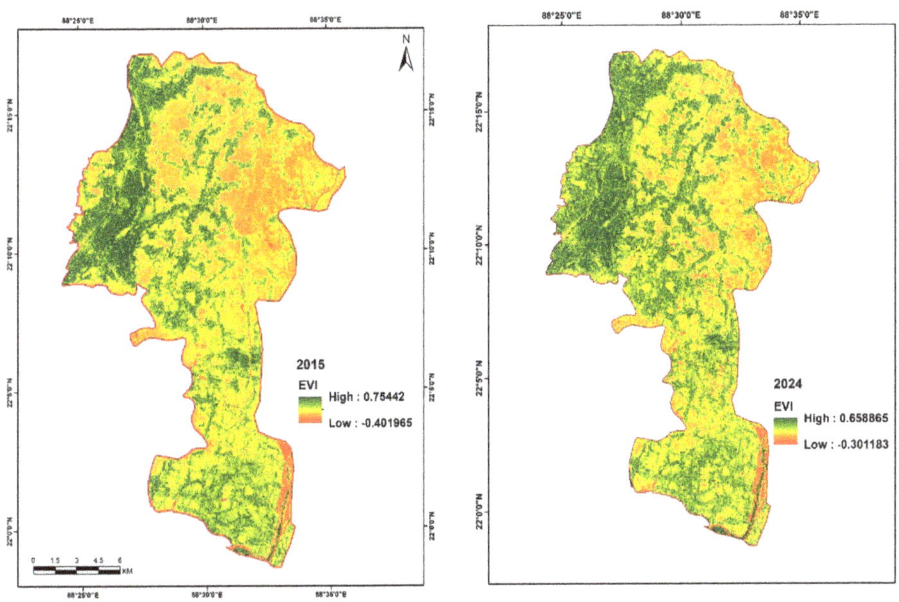

Fig. 9.5 EVI, 2015 and 2024

area as it displays positive values whereas the southern part of AOI from the figure it indicates that it is non-urban areas. During the year 2024 the NDBI indicates that the study area is almost urbanise in every part as it shows higher positive values comparing to the year 2015.

9.6.1.5 NDWI

The value of the NDWI varies from -1 to 1. Theoretically, the NDWI value more than 0 indicates the water bodies and NDWI value less than 0 represents non-water bodies. The Normalised Difference Water Index (NDWI) of the study area for a period 2015–2024 is analysed using GIS platform. The maximum value of NDWI has decreased from 0.33 to 0.19 during the time period of 2015–2024 indicating that the land surface wetness in the area of study is decreasing. The lower value of the study area increasing which signifies that the dry land has increased during the time period of 2015–2024 (Fig. 9.7).

9.6.2 Correlattion Analysis

The study established correlation among EVI & LST, NDWI & LST, and NDBI & LST that convey dependency of LST on different environmental conditioning factors (Fig. 9.8). Basesd on pearson's method about -0.388 indicate slightly negative correlation among

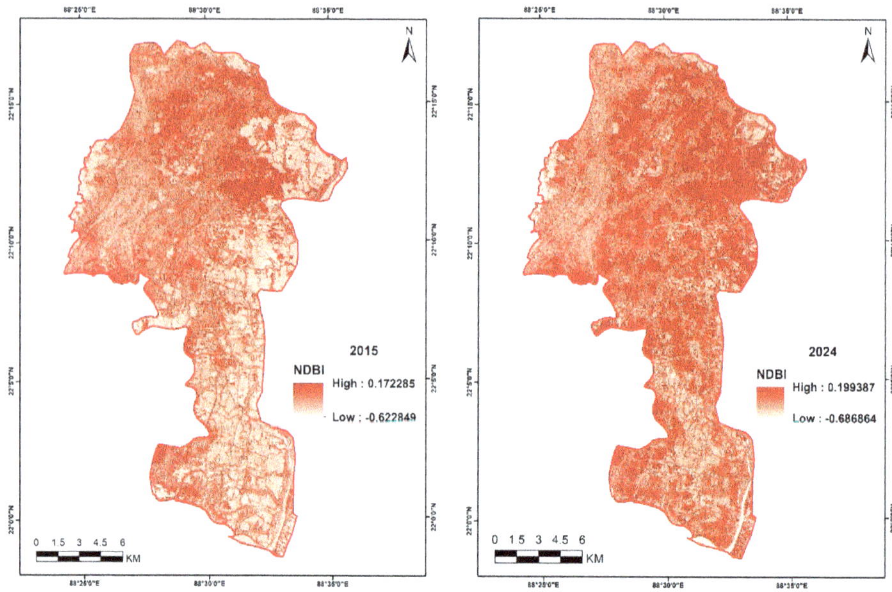

Fig. 9.6 NDBI, 2015 and 2024

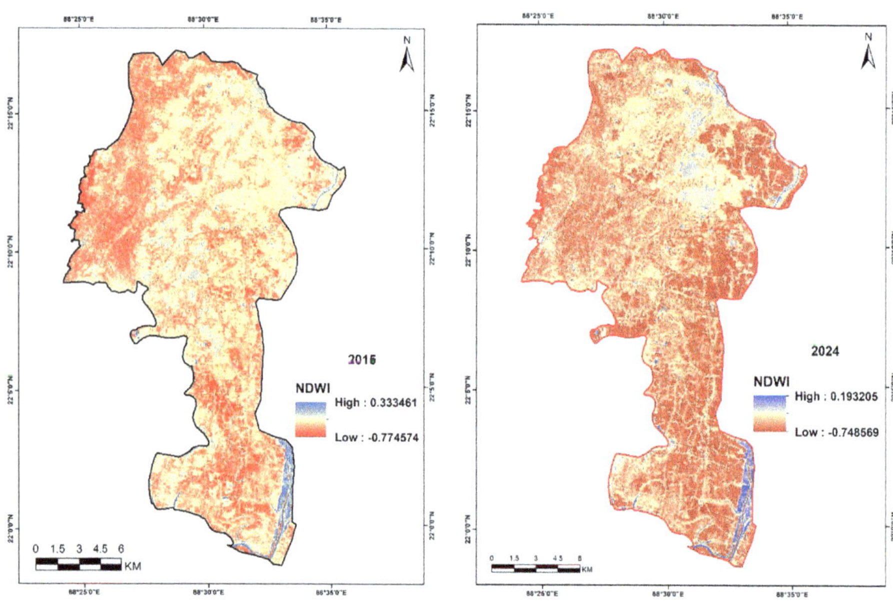

Fig. 9.7 NDWI, 2015 and 2024

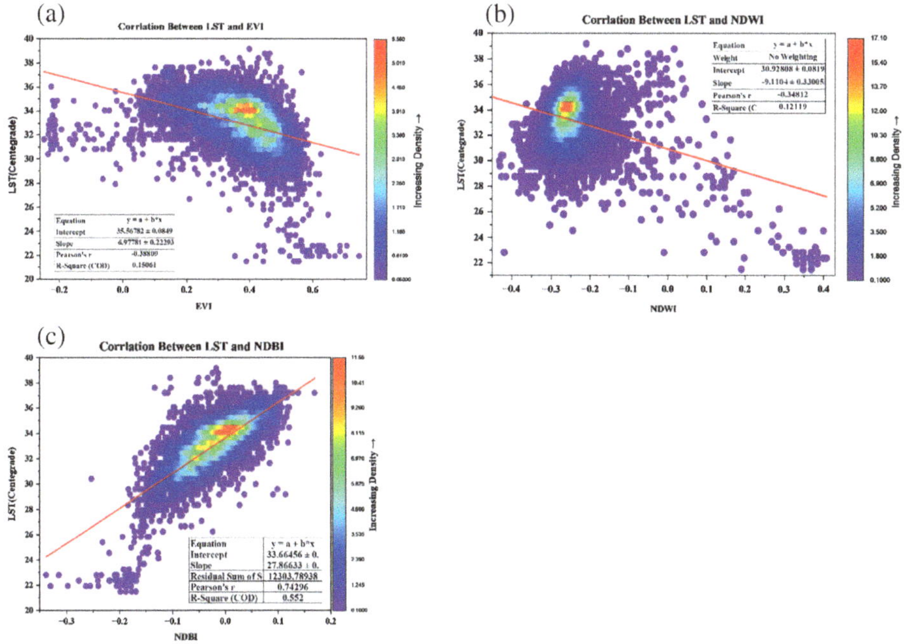

Fig. 9.8 Correlation analysis (LST & EVI, LST & NDWI, and LST & NDBI)

the LST and EVI in the study area. When EVI decreases the LST increase, the good vegetation health is control over LST, as its absorb heat.

The NDWI and LST shows marginal negative correlation based on the pearson's correlation method with r^2 value of -0.34812 in the area of interest. The value of NDWI increases when the value of LST decreases which indicate that water plays and important role controlling over LST by absorbing heat.

The correlation between the NDBI and LST of the study area show significant positive correlation with pearson's value 0.74286. The human activities or the urban expansion is directly related to the incrreasing temperature in the study area. The correlation result shows that the human activies or build-up lands has significant role in increasing temperature. It can be concluded that with the increase in build-up area the temperature relatively increases in the study area.

Kappa Coefficient for LU/LC Validation

Computing kappa coefficient techniques we got 0.81 in the year 2015, 0.85 in the year 2024 which indicates strongly agree with classified accuracy and actual accuracy.

9.7 Discussion

This research critically assesses the socio-spatial changes in the urbanizing areas of the Indian Sundarbans during the last two decades (2001–2021), with emphasis on the dynamic interaction between urban growth, LULC dynamics, and environmental conditions. The results highlight a complex pattern of development-induced ecological change, where physical landscape transformation is strongly interrelated with vegetation condition, surface hydrology, and thermal regimes.

Vegetation Cover and Environmental Health
Between 2015 and 2024, the overall vegetation cover of the study area increased from 78.85–85.12%, suggesting an apparent rise in green cover. Yet the EVI, which is an indicator of vegetation health and vigor, reduced between 0.75 and 0.65 over the same timeframe. This divergence indicates that while vegetated areas expanded, their physiological condition and ecological function may have deteriorated, possibly due to heat stress, increased pollution, or replacement by monoculture plantations or degraded secondary growth.

Built-up area Dynamics
Contrary to what could have been expected, the percentage of developed land in the area reduced from 97.32% in 2015 to 91.45% in 2024. This may not be a reduction in urban development, but rather a shift in the spatial extent and density of urban sprawl—possibly to low-density peri-urbanization or re-categorization of land-use in more recent satellite imagery. Urban development patterns may also be regulated by environmental processes such as tidal flooding, soil salinity, or cyclone exposure (Barui et al. 2025).

Agricultural and Hydrological Shift
Cropland experienced an initial increase from 83.67 to 89.23% between 2015 and 2024, subsequently a moderate decline to 88.56%. This trend indicates a short-term intensification of agriculture, perhaps due to improved irrigation or land reclamation, followed by minor abandonment or conversion due to urban encroachment, salinity intrusion, or declining soil fertility.

Similarly, the region occupied by water bodies and marshy lands increased from 88.10 to 93.22% and 75.34 to 79.45%, respectively, due to a shift in surface hydrology. It may be a consequence of greater rainfall variability, drainage deficiency, or coastal subsidence in low-lying land. Nonetheless, NDWI trends during the years 2015–2024 indicated lowering surface wetness, that while the spatial cover of water features rose, the total supply of moisture and hydrological integrity perhaps diminished, leading to a drier land surface.

Thermal Trends and Urban Heat Island Effects
The analysis of LST from 2015 to 2024 revealed a notable warming trend, with maximum values rising from 37.99 to 46.63 °C and minimum values increasing from 25.91 to 32.70 °C. These thermal elevations point to a growing UHI effect, likely exacerbated by impervious surfaces, reduced evapotranspiration, and land cover degradation (Weng et al. 2004).

There is high positive correlation between the NDBI and LST, which is 0.74286 for Pearson r^2, and this reflects the effect of urbanization in rising temperatures. In contrast, the EVI negatively correlates with LST ($r^2 = -0.388$), and this implies that though vegetative cover has a cooling effect, it is not strong enough to neutralize the effect of warming caused by urbanization.

Spatial Asymmetry and Urban Pressure
The spatial analysis further identified that the north section of the study area is more urbanized than the south. This spatial inequality is likely an indication of the proximity of the northern region to transportation lines, economic areas, or administrative zones, which boosts land conversion. Consequently, the urban-rural thermal gradient is steeper in the north, reinforcing the pattern that urban development is closely aligned with localized temperature escalation.

Synthesis of Findings
The study reveals that during the time interval of 2015–2024 the Sundarbans has undergone through substantial shifts in urbanization which impacts the LU/LC. The extensive forested areas of Sundarbans have significantly drop during this decade was found out by Chowdhury and Hafsa (2022) and Sardar and Samadder (2021).

EVI, NDBI and NDWI are the indices used for analysing the major changes of the study area for the different years. Through EVI we have observe that the vegetation health is declining in the past two decades. This decline of vegetation health in the area of study could be due to different factors such as changing of climate, salt content in the water or changes in the species composition in water Mitra et al. (2021). While the NDBI during the time period of 2015–2024 have increase which shows that urbanisation is increasing in the area. Laskar et al. (2024) states that build-up areas increases with increase in the value of NDBI that support my study. The NDWI of sundarbans have decreases in a decade of 2015–2024 which demostrate that the dryness of the land has increased and and the wetness of the land is declining. Saha et al. (2024) report that decreasing in wetland area is primarily due to the transform into build-up land or change in agricultural land.

The LST is has significantly increasing in the time period of 2015–2024 indicating that the urbanization is increasing where build-up areas were increasing. Recent studies have also highlight that the LST shows a significant increase in decreasing vegetation areas and increase in urbanisation where build-up land plays a crucial role (Laskar et al. 2024).

Sharma and Vashishtha (2023) report that the open surfaces of the land rapidly heated up and increase radiant energies because of decreasing vegetation cover which contribute to higher LST.

In the end, the goal cannot be to simply freeze the Sundarbans in time. Change is inevitable. The true objective of these policies is to guide the powerful forces of urbanization and development, steering them onto a path that is both sustainable for the environment and equitable for its people. We must remember that in the delicate balance of the delta, the health of the ecosystem and the well-being of its communities are deeply intertwined. It demands a firm commitment from leadership, a phased and coordinated approach, and, most importantly, the active partnership and wisdom of the communities who call the Sundarbans home. Their involvement is the key to building a resilient future for this unique region.

9.8 Limitation of the Study

Despite providing valuable insights into the socio-spatial transformations in the Indian Sundarbans, this research is subject to certain limitations. Firstly, the study primarily relied on secondary satellite data, which, despite being pre-processed, may have inherent spatial and radiometric inaccuracies. The classification of LULC was based on medium-resolution imagery (30 m), which may not fully capture fine-scale landscape changes or small-scale human interventions. Additionally, field validation was limited, which could affect the accuracy of certain land cover classifications, particularly in heterogeneous areas like urban fringes or marshlands. Climatic and hydrological data used in correlation analyses were constrained by temporal availability, potentially affecting the robustness of long-term trend assessments. Furthermore, socio-economic variables influencing land transformation were not deeply analyzed due to data unavailability at finer administrative levels. These limitations suggest that future studies could benefit from higher-resolution imagery, continuous ground-truthing, and the integration of more comprehensive socio-economic datasets.

9.9 Conclusion

The study of socio-spatial transformations in the urbanizing Indian Sundarbans, with a focus on Jaynagar, revealed significant land cover changes and environmental impacts over the past two decades. Between 2015 and 2024, vegetation cover increased from 78.45 to 85.12%, and water bodies expanded from 88.10 to 93.22%. However, the EVI declined from 0.75 in 2000 to 0.65 in 2023, suggesting that vegetation health deteriorated despite the increase in area. Built-up areas decreased slightly from 97.32 to 91.45%, but still showed a strong positive correlation with rising LST), which increased from a maximum of 37.99 °C in 2015 to 46.63 °C in 2024. Simultaneously, the NDWI decreased from 0.33 to 0.19, indicating a reduction in surface wetness and a rise in dry land areas. The Pearson correlation coefficient showed a significant positive relationship between NDBI and LST ($r = 0.74286$) and a negative relationship between EVI and LST ($r = -0.388$), confirming that urbanization and loss of vegetation health are key drivers of rising temperatures. These findings underline the urgent need for sustainable urban planning and adaptive land management strategies in the Sundarbans. The integration of geospatial technologies with environmental indicators provides a reliable foundation for policymakers and conservationists to develop resilient strategies in response to climate change and urban pressure. For future scope of the study to create policies that are truly fair, we need to understand how these changes—who gets land, what jobs are available, and who controls resources—impact men and women in different ways. That deeper look is essential.

References

Abdullah AYM, Masrur A, Adnan MSG, Baky MAA, Hassan QK, Dewan A (2019) Spatio-temporal patterns of land use/land cover change in the heterogeneous coastal region of Bangladesh between 1990 and 2017. Remote Sens 11(7):790

Abidin MR, Nur R, Mayzarah EM, Umar R (2021) Estimating and monitoring the land surface temperature (LST) using Landsat OLI 8 TIRS. Int J Environ Eng Educ 3(1):17–24

Armstrong RA (2019) Should Pearson's correlation coefficient be avoided? Ophthalmic Physiol Opt 39(5):316–327

Azmi N, Mustafa S, Mohd Yunos NH, Sakri WNAWM, Abdul Halim MN, Aadenan A (2020) Identification of planck's constant by light emitting diodes (LEDS). J Acad 8:33–38

Bakar SBA, Pradhan B, Lay US, Abdullahi S (2016) Spatial assessment of land surface temperature and land use/land cover in Langkawi Island. Earth and environmental science, IOP conference series

Barui I, Bhakta S, Ghosh K (2025) Storm surge-induced soil salinization and its impact on agriculture in the coastal area of the Indian Sundarban. Environ Dev 56:101250

Biswas S, Sarkar K, Das TK (2024) Decadal changes (1980–2021) of shoreline and mangrove cover in Sundarban Delta, India, using remote sensing and GIS. J Phys Oceanogr 54(8):1655–1674

Brenner H, Kliebsch U (1996) Dependence of weighted kappa coefficients on the number of categories. Epidemiology:199–202

Chakraborty A, Sen A, Biswas D (2024) Local institutional strategies and responses to climate change risks in the Indian Sundarbans: a political economic analysis. Nature And Space, Environment And Planning E. https://doi.org/10.1177/25148486241295536

Chander G, Markham BL, Helder DL (2009) Summary of current radiometric calibration coefficients for Landsat MSS, TM, ETM+, and EO-1 ALI sensors. Remote Sens Environ 113(5):893–903

Chatterjee S (2021) Decreasing climate change vulnerability through adaptation with special reference to migration: a study in the Indian Sundarbans. Kansas State University

Chauhan T, Bhadury P, Rodda SR, Thumaty KC, Jha CS, Ghosh S (2025) Resilience of Sundarban mangroves in South Asia to weather extremes and anthropogenic water pollution. Environ Res Commun. https://doi.org/10.1088/2515-7620/ada675

Chavez PS (1996) Image-based atmospheric corrections-revisited and improved. Photogramm Eng Remote Sens 62(9):1025–1035

Chowdhury MS, Hafsa B (2022) Multi-decadal land cover change analysis over Sundarbans mangrove Forest of Bangladesh: a GIS and remote sensing based approach. Glob Ecol Conserv 37:e02151. https://doi.org/10.1016/j.gecco.2022.e02151

Congalton RG, Green K (2019) Assessing the accuracy of remotely sensed data: principles and practices. CRC press

Crum SM, Jenerette GD (2017) Microclimate variation among urban land covers: the importance of vertical and horizontal structure in air and land surface temperature relationships. J Appl Meteorol Climatol 56(9):2531–2543. https://doi.org/10.1175/JAMC-D-17-0054.1

Das CS, Sarkar P (2025) The Transboundary Sundarban Mangroves: A Struggling Ecosystem, pp 80–103. https://doi.org/10.9734/bpi/mono/978-93-48859-73-0/ch5

Das K, Das K (2025) Climate resilience or impeding vulnerabilities in the light of mangroves of Indian Sundarban, pp 169–185. https://doi.org/10.9734/bpi/mono/978-93-48859-73-0/ch10

ESRI (2020) ArcGIS desktop: release 10.8 [computer software]. Environmental Systems Research Institute. https://www.esri.com

Foody GM (2020) Explaining the unsuitability of the kappa coefficient in the assessment and comparison of the accuracy of thematic maps obtained by image classification. Remote Sens Environ 239:111630

Gorelick N, Hancher M, Dixon M, Ilyushchenko S, Thau D, Moore R (2017) Google earth engine: planetary-scale geospatial analysis for everyone. Remote Sens Environ 202:18–27

Hashim BM, Al Maliki A, Sultan MA, Shahid S, Yaseen ZM (2022) Effect of land use land cover changes on land surface temperature during 1984–2020: a case study of Baghdad city using landsat image. Nat Hazards 112(2):1223–1246

Haug EG (2020) Finding the Planck length multiplied by the speed of light without any knowledge of G, c, or $\hbar$, using a Newton force spring. J Phys Commun 4(7):075001

Hazra S, Ghosh T, DasGupta R, Sen G (2002) Sea level and associated changes in the Sundarbans. Sci Cult 68(9/12):309–321

Jin M, Liang S (2006) An improved land surface emissivity parameter for land surface models using global remote sensing observations. J Clim 19(12):2867–2881

Kadaverugu R, Nandeshwar S, Biniwale R (2024) Wavelet local multiple correlation analysis of long-term AOD, LST, and NDVI time-series over different climatic zones of India. Theor Appl Climatol 155(10):9231–9246. https://doi.org/10.1007/s00704-024-05174-4

Katpatal YB, Kute A, Satapathy DR (2008) Surface-and air-temperature studies in relation to land use/land cover of Nagpur urban area using Landsat 5 TM data. J Urban Plan Dev 134(3):110–118

Kebede TA, Hailu BT, Suryabhagavan KV (2022) Evaluation of spectral built-up indices for impervious surface extraction using sentinel-2A MSI imageries: a case of Addis Ababa city. Ethiop Environ ChallS 8:100568. https://doi.org/10.1016/j.envc.2022.100568

Kumari B, Tayyab M, Shahfahad S, Mallick J, Khan MF, Rahman A (2018) Satellite-driven land surface temperature (LST) using Landsat 5, 7 (TM/ETM+ SLC) and Landsat 8 (OLI/TIRS) data and its association with built-up and green cover over urban Delhi, India. Remote Sens Earth Syst Sci 1:63–78

Lambin EF, Geist HJ, Lepers E (2003) Dynamics of land-use and land-cover change in tropical regions. Annu Rev Environ Resour 28(1):205–241

Laskar S, Mishra P, Behera B (2024) Impact of land use dynamics on land surface temperature: a multi-decadal analysis in Indian Sundarbans region. Environ Dev Sustain:1–29

Lévy-Leblond JM (2002) The meanings of Planck's constant. In: Conference proceedings-Italian physical society, vol 79, Editrice Compositori, 1999, pp 211–226

Liping C, Yujun S, Saeed S (2018) Monitoring and predicting land use and land cover changes using remote sensing and GIS techniques—a case study of a hilly area, Jiangle. China PloS one 13(7):e0200493

Maity DS, Srivastava DGL, Mane SP (2020) Comprehensive analysis of land surface temperature with different indices using Landsat–8 (OLI/TIRS) data in Kanpur Metropolis, India. Akshar Wangmay (*1*)

Mitra S, Basu S, Nandan M (2021) Analysis of vegetation health of the Sundarbans region-using remote sensing and GIS techniques. Brainwave: Multidiscip J 2(1):22–30

Mukhopadhyay A (2021) Climate change and the Sundarbans: challenges and adaptation strategies. Oxford University Press

Næsset E (1996) Use of the weighted kappa coefficient in classification error assessment of thematic maps. Int J Geogr Inf Syst 10(5):591–603

Office of the Registrar General & Census Commissioner, India (DCHB) (2014) Census of India 2011: West Bengal, series 20, part XII-A, district census handbook—south twenty FourParganas. Directorate of Census Operations, West Bengal. https://censusindia.gov.in

Ogunode A, Akombelwa M (2017) An algorithm to retrieve land surface temperature using Landsat-8 dataset. S Afr J Geomat 6(2):262–276

Paul S (2021) Mangrove ecosystems and socio-spatial dynamics in South Asia. Springer

Perera NGR, Samanthilaka KPPR (2014) Effect of street canyon materials on the urban Heat Island phenomenon in Colombo. In: International conference on 'cities, people and places'—ICCPP-2014, July 2017

Roy DP, Wulder MA, Loveland TR, Woodcock CE, Allen RG, Anderson MC et al (2014) Landsat-8: science and product vision for terrestrial global change research. Remote Sens Environ 145:154–172

Rwanga SS, Ndambuki JM (2017) Accuracy assessment of land use/land cover classification using remote sensing and GIS. Int J Geosci 8(04):611

Saha TK, Sajjad H, Roshani R, M. H., & Sharma, Y. (2024) Exploring the impact of land use/ land cover changes on the dynamics of Deepor wetland (a Ramsar site) in Assam, India using geospatial techniques and machine learning models. Model Earth Syst Environ 10(3):4043–4065

Sardar P, Samadder SR (2021) Understanding the dynamics of landscape of greater Sundarban area using multi-layer perceptron Markov chain and landscape statistics approach. Ecol Indic 121:106914. https://doi.org/10.1016/j.ecolind.2020.106914

Sarkar B, Islam A, Majumder A (2021) Seawater intrusion into groundwater and its impact on irrigation and agriculture: evidence from the coastal region of West Bengal. India RegNal Stud Mar Sci 44:101751

Sekertekin A, Bonafoni S (2020b) Land surface temperature retrieval from Landsat 5, 7, and 8 over rural areas: assessment of different retrieval algorithms and emissivity models and toolbox implementation. Remote Sens 12(2):294

Sekertekin A, Bonafoni S (2020a) Sensitivity analysis and validation of daytime and nighttime land surface temperature retrievals from Landsat 8 using different algorithms and emissivity models. Remote Sens 12(17):2776

Sen H (2019) The Sundarbans: a disaster-prone eco-region, vol 10. Springer, Berlin, pp 978–973

Sharma A, Vashishtha D (2023) Spatio-temporal assessment of land use land cover changes and their impact on variations of land surface temperature in Aligarh municipality. J Indian Soc Remote Sens 51(4):799–827

Sim J, Wright CC (2005) The kappa statistic in reliability studies: use, interpretation, and sample size requirements. Phys Ther 85(3):257–268

Su Y, Davidson JH (2018) A three-dimensional phonon energy transport model based on the non-dimensional lattice Boltzmann method. Int J Heat Mass Transf 127:303–318

Tabatabai M, Bailey S, Bursac Z, Tabatabai H, Wilus D, Singh KP (2021) An introduction to new robust linear and monotonic correlation coefficients. BMC Bioinform 22(1):170. https://doi.org/10.1186/s12859-021-04098-4

Tenhunen S, Uddin MJ, Roy D (2024) Confronting climate change in the Sundarbans in coastal India and Bangladesh. Wisdom Lett 2024(2)

Thakur S, Maity D, Mondal I, Basumatary G, Ghosh PB, Das P, De TK (2021) Assessment of changes in land use, land cover, and land surface temperature in the mangrove forest of Sundarbans, northeast coast of India. Environ Dev Sustain 23:1917–1943

Tian Y (2021) Photonic manipulation and Photothermal Conversion. Northeastern University

Titumir RAM (2022) Introduction: the Sundarbans as our mind. In: Sundarbans and its ecosystem services: traditional knowledge, customary sustainable use and community based innovation. Singapore, Springer Nature Singapore, pp 1–20

Town and Country Planning Organisation (TCPO) (2005) Urban and regional development plans formulation and implementation guidelines (URDPFI). Ministry of Urban Development, Government of India

Verma P, Raghubanshi A, Srivastava PK, Raghubanshi AS (2020) Appraisal of kappa-based metrics and disagreement indices of accuracy assessment for parametric and nonparametric techniques used in LULC classification and change detection. Model Earth Syst Environ 6:1045–1059

Weng Q, Lu D, Schubring J (2004) Estimation of land surface temperature–vegetation abundance relationship for urban heat Island studies. Remote Sens Environ 89(4):467–483

World Bank (2014) Landscape narrative of the Sundarban. World Bank Group. https://documents.worldbank.org/curated/en/539771546853079693/pdf/Sundarban-Joint-Landscape-Narrative.pdf

Yilmaz AE, Demirhan H (2023) Weighted kappa measures for ordinal multi-class classification performance. Appl Soft Comput 134:110020

Zhang F, Zhou G (2019) Estimation of vegetation water content using hyperspectral vegetation indices: a comparison of crop water indicators in response to water stress treatments for summer maize. BMC Ecol 19(1):18. https://doi.org/10.1186/s12898-019-0233-0

Zhang W-Y, Wei Z-W, Wang B-H, Han X-P (2016) Measuring mixing patterns in complex networks by spearman rank correlation coefficient. Phys A: Stat Mech Its Appl 451:440–450. https://doi.org/10.1016/j.physa.2016.01.056

Zheng Y, Tang L, Wang H (2021) An improved approach for monitoring urban built-up areas by combining NPP-VIIRS nighttime light, NDVI, NDWI, and NDBI. J Clean Prod 328:129488. https://doi.org/10.1016/j.jclepro.2021.129488

Impacts of Urban Growth and Industrial Expansion on Indonesia's Delta Ecosystems

10

Annisa Utami Rauf and Ratna Dwi Puji Astuti

Abstract

This chapter examines the complex interplay between urbanization, industrialization, public health, and climate change in shaping the sustainability of Indonesia's delta regions. Rapid urban expansion, particularly in Java Island, has led to uncontrolled sprawl, disrupting spatial planning and increasing flood risks due to the loss of natural water absorption areas. The delta region, as one of the important natural resources, has suffered significant degradation, partly due to extensive industrial activities. Several areas in the provinces of Central Kalimantan, Riau, and Southeast Sulawesi, which serve as mining zones and have fostered urban growth around industrial sites, are also facing environmental challenges particularly the limited access to clean water and sanitation, which poses potential health risks. Industrial growth has intensified environmental degradation, contributing to rising carbon emissions, pollution, and inefficient waste management. These challenges, exacerbated by climate change, threaten ecological balance and heighten public health risks, including waterborne diseases and bacterial infections. To address these complexities, a multidisciplinary approach is essential. Integrated adaptive urban strategies and community-based solutions such as constructed wetlands and sustainable waste management play a crucial role in enhancing resilience. A holistic framework that combines environmental governance,

A. U. Rauf (✉)
Department of Health Behavior, Environment and Social Medicine, Faculty of Medicine, Public Health and Nursing, Universitas Gadjah Mada, Yogyakarta, Indonesia
e-mail: annisaur@ugm.ac.id

R. D. P. Astuti
Department of Environmental Health, Faculty of Public Health, Universitas Airlangga, Surabaya, Indonesia
e-mail: ratna.dwi@fkm.unair.ac.id

C. Singha et al. (eds.), *Sustainable Management of Delta Ecosystems Resilience*,
Deltas of the World, https://doi.org/10.1007/978-3-032-16261-8_10

public health initiatives, and climate adaptation measures is key to ensuring the long-term sustainability of Indonesia's delta regions amid ongoing environmental and socio-economic transformations.

Abbreviations

AHP	Analytic hierarchy process
BGI	Integrated blue-green infrastructure
CBMM	Community-based mangrove management
GDP	Gross domestic product
GGC	Green growth compact
ICZM	Integrated coastal zone management
IDR	Indonesian rupiah
JMA	Jakarta metropolitan area
LEDS	Low emissions development strategy
MP	Microplastic
NGO	Non-governmental organization
NPOA	National plan of action
SDMs	Species distribution models
SUHI	Surface urban heat islands

10.1 Introduction

Urbanization and industrialization have dramatically transformed delta regions worldwide, particularly in developing countries (Ahmed et al. 2025; Hua et al. 2024; Najimi et al. 2025). These dynamic processes drive economic growth and urban expansion but simultaneously pose significant environmental and socio-economic challenges. The rapid expansion of urban areas and industries in delta regions exacerbates environmental degradation, increases pollution, and disrupts ecological balance (Moghimi Dehkordi et al. 2024). Furthermore, as cities expand, the pressures on water resources, air quality, and land use intensify, leading to heightened flood risks, biodiversity loss, and socio-cultural shifts. Deltaic wetlands play a crucial role in capturing and storing large amounts of atmospheric carbon dioxide, contributing significantly to carbon sequestration. This process not only helps mitigate climate change but also supports the stability and health of these ecosystems, making them essential for global climate regulation (Wu et al. 2024; Zhi et al. 2023).

Across the globe, delta ecosystems have been profoundly affected by urban and industrial development. In China's Yangtze River Delta, for instance, economic, spatial, and demographic factors significantly influence air pollution levels, while industrial agglomeration fosters urban expansion, sometimes at the cost of environmental sustainability (She et al. 2021). Similarly, in the Pearl River Delta, industrial clustering has played a vital role in shaping urban growth, illustrating both the opportunities and challenges associated with rapid industrialization (Gong et al. 2023). However, without effective policies and governance, these urban transformations can lead to long-term environmental and public health concerns (Masud et al. 2024; Rusprayunita et al. 2025). From an urbanization perspective, the negative effects of waste accumulation and pollutants from human activities will degrade the quality of deltas, disrupt aquatic ecosystems, and pose risks to the food chain (Häder et al. 2020; Patel et al. 2024).

Indonesia's delta ecosystems are vital for maintaining ecological stability and supporting fisheries, agriculture, and freshwater resources. However, rapid industrial growth particularly around river mouths and coastal zones has emerged as a major cause of environmental degradation. Industrial operations are often concentrated in delta areas due to easy access to water for processing and waste disposal, leading to increased pollution from heavy metals, organic matter, and microplastic (MP) that disrupt aquatic ecosystems. Prominent deltas area such as the Mahakam, Citarum, Musi, and Siak serve as crucial ecological and economic corridors, yet unregulated industrial expansion has caused sedimentation changes, habitat loss, and declining water quality. The discharge of untreated industrial effluents, coupled with weak enforcement of environmental regulations, has further accelerated eutrophication and diminished the resilience of mangrove and estuarine systems.

The transformation of Indonesia's coastline between 1990 and 2018 underscores the growing influence of industrialization and human activities on the country's delta and coastal ecosystems. Over 28-year period, while the total coastline length increased slightly by about 777 km, the proportion of natural coastlines declined by nearly 6000 km, accompanied by an increase of more than 6700 km in artificial coastlines (Sui et al. 2020). This shift, most prominent along the southern coasts of Sumatra and Java, the southeastern coast of Kalimantan, and the northern coast of the Maluku Islands, reflects the rapid expansion of industrial zones, ports, and urban coastal infrastructure. The physical transformations indicate a broader ecological imbalance, as industrial and urban development encroach upon natural buffers such as mangroves and wetlands.

Addressing these issues requires comprehensive management strategies, including the implementation of strict regulations, the development and adoption of environmentally friendly technologies, and increased public awareness of sustainable plastic waste management. Moreover, given Indonesia's high plastic production and consumption, stronger policies such as restrictions on single-use plastics and the optimization of recycling systems are crucial. As the fourth most populous country, Indonesia ranks as the world's second-largest plastic polluter after China, generating 3.2 million tonnes of unmanaged plastic waste annually, with around 1.29 million tonnes ending up in the deltas and ocean (Ministry of Environment and Forestry, Republic of Indonesia 2020). Industrial zones, often located near rivers and coastal areas, are major sources of waste runoff. This runoff contaminates water bodies, leading to ecological degradation and posing risks to human health (Jagemma et al. 2023; Kiran and Ramaraju 2022; Ruzieva and Ruziev 2024). To mitigate these environmental and public health threats, it is essential to integrate waste management improvements with public health initiatives, including stricter industrial waste regulations, enhanced sanitation infrastructure, pollution control measures, and community-based education programs in Indonesia (Ibrahim et al. 2024; Lasut, Jensen, and Shivakoti 2008; Ministry of Environment and Forestry, Republic of Indonesia 2020; Susilawati et al. 2022).

In this review, the effects of urbanization and industrialization on Indonesia's delta ecosystems are examined, with particular attention paid to the main causes of ecological degradation. Its comprehensive perspective, which links industrial expansion, environmental consequences, and public health threats, is what makes this review related to the field of environmental and public health sustainability. By combining the available information, this analysis seeks to pinpoint the main obstacles and suggest solutions for sustainable delta management and policy change.

10.2 Urban Expansion and Its Consequences

10.2.1 Patterns of Urban Grown in Indonesia's Delta Regions

Indonesia has undergone rapid urbanization, with its urban population increasing from 32.8 million in 1980 to 118.3 million in 2010. This expansion has driven the growth of urban areas, particularly on the island of Java, encompassing major metropolitan centers such as Jakarta, Bandung, Surabaya, Semarang, Malang, Surakarta, and Yogyakarta (Mardiansjah et al. 2021). Urbanization in Indonesia is marked by a high concentration of the urban population in a few major cities, especially the Jakarta Metropolitan Area (JMA), highlighting its integration into the global economy and the disparities between large and smaller cities (Firman 2004).

Jakarta, as Indonesia's economic hub, has experienced substantial urban expansion, particularly in its peri-urban and delta regions (Farid et al. 2022). This expansion has manifested through various urban growth patterns, including infilling, edge expansion, and outlying development, contributing to increased urban density and spatial reconfiguration. While this growth has facilitated economic development and infrastructure expansion, it has also significantly altered the city's landscape, leading to notable environmental challenges. The high contamination levels in Jakarta's bay area are largely attributed to untreated sewage discharge, industrial runoff, and inadequate wastewater management, leading to the accumulation of persistent organic pollutants (Dsikowitzky et al. 2020; Dwiyitno et al. 2016). The presence of these pollutants in delta sediments not only disrupts marine ecosystems but also poses significant risks to coastal communities, exacerbating

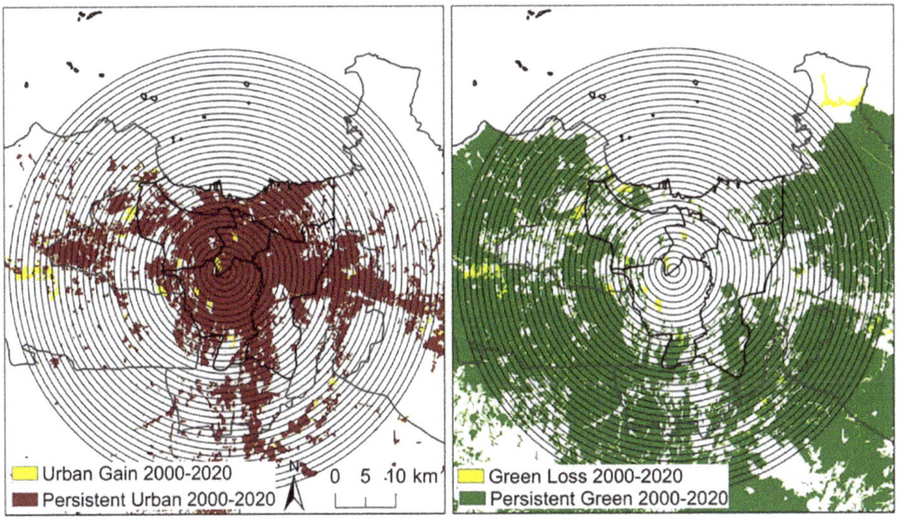

Fig. 10.1 Land cover transformation in Jakarta: **a** expansion of built-up areas (left) and **b** reduction of green spaces (right). Note: From 2000 to 2020, urban development extended outward from the city center to surrounding areas, while green spaces within the city core declined. Adapted from (Sarker et al. 2024)

water quality degradation and long-term environmental instability. Figure 10.1 shows that Jakarta's built-up area expanded faster than its built-up volume between 2000 and 2020. The largest growth occurred in East and West Jakarta, while Central Jakarta showed minimal change. In contrast, green space declined sharply by about 31%, shrinking from 148 km^2 to 102 km^2 during the same period.

In contrast, Surakarta has experienced uneven urban growth, with rapid expansion in its peripheral areas and decline in older city centers, creating socio-spatial inequalities. As part of the broader Joglosemar Urban Region, which includes Semarang and Yogyakarta, urban sprawl has led to significant vegetation loss (Dewa et al. 2024). Likewise, Yogyakarta has experienced fragmented urbanization on its outskirts, gradually developing into stable urban zones, and contributing to environmental issues such as habitat loss, water pollution, and increased flood risks (Divigalpitiya and Handayani 2015). In Pekalongan Regency, located on the other side of Java Island, microplastic (MP) levels in water ranged between 45.2 and 99.1 particles/L, with sediment concentrations between 0.77 and 1.01 particles/g. The most abundant types of MPs were fragments and films, making up 30.1% and 25.4% of the total, respectively. The growing urbanization in Pekalongan, coupled with increased industrialization and the widespread use of single-use plastics, has led to a surge in improperly managed plastic waste (Ismanto et al. 2023). In other areas, an increase in MPs has also been recorded due to human activities and behaviors, as well as facilities and policies that are not stringent enough (Henny et al.

2022; Mulyasari, Mukono, and Sincihu 2023; Pramaningsih et al. 2023; Sari et al. 2021). This has led to rising MP levels that threaten delta ecosystems, particularly on the highly populated island of Java.

There are about 2.5 million people living in Bandung, Surabaya, and Medan, followed by Makassar, Semarang, and Palembang, each of which has about 1.5 million inhabitants. The population of emerging metropolitan regions like Batam, Padang, Bandar Lampung, and Pekanbaru is getting close to one million, while cities like Malang, Samarinda, Balikpapan, and Banjarmasin are anticipated to do so shortly. The rapid urbanization of Indonesia outside of its main urban centers is reflected in these "medium-sized metropolitan cities." The trend of urban expansion is exemplified by this continuous growth, which results in significant land conversion, increased strain on natural resources and infrastructure, and growing difficulties in attaining sustainable urban management.

10.2.2 Spatial Planning Challenges and Land-Use Conflicts

Land use changes and conflicts have become significant concerns in newly developed regions of Indonesia, especially since the implementation of regional autonomy policies that emphasize decentralized land use governance. In South Sulawesi Province, two main land use issues have emerged, encroachment on protected forest lands in rural areas and the expansion of urban development into rural zones, leading to various land use conflicts (Baja et al. 2019). Decentralization and local autonomy in Indonesia have created challenges in spatial planning, including limited resources, regional egoism, and violations of regulations. Strengthening government management, enforcing laws, and improving communication between central and local governments are key to addressing these issues and enhancing spatial planning effectiveness (Baja et al. 2019; Kuncorowati et al. 2024).

One of the key drivers of land-use conflicts in Indonesia is its imbalance in spatial planning. Jakarta, Southeast Asia's most disaster-prone city, faces escalating land-use disputes as worsening annual floods are fueled by land subsidence, declining green spaces, and clogged waterways (Tjoei et al. 2020). Although northern mangrove forests could help mitigate flooding, around 10–20% of urban lakes have disappeared, and green space has shrunk from 40% in 1985 to just 9% in 2002, exacerbating surface runoff and land subsidence. Land reclamation has expanded urban areas but reduced sea space, transforming once-communal land into private property. Control over these reclaimed areas is concentrated in the hands of specific groups, restricting public access. There are growing concerns that developers will prioritize affluent buyers, turning these areas into exclusive luxury districts for profit (Colven 2020; Kawata 2022).

The most common urban issues is the emergence of illegal settlements, primarily inhabited by migrants with limited financial resources. These individuals move to cities in search of better opportunities but, due to restricted access to proper housing, end up

Fig. 10.2 Urban settlements formed by rural migrants seeking employment, often contributing to the degradation of the Ciliwung River ecosystem. Adapted from (Kawata 2022)

establishing informal settlements in unsuitable areas, including along riverbanks (Purwanto et al. 2023). These settlements not only create social and economic challenges but also have severe environmental consequences. Built without proper planning, they contribute to river pollution from domestic waste, sedimentation, and the narrowing of waterways. As a result, the river's capacity to accommodate water decreases, heightening the risk of flooding. This phenomenon is prevalent across Indonesia, particularly along the Ciliwung River (West Java), Sikambing River (North Sumatra), and Kuin River (South Kalimantan), Siak River (Riau) (Angriani et al. 2018; Fachrudin and Lubis 2016; Kawata 2022). When these settlements are subjected to eviction, it often leads to conflicts between the community and authorities. Figure 10.2 illustrates slum settlements around riverbanks, which alter land use functions.

10.3 Public Health Challenges in Delta Regions

As the population in coastal urban areas continues to grow, an increasing volume of land-based pollutants is being transported into coastal waters, negatively impacting tropical ecosystems. Previous studies on anthropogenic pollutants in Indonesian coastal waters and ecosystems from 1986 to 2021 have identified the most commonly studied contaminants, including nutrients, heavy metals, organic pollutants, and plastic debris. It was reported that concentrations of nutrients, heavy metals, and organic pollutants have exceeded standard limits, pointing to poor water quality in some regions of Indonesian coastal waters. Coral reef ecosystems are particularly vulnerable to these anthropogenic disturbances (Adyasari et al. 2021). Another example, the expansion of residential areas in Eastern Indonesia has contributed to the degradation of riparian zones and a marked increase in river pollution. Simulations indicate that this pollution represents an ongoing threat to the river ecosystem, particularly impacting the mangrove habitats at the mouth of the Wanggu River, thereby exacerbating pollution levels in Kendari Bay, Southeast Sulawesi (Jurumai et al. 2023).

A significant portion of Indonesia's coastal waters is in poor condition, largely due to anthropogenic pollutants such as nutrients, heavy metals, organic pollutants, and plastic debris (Asmal et al. 2022; Iskandar et al. 2021). These issues are further exacerbated by rapid urbanization and industrial expansion, which contribute to increased wastewater discharge, untreated industrial effluents, and improper waste management. A study indicated that 82% of field assessments have recorded excessive nutrient levels, while 54% and 50% exceed environmental standards for heavy metals and organic pollutants, respectively (Adyasari et al. 2021). Among marine ecosystems, coral reefs are the most affected by these pollutants, followed by mangroves and seagrass. In addition to environmental pollution, inadequate sanitation has had severe economic and public health consequences, contributing to GDP losses of up to 58 trillion IDR (Lisafitri et al. 2021; Wulandari et al. 2024). While short-term solutions may provide temporary relief, they often lead to long-term issues. Limited access to proper sanitation services threatens both human health and the surrounding environment, making the implementation of sustainable sanitation systems essential.

Coastal communities are particularly vulnerable due to persistent hygiene challenges. Many of these areas struggle with limited access to clean water and proper sanitation facilities, further exacerbated by low awareness of personal hygiene practices such as handwashing (Hussainzad and Gou 2024; Susilawati et al. 2022). Informal settlements in these regions are often unplanned and predominantly inhabited by low-income populations, reinforcing cycles of poor hygiene and environmental degradation. These findings align with previous studies, emphasizing the urgent need for integrated solutions to address both pollution and sanitation in Indonesia's coastal regions (Adyasari et al. 2021; Danielaini et al. 2019; Koagouw et al. 2021).

Indonesia's vast delta regions, including those in North Jakarta and Jambi, face persistent challenges related to waterborne diseases due to poor sanitation, flooding, and environmental degradation. In densely populated urban deltas, inadequate wastewater management and open defecation contribute to the contamination of water sources with fecal pathogens, increasing the prevalence of diseases such as cholera, dysentery, and typhoid (Merten et al. 2021; Prana et al. 2024). Environmental and economic shifts have worsened the health risks associated with waterborne diseases. Historically, flooding was a manageable seasonal occurrence, but the expansion of agro-industrial plantations has altered the region's hydrological balance. Large-scale deforestation and land conversion have reduced natural water absorption, increasing flood severity and water stagnation (Cameron et al. 2021; Merten et al. 2021).

Climate-related factors such as heavy rainfall, flooding, and rising sea levels further contribute to the increased incidence of illnesses like cholera, hepatitis A and E, and Vibrio infections (Danielaini et al. 2019). Insufficient wastewater management and limited access to clean drinking water heighten these health hazards, especially in densely populated coastal and delta regions in Indonesia (Adyasari et al. 2019, 2021). Waterborne diseases, particularly diarrhea, are prevalent in Indonesia. Research indicates that upstream river bathing can account for approximately 7.5% of all diarrhea-related deaths in these areas (Garg et al. 2018). Similarly, the Mekong Delta in Vietnam, which shares environmental characteristics with Indonesia's delta regions, experiences high rates of diarrhea. Studies have found significant correlations between increased diarrhea incidence and climatic factors such as elevated temperatures, high relative humidity, and substantial cumulative rainfall (Phung et al. 2015). In Jakarta, climate change has adversely affected the urban water environment, leading to a higher incidence of waterborne infections (Fukushi 2019).

10.4 Strategies for Sustainable Development

10.4.1 Integrated Spatial Planning and Green Infrastructure

The incorporation of green infrastructure into spatial planning in Indonesia's delta regions offers significant environmental and economic benefits, including improved air quality, reduced energy consumption, and increased property values. However, its implementation remains constrained by several key challenges, such as a shortage of skilled human resources, inadequate inter-agency coordination, and the absence of a regulatory framework incorporating incentives and disincentives to ensure compliance. Financial limitations for initiating spatial planning in autonomous regions, regional fragmentation, and frequent violations of spatial planning policies further impede progress. Despite these obstacles, there is an increasing acknowledgment of the need for structured management and professional expertise in spatial planning. In response, local governments are prioritizing the recruitment of professionals to formulate more systematic and effective planning policies that align with sustainable development objectives. Strengthening regulatory enforcement and ensuring adherence to spatial planning frameworks are essential to mitigating the negative impacts of unregulated urban expansion and environmental degradation. Adopting a systematic approach to regional revenue management can enhance financial support for green infrastructure initiatives, fostering long-term sustainability (Maryati and Humaira 2017; Monteiro et al. 2022). A study on the Siak River near the

Sari Residence housing complex in Pekanbaru City in Riau Province, found that the density of coliform bacteria ranged from 11,267 to 15,650 units per 100 mL, exceeding the standard limit for drinking water quality. Another investigation examined the impact of reforestation on water quality in Maredan Village, Tualang District, an area with intensive industrial activities where the study reported that after reforestation efforts carried out between 2014 and 2016, the river's condition became more stable and showed improvement compared to the period before rehabilitation in 2003–2005 (Basuki et al. 2024). Through integrated governance and strategic policy interventions will facilitate the effective incorporation of green infrastructure into Indonesia's spatial planning framework, enhancing resilience and promoting sustainable development in delta regions.

Effective spatial planning is essential for achieving a balance between environmental sustainability and economic development, particularly in rapidly urbanizing and ecologically sensitive regions such as Indonesia. However, significant gaps remain in integrating ecological considerations into spatial planning, particularly in marine conservation, urban climate resilience, and coastal land-use management (Breen et al. 2016; Kaschner et al. 2006). One critical gap is the absence of comprehensive distribution maps for cetacean habitats at eco-regional and seascape scales, despite directives from the National Plan of Action (NPOA) for Cetaceans and the Coral Triangle Initiative (Sahri et al. 2021). Understanding cetacean habitat preferences and distribution patterns is fundamental to mitigating human-wildlife conflicts and implementing spatially explicit conservation measures. The application of species distribution models (SDMs) in conjunction with human activity mapping presents an opportunity to enhance marine spatial planning, yet such an approach has not been systematically applied in Indonesia.

Urbanization further exacerbates environmental challenges, particularly through the intensification of Surface Urban Heat Islands (SUHI). Spatial analyses have revealed substantial intra-city variations in heat concentration, highlighting the need for targeted mitigation strategies such as urban greening (Fajary et al. 2024). Expanding these studies to a broader set of cities could provide deeper insights into spatial patterns of urban heat and inform more effective urban climate adaptation policies. Large-scale land-use transformations in northern Java, where 55,220 hectares of flood-prone areas—including protected zones, fishponds, and agricultural land—are designated for industrial and residential expansion, heighten exposure to coastal inundation. Without the integration of climate adaptation measures, these changes could result in economic losses exceeding $246.6 billion by 2030 (Suroso and Firman 2018). The current spatial planning framework has not adequately accounted for climate-induced hazards, underscoring the need for resilience-based approaches to mitigate future risks.

10.4.2 Community Participation

Community participation plays a crucial role in shaping industrialization and urbanization in delta regions, influencing economic transitions, environmental sustainability, and governance. The shift from rural to urban economies, as observed in areas such as Temanggung Regency, is primarily driven by local workforce dynamics and resource availability, resulting in in situ urbanization that mitigates large-scale migration (Wijaya and Buchori 2023). The effective urban planning requires inclusive governance frameworks, as seen in Jakarta's Kampung Deret program, where limited transparency in fund allocation has raised concerns regarding accountability (Lestari and Kurniawan 2018; Prana et al. 2024). Similarly, the lack of community engagement in Semarang's transportation policies underscores the necessity of participatory decision-making to ensure socially equitable urban development (Nalle et al. 2023).

In the context of environmental management, community-based approaches have been instrumental in coastal and mangrove conservation. For example, Community-Based Mangrove Management (CBMM) initiatives in Pantai Bahagia Village demonstrate that the success of restoration projects largely depends on the level of local engagement and external institutional support (Majesty and Fadmastuti 2018). In East Java, integrated coastal ecosystem management prioritizes collaborative strategies involving the government, private sector, and communities to rehabilitate mangroves, coral reefs, estuaries, and seagrass ecosystems (Damastuti and De Groot 2017). Despite these efforts, challenges such as inadequate capacity-building initiatives, as seen in the mangrove edu-tourism project in Teluk Awur, Jepara, hinder active community participation, emphasizing the need for structured education and awareness programs (Yamindago 2015).

The legal and institutional landscape significantly impacts community participation in resource management. Weak stakeholder coordination and limited access to decision-making processes impede effective watershed governance in Indonesia (Narendra et al. 2021). Comparative studies highlight the varying effectiveness of community participation models across regions. While in Iskandar Malaysia, operational dimensions of community engagement are prioritized over structural and cultural aspects, the success of community-led policing in mangrove management in Cilacap, Central Java, underscores the importance of integrating local communities into legal and environmental protection frameworks (Purwendah et al. 2024; Zanudin et al. 2021). Strengthening public awareness, fostering participatory governance, and ensuring institutional support are imperative for enhancing sustainable industrial and urbanization processes in delta regions.

Table 10.1 presents various approaches to managing industrial and urbanization challenges in Indonesia's delta regions, highlighting key locations, outcomes, and relevant sources. These approaches encompass integrated coastal zone management, stakeholder collaborations, ecosystem rehabilitation, governance frameworks, and sustainable urban planning strategies. By examining case studies from different regions in Indonesia, providing insights into the effectiveness of community participation, institutional support, and policy interventions in addressing environmental and socio-economic challenges.

10.5 Conclusion and Future Directions

The rapid urbanization and industrialization of delta regions in Indonesia have profoundly altered ecological systems and socio-economic structures. While these developments contribute to economic expansion, they simultaneously exacerbate environmental stressors, including pollution, land-use conflicts, and biodiversity loss. Coastal areas are particularly vulnerable due to their exposure to both natural and anthropogenic pressures, making them highly susceptible to degradation. The interplay between human-induced stressors and climate change necessitates an integrated approach to sustainable management. Addressing these challenges requires evidence-based policies, interdisciplinary research, and the incorporation of ecological and socio-economic resilience frameworks to ensure long-term environmental stability and human well-being.

Indonesia's delta regions face multifaceted challenges driven by industrial activities, urban expansion, and climate change. Persistent pollution from industrial discharge and domestic waste contributes to declining water quality and ecosystem degradation. Land subsidence, exacerbated by excessive groundwater extraction, heightens vulnerability to flooding and coastal erosion. Coastal areas, as the most vulnerable regions, are disproportionately affected by these environmental stressors. Weak enforcement of environmental regulations further accelerates deltaic degradation, as unsustainable land-use practices and unchecked industrial expansion continue to strain fragile ecosystems.

Table 10.1 Approaches to managing industrial and urbanization challenges in Indonesia's Delta Regions

Approaches	Locations/city	Result	Source
Integrated Coastal Zone Management (ICZM)	Seribu Islands, East Bintan, Jakarta Bay and Banten Province	Implementing integrated coastal zone management strategies is crucial to preserving coastal ecosystems and ensuring the well-being of communities that depend on them. The government has prioritized land-based development, often neglecting comprehensive coastal management	(Farhan and Lim 2012; Irmadhiany et al. 2024)
Stakeholder collaborations	Ambon, Maluku Province	This approach emphasizes collaborative efforts across national boundaries to manage marine resources holistically, considering entire ecosystems rather than isolated components.	(Kobayashi 2023)
Rehabilitation of mangrove ecosystem	Mahakam Delta, East Kalimantan	Mangrove restoration in the Mahakam Delta faces key policy challenges, including stakeholder disagreements on priorities and significant power imbalances in decision-making	(Powell and Osbeck 2010)

(continued)

Table 10.1 (continued)

Approaches	Locations/city	Result	Source
Governance, institutions, social–ecological systems	National data	The subsections outline key policy recommendations for governing mangroves, fisheries, watersheds, marine conservation, coastal livelihoods, and pollution. Recognizing overlaps and trade-offs, the analysis proposes integrated, adaptive, and inclusive governance strategies	(Glaser et al. 2022)
Community-based mangrove management (CBMM)	Central Java	Effective community-based mangrove management (CBMM) enhances resource use and local livelihoods through participation and external support. Strengthening institutions, diversifying incomes, and continuous monitoring can further improve CBMM for sustainable mangrove management	(Damastuti and De Groot 2017)
Green Growth Compact (GGC) initiative	Mahakam Delta, East Kalimantan	Stakeholder analysis in Mahakam Delta identifies three key groups: government/NGOs, oil and gas companies, and shrimp farmers. Findings support the Green Growth Compact (GGC) initiative for better collaborative mangrove management	(Wahyuni et al. 2022)

(continued)

Table 10.1 (continued)

Approaches	Locations/city	Result	Source
Integrated water management	Jakarta	Effective flood risk management in Jakarta requires integration with broader water management strategies, tackling challenges from urbanization and climate change	(Lamond 2014)
Institutional capacity building	Jakarta	Supporting factors for Urban-LEDS implementation include Indonesia's global climate commitments, regional policies for low-emission cities, and multi-stakeholder collaboration. However, the main challenge is the lack of regulations limiting motorized vehicle overuse	(Zalsabila et al. 2020)
Land use planning	Jakarta	Jakarta's neoliberal urban development fosters competition but deepens inequalities and strains land access. Governance modes are challenged as informal power structures emerge alongside official planning. Case studies in Budidharma and Bintara highlight negotiations, contestations, and the tension between individualization and collective mobilization in urban space	(Dietrich 2015)

(continued)

Table 10.1 (continued)

Approaches	Locations/city	Result	Source
Wastewater management	Manado, North Sulawesi	Limited law enforcement and lack of community initiatives exacerbate pollution. An integrated wastewater management plan is urgently needed to address environmental, socio-economic, and institutional challenges	(Lasut et al. 2008)
Blue-green infrastructure implementation	Surabaya, East Java	Integrated blue-green infrastructure (BGI) is explored as a sustainable flood management solution. Using geospatial analysis and the analytic hierarchy process (AHP), the study identifies optimal BGI nodes and corridors. The eastern and western regions, including agricultural and mangrove areas, are prime candidates, along with public land under city government jurisdiction	(Harlis and Seo 2024)

(continued)

Table 10.1 (continued)

Approaches	Locations/city	Result	Source
Participatory approach	South Tanggerang	In Kampung Nambo, South Tangerang, a participatory approach involving residents, scholars, NGOs, and local government developed a master plan to tackle urban pollution, promoting sustainable development through education and awareness	(Dewi et al. 2019)

References

Adyasari D, Hassenrück C, Oehler T, Sabdaningsih A, Moosdorf N (2019) Microbial community structure associated with submarine groundwater discharge in Northern Java (Indonesia). Sci Total Environ 689:590–601. https://doi.org/10.1016/j.scitotenv.2019.06.193

Adyasari D, Pratama MA, Teguh NA, Sabdaningsih A, Kusumaningtyas MA, Dimova N (2021) Anthropogenic impact on indonesian coastal water and ecosystems: current status and future opportunities. Mar Pollut Bull 171:112689. https://doi.org/10.1016/j.marpolbul.2021.112689

Ahmed I, Mostafa Monwar M, Amran MIUA, Iqbal MM, Isha IS, Billah MM, Islam MR, Mamun AA, Bhuiyan MKA (2025) Are industrial footprints driving potentially toxic element contamination in the Karnaphuli River Estuary? J Trace Elem Miner 12:100220. https://doi.org/10.1016/j.jtemin.2025.100220

Al Masud A, Islam AR, Al Mamun A, Alam GM, Arabameri A, Bari AM, Pal SC, Rakib MR, Senapathi V, Bodrud-Doza M, Idris AM (2024) A new approach from public behavioral attitudes and perceptions towards microplastics: influencing factors, and policy proposals. Ocean Coast Manag 253:107144. https://doi.org/10.1016/j.ocecoaman.2024.107144

Angriani P, Sumarmi INR, Bachri S (2018) River management: the importance of the roles of the public sector and community in river preservation in Banjarmasin (a case study of the Kuin River, Banjarmasin, South Kalimantan—Indonesia). Sustain Cities Soc 43:11–20. https://doi.org/10.1016/j.scs.2018.08.004

Asmal I, Syarif E, Amin S, Walenna MA (2022) The impact of the environment and people's attitudes on greywater management in slum coastal settlements. Civ Eng J 8(12):2734–2748. https://doi.org/10.28991/CEJ-2022-08-12-05

Baja S, Pulubuhu DA, Neswati R, Arif S, Nurmiaty (2019) Land use conflict with a particular reference to spatial planning implementation in South Sulawesi. In: IOP conference series: earth and environmental science, vol 279, pp 012006. https://doi.org/10.1088/1755-1315/279/1/012006

Basuki TM, Indrawati DR, Nugroho HY, Pramono IB, Setiawan O, Nugroho NP, Nada FM, Nandini R, Savitri E, Adi RN, Purwanto P (2024) Water pollution of some major rivers in Indonesia: the status, institution, regulation, and recommendation for its mitigation. Pol J Environ Stud 33(4), 3515–3530. https://doi.org/10.15244/pjoes/178532

Breen P, Brown S, Reid D, Rogan E (2016) Modelling cetacean distribution and mapping overlap with fisheries in the northeast Atlantic. Ocean Coast Manag 134:140–149. https://doi.org/10.1016/j.ocecoaman.2016.09.004

Cameron L, Chase C, Haque S, Joseph G, Pinto R, Wang Q (2021) Childhood stunting and cognitive effects of water and sanitation in Indonesia. Econ Hum Biol 40:100944. https://doi.org/10.1016/j.ehb.2020.100944

Colven E (2020) Subterranean infrastructures in a sinking city: the politics of visibility in Jakarta. Crit Asian Stud 52(3):311–331. https://doi.org/10.1080/14672715.2020.1793210

Damastuti E, De Groot R (2017) Effectiveness of community-based mangrove management for sustainable resource use and livelihood support: a case study of four villages in Central Java, Indonesia. J Environ Manage 203:510–521. https://doi.org/10.1016/j.jenvman.2017.07.025

Danielaini TT, Maheshwari B, Hagare D (2019) An assessment of household water insecurity in a rapidly developing coastal metropolitan region of Indonesia. Sustain Cities Soc 46:101382. https://doi.org/10.1016/j.scs.2018.12.010

Dehkordi M, Mohsen ZP, Nodeh KS, Dehkordi HS, Khorjestan RR, Ghaffarzadeh M (2024) Soil, air, and water pollution from mining and industrial activities: sources of pollution, environmental impacts, and prevention and control methods. Results Eng 23:102729. https://doi.org/10.1016/j.rineng.2024.102729

Dewa DD, Buchori I, Rudiarto I, Sejati AW, Liu Y (2024) Spatiotemporal variations of urban spatial dispersion and its vegetation degradation in Joglosemar's urban growth corridors, Indonesia. Aust Geogr 55(2):179–201. https://doi.org/10.1080/00049182.2024.2331102

Dewi O, Atmoko T, Ha Alrasyid D, Priyandana MD, Siska D (2019) Tackling environmental pollution through participatory approach in Nambo, South Tangerang, Indonesia. IOP Conf Ser: Earth Environ Scie 291(1):012003. https://doi.org/10.1088/1755-1315/291/1/012003

Dietrich J (2015) Contestation Ou Négociation à Jakarta ? Deux Quartiers Face Aux Projets Urbains de La Métropole. Cybergeo. https://doi.org/10.4000/cybergeo.27370

Divigalpitiya P, Nurul Handayani K (2015) Measuring the urban expansion process of Yogyakarta City in Indonesia: urban expansion process and spatial and temporal characteristics of growing cities. Int Rev Spat Plan Sustain Dev 3(4):18–32. https://doi.org/10.14246/irspsd.3.4_18

Dsikowitzky L, Crawford SE, Nordhaus I, Lindner F, Dwiyitno HE, Irianto FA, Schwarzbauer J (2020) Analysis and environmental risk assessment of priority and emerging organic pollutants in sediments from the tropical coastal megacity jakarta, Indonesia. Reg Stud Mar Sci 34:101021. https://doi.org/10.1016/j.rsma.2019.101021

Dwiyitno LD, Nordhaus I, Andarwulan N, Irianto HE, Lioe HN, Ariyani F, Kleinertz S, Schwarzbauer J (2016) Accumulation Patterns of lipophilic organic contaminants in surface sediments and in economic important mussel and fish species from Jakarta Bay, Indonesia. Mar Pollut Bull 110(2):767–777. https://doi.org/10.1016/j.marpolbul.2016.01.034

Fachrudin HT, Lubis MD (2016) Planning for riverside area as water tourism destination to improve quality of life local residents, case study: Batuan—Sikambing River, Medan, Indonesia. Procedia Soc Behav Sci 234:434–441. https://doi.org/10.1016/j.sbspro.2016.10.261

Fajary FR, Lee HS, Kubota T, Bhanage V, Pradana RP, Nimiya H, Putra ID (2024) Comprehensive spatiotemporal evaluation of urban growth, surface urban heat island, and urban thermal conditions on java Island of Indonesia and implications for urban planning. Heliyon 10(13):e33708. https://doi.org/10.1016/j.heliyon.2024.e33708

Farhan AR, Lim S (2012) Vulnerability assessment of ecological conditions in Seribu Islands, Indonesia. Ocean Coast Manag 65:1–14. https://doi.org/10.1016/j.ocecoaman.2012.04.015

Farid M, Pratama MI, Kuntoro AA, Adityawan MB, Rohmat FIW, Moe IR (2022) Flood prediction due to land cover change in the Ciliwung River Basin. Int J Techn 13(2):356. https://doi.org/10.14716/ijtech.v13i2.4662

Firman T (2004) Demographic and spatial patterns of Indonesia's recent urbanisation. Popul Space Place 10(6):421–434. https://doi.org/10.1002/psp.339

Fukushi, K (2019) Health risk assessment for planning of a resilient city in the changing regional environment. In: Watanabe T, Watanabe C (eds) Health in ecological perspectives in the Anthropocene. Springer Singapore, Singapore, pp. 109–15

Garg T, Hamilton SE, Hochard JP, Kresch EP, Talbot J (2018) (Not so) Gently down the stream: river pollution and health in Indonesia. J Environ Econ Manag 92:35–53. https://doi.org/10.1016/j.jeem.2018.08.011

Glaser M, Adrianto L, Breckwoldt A, Buhari N, Deswandi R, Ferse S, Gorris P, Paragay SH, Glaeser B, Mohammad N, Máñez KS (2022) The governance of coastal and marine social–ecological systems. In: Science for the protection of Indonesian Coastal Ecosystems (SPICE), pp 407–43

Gong J, Hongyan D, Sun Y, Zhan Y (2023) Simulation and prediction of land use in urban agglomerations based on the PLUS model: a case study of the Pearl River Delta, China. Front Environ Sci 11:1306187. https://doi.org/10.3389/fenvs.2023.1306187

Harahap RH, Mulya MB, Andayani LS (2022) Behavior model of community-based sanitation management in coastal areas: confirmatory factor analysis. Heliyon 8(11):e11756. https://doi.org/10.1016/j.heliyon.2022.e11756

Häder D-P, Banaszak AT, Villafañe VE, Narvarte MA, González RA, Walter Helbling E (2020) Anthropogenic pollution of aquatic ecosystems: emerging problems with global implications. Sci Total Environ 713:136586. https://doi.org/10.1016/j.scitotenv.2020.136586

Harlis TA, Seo SB (2024) Land suitability analysis for blue-green infrastructure implementation in an urban stormwater management system in Surabaya, Indonesia. Water Sci Technol 90(5):1501–1519. https://doi.org/10.2166/wst.2024.265

Henny C, Rohaningsih D, Suryono T, Santoso AB, Waluyo A (2022) Microplastic pollution in the surface water of lake Singkarak, Indonesia. IOP Conf Ser: Earth Environ Sci 1118(1):012050. https://doi.org/10.1088/1755-1315/1118/1/012050

Hua HH, Cremin E, Van Huynh D, Long G, Renaud FG (2024) Impacts of aquaculture practices on the sustainability of social-ecological systems in coastal zones of the Mekong Delta. Ocean Coast Manag 258:107392. https://doi.org/10.1016/j.ocecoaman.2024.107392

Hussainzad EA, Gou Z (2024) Climate risk and vulnerability assessment in informal settlements of the global south: a critical review. Land 13(9):1357. https://doi.org/10.3390/land13091357

Ibrahim R, Akil A, Zubair A, Samudro G, Samudro H, Mangkoedihardjo S (2024) Improving Hasanuddin University lake water quality by controlling contamination sources and biological monitoring systems. J Adv Res Fluid Mech Therm Sci 114(1):166–177. https://doi.org/10.37934/arfmts.114.1.166177

Irmadhiany M, Nikijuluw V, Putra KS, Prasetia R, Burhanuddin B, Wibisono E, Hakim A, Tomasow J, Armanto T (2024) Integrated Coastal Management (ICM) with a reference on protection-production approach: an initial lesson of TWP East Bintan, Kepulauan Riau Province. In: Adrianto L, Damar A, Ramadhan YW, Tarman K, Ferse S, Castrence Gonzales R, Maulidiani, Nitayapat N, Akmal SG, Nhi Cong LT, Narcise CIS (eds) BIO web of conferences, vol 92, pp 01025. https://doi.org/10.1051/bioconf/20249201025

Iskandar MR, Surinati D, Cordova MR, Siong K (2021) Pathways of floating marine Debris in Jakarta Bay, Indonesia. Mar Pollut Bull 169:112511. https://doi.org/10.1016/j.marpolbul.2021.112511

Ismanto A, Hadibarata T, Kristanti RA, Sugianto DN, Widada S, Atmodjo W, Satriadi A, Anindita MA, Al-Mohaimeed AM, Abbasi AM (2023) A Novel report on the occurrence of microplastics in Pekalongan River Estuary, Java Island, Indonesia. Mar Pollut Bull 196:115563. https://doi.org/10.1016/j.marpolbul.2023.115563

Jagemma M, Worku H, Gemeda FT, Gemeda BS (2023) Variability in space effects of industrial wastewater on river physicochemical characteristics in a sample of towns of Oromia. Int J Environ Sci Technol 20(12):13299–13306. https://doi.org/10.1007/s13762-023-04874-0

Jurumai LP, Abidin AA, Soetrisno ET, Wibowo D (2023) Impact of population growth and housing development on the riverine environment: identifying environmental threat and solution in the Wanggu River, Indonesia. Ecol Model 486:110540. https://doi.org/10.1016/j.ecolmodel.2023.110540

Kaschner K, Watson R, Trites A, Pauly D (2006) Mapping world-wide distributions of marine mammal species using a relative environmental suitability (RES) model. Mar Ecol Prog Ser 316:285–310. https://doi.org/10.3354/meps316285

Kawata Y (2022) Green infrastructures in Megacity Jakarta: current status and possibilities of Mangroves for Flood Damage Mitigation. In: Nakamura F (ed) Green infrastructure and climate change adaptation, ecological research monographs. Springer Nature Singapore, Singapore, pp 335–51

Kiran DA, Ramaraju HK (2022) Evaluation and correlation analysis of surface water in and around Mangalore Coastal Belt, Karnataka, India. Asian J Water Environ Pollut 19(1):27–32. https://doi.org/10.3233/AJW220005

Koagouw W, Arifin Z, Olivier GWJ, Ciocan C (2021) High concentrations of paracetamol in effluent dominated waters of Jakarta Bay, Indonesia. Mar Pollut Bull 169:112558. https://doi.org/10.1016/j.marpolbul.2021.112558

Kobayashi M (2023) Promoting sustainable fisheries and aquaculture—enabling policies, innovation and challenges for achieving sustainable blue economies in Ambon, Maluku Province, Indonesia. IOP Conf Ser: Earth Environ Sci 1207(1):012006. https://doi.org/10.1088/1755-1315/1207/1/012006

Kuncorowati PW, Handayani IG, Subekti R (2024) Enforcement of spatial planning laws in controlling spatial use permits in Yogyakarta. Jurnal Civics: Media Kajian Kewarganegaraan 21(1):54–62. https://doi.org/10.21831/jc.v21i1.70161.

Lamond JE (2014) Balancing flood risk and water scarcity of the Asian Delta Regions. In: Booth CA, Charlesworth SM (eds) Water Resources in the Built Environment, pp 389–400. Wiley

Lasut MT, Jensen KR, Shivakoti G (2008) Analysis of constraints and potentials for wastewater management in the coastal city of Manado, North Sulawesi, Indonesia. J Environ Manage 88(4):1141–1150. https://doi.org/10.1016/j.jenvman.2007.06.011

Lestari DEP, Kurniawan T (2018) Community participation in upgrading slum area: study of the implementation of 'Kampung Deret' in Petogogan, Jakarta. IOP Conf Ser: Earth Environ Sci 179:012032. https://doi.org/10.1088/1755-1315/179/1/012032

Lisafitri Y, Setiawati E, Fajar M, Syafrizal M (2021) Identification of sanitation and public health's condition in densely populated settlement at coastal area (Case Study: Kangkung, Bandar Lampung). IOP Conf Ser: Earth Environ Sci 830(1):012089. https://doi.org/10.1088/1755-1315/830/1/012089

Majesty KI, Fadmastuti M (2018) Degree of community participation in Mangrove resources management as livelihood support in West Java, Indonesia. In: Herdiansyah H (ed) E3S Web of conferences, vol 74, pp 10005. https://doi.org/10.1051/e3sconf/20187410005

Mardiansjah FH, Rahayu P, Rukmana D (2021) New Patterns of urbanization in Indonesia: emergence of non-statutory towns and new extended urban regions. Environ Urban ASIA 12(1):11–26. https://doi.org/10.1177/0975425321990384

Maryati S, Humaira AN (2017) Implementation of green infrastructure concept in Citarum watershed. Semarang, Indonesia, pp 020031

Merten J, Nielsen JØ, Rosyani ES, Faust H (2021) From rising water to floods: disentangling the production of flooding as a hazard in Sumatra, Indonesia. Geoforum 118:56–65. https://doi.org/10.1016/j.geoforum.2020.11.005

Ministry of Environment and Forestry, Republic of Indonesia (2020) National Plastic Waste Reduction Strategic Actions for Indonesia

Monteiro R, Ferreira JC, Antunes P (2022) Green infrastructure planning principles: identification of priorities using analytic hierarchy process. Sustainability 14(9):5170. https://doi.org/10.3390/su14095170

Mulyasari TM, Mukono J, Sincihu Y (2023) The presence of microplastics in the indonesian environment and its effects on health. J Public Health Afr 14(2):8. https://doi.org/10.4081/jphia.2023.2565

Najimi C, Bechar S, Cherkaoui E, Khamar M, Nounah A (2025) Potentially toxic elements accumulation in the surface sediment of the Bouregreg Estuary, Morocco: implications for environmental changes and human activities over the last decades. Mar Pollut Bull 213:117708. https://doi.org/10.1016/j.marpolbul.2025.117708

Nalle VI, Syaputri MD, Krisnanto W, Tjandra OC (2023) Public participation in bus transit policymaking: the case of Semarang, Indonesia. Int J Transp Dev Integr 7(3):235–45. https://doi.org/10.18280/ijtdi.070307

Narendra BH, Siregar CA, Dharmawan IW, Sukmana A, Pratiwi, Pramono IB, Basuki TM, Nugroho HY, Supangat AB, Purwanto, Setiawan O (2021) A review on sustainability of watershed management in Indonesia. Sustainability 13(19):11125. https://doi.org/10.3390/su131911125.

Patel D, Verma R, Hait M (2024) Macrophytes as a sustainable biomonitoring tool sustainable of aquatic pollution. In: Izah SC, Ogwu MC, Hamidifar H (eds) Biomonitoring of pollutants in the global south. Springer Nature Singapore, Singapore, pp 391–420

Phung D, Huang C, Rutherford S, Chu C, Wang X, Nguyen M, Nguyen NH, Manh CD, Nguyen TH (2015) Association between climate factors and diarrhoea in a Mekong Delta Area. Int J Biometeorol 59(9):1321–1331. https://doi.org/10.1007/s00484-014-0942-1

Powell N, Osbeck M (2010) Approaches for understanding and embedding stakeholder realities in Mangrove rehabilitation processes in Southeast Asia: lessons learnt from Mahakam Delta, East Kalimantan. Sustain Dev 18(5):260–270. https://doi.org/10.1002/sd.477

Pramaningsih V, Isworo S, Purwanto AD, Kurniawan D (2023) Microplastic contaminant in Indonesia: a review on debris, exposure, health risk and future perspective. IOP Conf Ser: Earth Environ Sci 1282(1):012004. https://doi.org/10.1088/1755-1315/1282/1/012004

Prana AM, Dionisio R, Curl A, Hart D, Gomez C, Apriyanto H, Prasetya H (2024) Informal adaptation to flooding in north Jakarta, Indonesia. Prog Plan 186:100851. https://doi.org/10.1016/j.progress.2024.100851

Purwanto SL, Yonariza FA, Sofiana EI, Ferdiansah MR (2023) Land cover change assessment using random forest and CA Markov from remote sensing images in the protected forest of south Malang, Indonesia. Remote Sens Appl: Soc Environ 32:101061. https://doi.org/10.1016/j.rsase.2023.101061

Purwendah EK, Sasongko NA, Susanto H, Mawardi R, Cahyono T, Susilawati HL, Wahyuni T, Juhandi D, Rahman T, Gustina A, Triana IDS, Pudyastiwi E, Kusumaningsih O, Martini T (2024) The influence of legal compliance in farmer group on the growth and development of sustainable mangrove ecosystem. Glob J Environ Sci Manag 10(3). https://doi.org/10.22034/gjesm.2024.03.26

Rusprayunita NR, Puspitasari S, Mahiroh H, Setyani ER, Citraningtyas VI, Wahyuningsih WS, Rauf AU (2025) Water pollution reduction for sustainable urban development. In: Parray JA, Shameem N, Haghi AK (eds) Sustainable urban environment and waste management, urban sustainability. Springer Nature Singapore, Singapore, pp 1–21

Ruzieva M, Ruziev U (2024) Assessing environmental problems: primary indicators and their role in ecosystem management. In: Yekimov S, Tsipko V (eds) E3S Web of Conferences, vol 587, pp 02012. https://doi.org/10.1051/e3sconf/202458702012

Sahri A, Putra MIH, Mustika PLK, Kreb D, Murk AJ (2021) Cetacean habitat modelling to inform conservation management, marine spatial planning, and as a basis for anthropogenic threat mitigation in Indonesia. Ocean Coast Manag 205:105555. https://doi.org/10.1016/j.ocecoaman.2021.105555

Sari G, Kasasiah A, Utami MR, Trihadiningrum Y (2021) Microplastics contamination in the aquatic environment of Indonesia: a comprehensive review. J Ecol Eng 22(10):127–140. https://doi.org/10.12911/22998993/142118

Sarker T, Fan P, Messina JP, Mujahid N, Aldrian E, Chen J (2024) Impact of urban built-up volume on urban environment: a case of Jakarta. Sustain Cities Soc 105(346). https://doi.org/10.1016/j.scs.2024.105346

She Q, Cao S, Zhang S, Zhang J, Zhu H, Bao J, Meng X, Liu M, Liu Y (2021) The impacts of comprehensive urbanization on pm2.5 concentrations in the Yangtze River Delta, China. Ecol Indic 132:108337. https://doi.org/10.1016/j.ecolind.2021.108337

Sui L, Wang J, Yang X, Wang Z (2020) Spatial-temporal characteristics of coastline changes in Indonesia from 1990 to 2018. Sustainability 12(8):3242. https://doi.org/10.3390/su12083242

Suroso DSA, Firman T (2018) The role of spatial planning in reducing exposure towards impacts of global sea level rise case study: northern Coast of Java, Indonesia. Ocean Coast Manag 153:84–97. https://doi.org/10.1016/j.ocecoaman.2017.12.007

Tjoei A, Kesuma M (2020) North Coast Jakarta Land Reclamation as a Solution of Land Requirements. In: IOP Conference series: materials science and engineering, vol 852, pp 012172. https://doi.org/10.1088/1757-899X/852/1/012172

Wahyuni T, Karmilasanti, Indriyanti SY, Abdurachman (2022) Involvement and roles of stakeholders in Mahakam Delta management to support mitigation and adaptation effort of climate change in East Kalimantan. In: IOP Conference series: earth and environmental science, vol. 976, pp 012028. https://doi.org/10.1088/1755-1315/976/1/012028

Wijaya HB, Buchori I (2023) The origin of industrial workers and rural in situ urbanization in Temanggung Regency, Indonesia. Int J Urban Sci 27(1):112–128. https://doi.org/10.1080/12265934.2022.2097119

Wu S, Guo Z, Askar A, Li X, Yang H, Li H, Saria AE (2024) Dynamic land cover and ecosystem service changes in global coastal deltas under future climate scenarios. Ocean Coast Manag 258:107384. https://doi.org/10.1016/j.ocecoaman.2024.107384

Wulandari R, Iswara AP, Qadafi M, Prayogo W, Astuti RDP, Utami RR, Jayanti M, Dion Awfa I, Suryawan WK, Fitria L, Andhikaputra G (2024) Water pollution and sanitation in Indonesia: a review on water quality, health and environmental impacts, management, and future challenges. Environ Sci Pollut Res 31(58):65967–65992. https://doi.org/10.1007/s11356-024-35567-x

Yamindago R-I-A (2015) Restoring coastal ecosystems—a case study Malang and Gresik Regency, Indonesia. J Coast Conserv 19(2):119–130. https://doi.org/10.1007/s11852-015-0373-0

Zalsabila F, Cangara AR, Burhanuddin A, Baharuddin A, Marifat IDN (2020) The impact of UN-Habitat and ICLEI cooperation through urban-LEDS program towards the development of low-emission cities in Indonesia. IOP Conf Ser: Earth Environ Sci 575(1):012244. https://doi.org/10.1088/1755-1315/575/1/012244

Zanudin K, Ngah I, Misnan SH (2021) Public planner's perception of the community participation in planning and operational decision-making in Iskandar Malaysia. Eng J 25(1):275–285. https://doi.org/10.4186/ej.2021.25.1.275

Zhi L, Li X, Bai J, Shao D, Cui B, Yonglin M, Tiantian Ma X, Xie HB, Abdullahi U (2023) Seawall-induced impacts on large river delta wetlands and blue carbon storage under sea level rise. Sci Total Environ 859:159891. https://doi.org/10.1016/j.scitotenv.2022.159891

Expression of Bioremediation in Sustainable Agriculture System

11

Balwant Singh

Abstract

Integrate bioremediation into the sustainable agriculture systems for innovation in mitigating environmental pollution and enhancing agricultural productivity. This is mainly with respect to the challenges brought by climate variability. Bioremediation harnesses microorganisms, fungi, and plants to metabolically degrade, detoxify, or immobilize heavy metals and organic contaminants such as pesticides, thereby restoring the quality of soil and water. Within the context of delta regions, where agriculture is highly vulnerable to salinity intrusion, flooding, and pollutant accumulation, bioremediation emerges as a vital strategy to strengthen delta resilience by maintaining ecosystem balance and securing long-term agricultural sustainability. This chapter discusses the mechanisms, applications, and potential of bioremediation strategies in agricultural systems, focusing on the role that bioremediation plays in addressing soil and water contamination while promoting crop health and productivity. It further examines the synergy between bioremediation and geospatial technologies, which enable precise monitoring and management of contaminated sites. Case studies and examples are provided to illustrate successful bioremediation practices that support sustainable agricultural development. The chapter concludes by summarizing the challenges, limitations, and future prospects of integrating bioremediation into climate-resilient agriculture systems, underlining its importance in attaining global food security, environmental sustainability, and the resilience of vulnerable delta systems.

B. Singh (✉)
Department of Life Science, Dr. C. V. Raman University, Vaishali, Bihar, India
e-mail: balwantsingh1642@gmail.com

C. Singha et al. (eds.), *Sustainable Management of Delta Ecosystems Resilience*, Deltas of the World, https://doi.org/10.1007/978-3-032-16261-8_11

Abbreviations

GIS	Geographic Information Systems
GPS	Global Positioning Systems
RS	Remote Sensing
PH	Petroleum Hydrocarbons
PCB	Polychlorinated Biphenyls
PAH	Polycyclic Aromatic Hydrocarbons
PGPR	Plant Growth-Promoting Rhizobacteria
PAH	Polycyclic Aromatic Hydrocarbons
AI	Artificial Intelligence
IoT	Internet of Things
UAV	Unmanned Aerial Vehicle
SDG	Sustainable Development Goals
GMO	Genetically Modified organisms

11.1 Introduction

Sustainable agriculture aims to weigh the demand for food production against environmental conservation such that farming systems do not hamper the potential of future generations to satisfy their own needs. One of the most important means of achieving balance is through preserving soil health, which is integral to productive agroecosystems. Conventional farming approaches, however, such as high application rates of chemical fertilizers and pesticides, have resulted in soil degradation and environmental contamination that present serious barriers to sustainability (Barwant et al. 2025a, b).

Bioremediation, or the application of microorganisms to break down or detoxify environmental pollutants, presents a potential solution to these issues. By taking advantage of microbes' natural metabolic functions, bioremediation can reestablish soil health, increase nutrient cycling, and minimize chemical input dependency. This mechanism is in accordance with the values of sustainable agriculture through its encouragement of ecological balance and diminished environmental impacts (Barwant et al. 2025c, d).

Current research in microbes has further widened the scope of agricultural applications of bioremediation. For example, research has proven the efficiency of some microbial strains in degrading toxic pollutants in agricultural land, thus preventing their harmful impacts on humans and animals (Saravanan et al. 2023). In addition, new methods, including the creation of crop varieties that enter into improved alliances with soil microorganisms, have been demonstrated to enhance nutrient acquisition and decrease the demand for chemical fertilizers (Sample 2025).

While these advances have been made, the application of bioremediation in mainstream agriculture is hampered by various issues. These are variability in microbial performance in field conditions, regulatory challenges, and lack of awareness among farmers on the advantages of microbial solutions. Overcoming these issues calls for a collaborative effort from researchers, policymakers, and industry players to design facilitating policies, conduct education and training, and undertake research and development.

In this chapter, we will look at the status of bioremediation in sustainable agriculture today, highlighting both the successes and the challenges of the future. We will review the contributions of research and innovation to the development of microbial technologies and discuss the policy frameworks that will enable their uptake. Through an understanding of the potential and the limitations of bioremediation, we can more effectively position this technology as a pillar of climate-resilient agricultural systems.

11.1.1 Concept of Bioremediation in Agriculture

Bioremediation is the action of utilizing living cells, in large part microorganisms, to disintegrate or detoxify pollution in the environment, leading the polluted land to return to normal. Agriculture defines bioremediation as utilizing soil bacteria for the breakdown of toxic materials such as pesticides, heavy metals, and other waste products produced with intensive use of land for agricultural purposes. This natural attenuation process not only decontaminates the soil but also increases its fertility and structure, resulting in better crop yields and ecosystem health. Recent research has shown the effectiveness of certain microbial strains in breaking down environmental pollutants, thus reducing their toxic impact on humans and animals (Saravanan et al. 2023).

11.1.2 Importance of Sustainable Agricultural Systems

Sustainable agriculture seeks to provide food and fiber demands of the present without compromising the future ability of generations to produce what they need. Such practice focuses on methods that are environmentally sound, economically effective, and socially just. Some of the crucial elements involve soil conservation, water resource conservation, minimizing chemical usage, and enhancing biodiversity. The incorporation of bioremediation in sustainable agricultural systems resolves soil pollution problems, decreases reliance on chemical fertilizers, and promotes ecological equilibrium. For example, the creation of crop varieties that develop improved relationships with soil microorganisms has been proven to enhance nutrient acquisition and lower the application of chemical fertilizers, promoting sustainable agriculture (Singh et al. 2022; Chauhan et al. 2024; Sample 2025).

11.1.3 Role of Geospatial Technologies in Climate-Resilient Agriculture

Geospatial technologies, including GIS, RS, GPS, are crucial for enhancing agricultural practices to become more climate-resilient. The technologies provide accurate mapping and analysis of environmental parameters, enabling well-informed decision-making in fields such as precision irrigation, crop monitoring, and resource management. For instance, artificial intelligence-driven software, if coupled with geospatial data, can help in high-level weather forecasting and soil moisture measurement, benefiting farmers in the most efficient use of water and making their crops climate variability-resistant (Reuters 2025). Moreover, geospatial analysis helps determine where rainwater can be best harvested, supporting agriculture in sustainable water management (Cropway 2024). By applying these technologies, farmers are able to increase productivity and reduce the negative impacts of climate change and, in so doing, enhance world food security (Chauhan et al. 2024).

11.1.4 Research Gap and Novelty

While substantive advances have occurred in the knowledge of bioremediation processes, their incorporation into sustainable agriculture systems has remained largely underdeveloped, especially in the context of climate variability and deltaic ecosystem resilience. Past research has mainly addressed laboratory-scale or site-specific use, with scant attention being paid to the scale-up of bioremediation approaches for large agroecosystems, nor the application of advanced monitoring methods such as UAV surveys and geospatial technologies. In addition, although the potential of microorganisms, fungi, and vegetation in cleansing pollutants has been amply documented, no complete frameworks exist to showcase how these biological approaches can be incorporated systematically into climate-resilient agriculture for improved crop productivity, water quality, and soil health.

The originality of this chapter is that it takes a systems perspective in positioning bioremediation as not only a remediation method but a fundamental aspect of sustainable agriculture. By interrelating bioremediation with ecosystem services, technological advancements, and climate-resilience interventions, the chapter offers an original approach to how biological processes can be involved in long-term agricultural sustainability and food security, particularly in at-risk deltaic areas.

11.2 Fundamentals of Bioremediation

Bioremediation is an environmentally friendly method that utilizes biological agents to detoxify and restore environments that have been polluted by harmful substances. This method utilizes the natural metabolic activities of microorganisms, plants, and fungi to break down, convert, or stabilize toxic substances, thus reducing environmental pollution (Garbisu and Alkorta 2001).

11.2.1 Mechanisms of Bioremediation

Bioremediation is a cost-effective and environmentally friendly method that employs biological organisms, primarily microorganisms, fungi, and plants, to detoxify and eliminate contaminants from the environment. It is an important component of sustainable agriculture in that it prevents soil and water pollution, protects crops, and restores ecosystem equilibrium. The six primary mechanisms of bioremediation: biodegradation, biotransformation, bioaccumulation, biomineralization, biosorption, and bioleaching, each of which plays a distinct role in contaminant removal (Fig. 11.1).

11.2.1.1 Biodegradation

Biodegradation is the enzymatic degradation of organic contaminants by microorganisms into less complex, non-toxic substances. Bacteria, fungi, and actinomycetes produce hydrolytic and oxidative enzymes to break down complex hydrocarbons, pesticides, and industrial compounds (Singh and Ward 2021). The process may be aerobic, in which oxygen-requiring enzymes catalyze the breakdown, or anaerobic, in which electron acceptors like nitrate, sulfate, or carbon dioxide support decomposition. Biodegradation in agriculture is necessary to degrade pesticide residues, petroleum hydrocarbons, and synthetic fertilizers, hence lowering environmental toxicity. Microorganisms degrade organic pollutants enzymatically into less toxic substances, finally into carbon dioxide, water, and biomass. It is central to the remediation of hydrocarbons and pesticides (Vidali 2001).

11.2.1.2 Biotransformation

Biotransformation entails microbial transformation of contaminants to less toxic or more extractable forms (Juwarkar et al. 2010). In contrast to complete mineralization, in which pollutants are completely degraded to CO_2 and water, biotransformation alters the chemical composition of contaminants. For instance, certain bacteria are capable of transforming toxic chromium (Cr^{6+}) into less toxic Cr^{3+}, which is less mobile and bioavailable. This process is especially useful for the cleanup of heavy metal contamination in soils and water of agricultural fields. Plants and microbes modify the chemical makeup of contaminants to make them less toxic or more susceptible to further breakdown (Garbisu and Alkorta 2001).

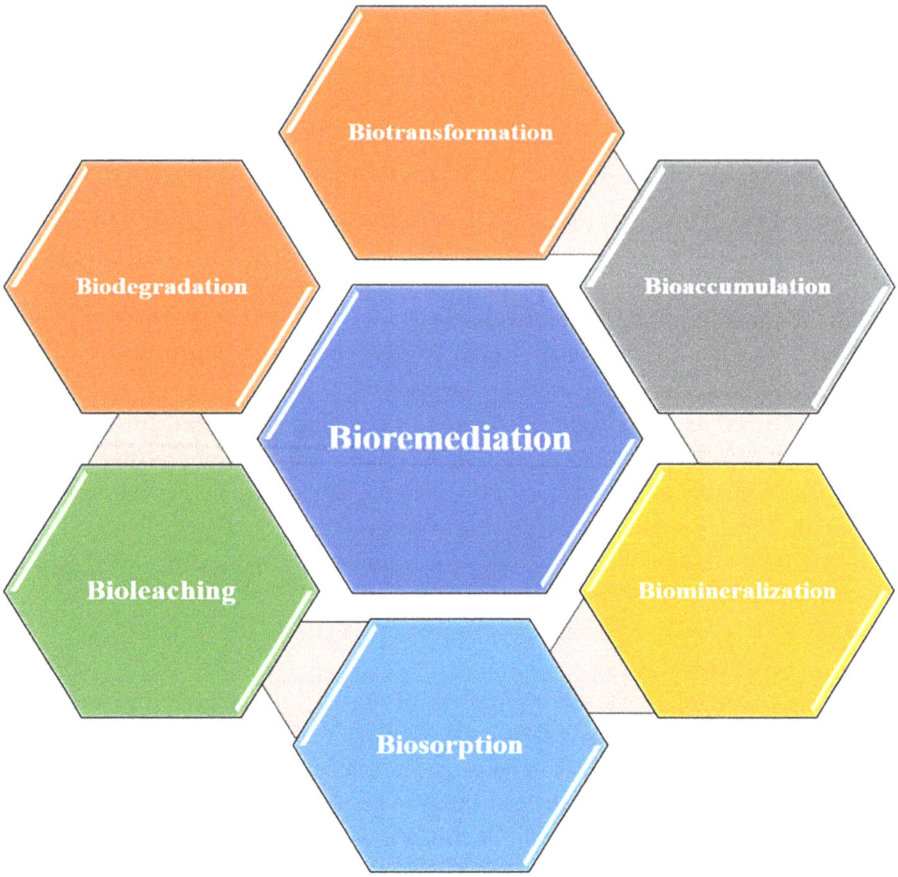

Fig. 11.1 Different mechanism involves in bioremediation

11.2.1.3 Bioaccumulation

Bioaccumulation refers to the uptake and storage of pollutants, especially heavy metals, by living organisms from their environment. Vegetation, phytoplankton, fungi, and microorganisms can take up contaminants like arsenic, cadmium, and lead into their tissues (Gadd 2010). Hyperaccumulator plants, such as Brassica juncea (Indian mustard) and Helianthus annuus (sunflower), are often employed in phytoremediation of metal-polluted agricultural fields. This approach bars the incorporation of harmful elements into the food chain while allowing metal recovery for industrial recycling. Some organisms take up and concentrate contaminants from their surroundings into their tissues. This process is especially applicable to heavy metals and radionuclides (Gadd 2004).

11.2.1.4 Biomineralization

Biomineralization is the biological conversion of metals to stable mineralized phases. Microorganisms like Pseudomonas and Bacillus species enable the precipitation of heavy metals as insoluble phosphates, sulfides, or carbonates, thus limiting their bioavailability (Ngwenya et al. 2010). In agriculture, biomineralization aids soil stabilization and inhibits the leaching of toxic constituents into groundwater. The process is essential for rehabilitating mine-contaminated soils and industrial waste areas.

11.2.1.5 Biosorption

Biosorption is a physicochemical, passive process whereby dead or live biomass adsorbs heavy metals and organic pollutants onto its surface (Vijayaraghavan and Yun 2008). Bacteria, fungi, and algae contain cell wall structures such as chitin, cellulose, and peptidoglycans, which have functional groups (e.g., hydroxyl, carboxyl, and amino groups) that are able to bind metal ions. The method is commonly applied in the treatment of wastewater from agricultural and industrial sources because it is efficient and economical.

11.2.1.6 Bioleaching

Bioleaching employs the metabolic processes of microorganisms to dissolve metals from solid substrates, rendering them bioavailable for recovery or removal (Rawlings and Johnson 2007). Acidophilic microorganisms like Acidithiobacillus ferrooxidans oxidize metal sulfides, liberating metals into solution. In sustainable agriculture, bioleaching is used to demetalize metal-polluted soils, increase nutrient recycling, and recover valuable metals from agricultural waste products. The introduction of nutrients or electron acceptors/donors promotes indigenous microbial activity, speeding the degradation process (Azubuike et al. 2016) (Table 11.1).

11.2.2 Types of Bioremediations

Bioremediation encompasses various strategies, notably-

11.2.2.1 Microbial Remediation

Utilizes bacteria, fungi, and archaea to degrade organic pollutants. For example, *Pseudomonas* species are renowned for their capacity to decompose hydrocarbons (Das and Chandran 2011).

11.2.2.2 Phytoremediation

Involves the use of plants to extract, stabilize, or degrade contaminants. Mechanisms include phytoextraction, where plants like *Brassica juncea* accumulate heavy metals in their biomass, and phytodegradation, wherein plants metabolize organic pollutants (Ali et al. 2013).

Table 11.1 Mechanisms of bioremediation

Mechanism	Description	Example pollutants	References
Biodegradation	Breakdown of organic pollutants by microbial enzymatic action into less toxic or non-toxic forms	Hydrocarbons, pesticides, chlorinated solvents	Vidali (2001), Das and Chandran (2011)
Bioaccumulation	Absorption and accumulation of pollutants (especially heavy metals) by microorganisms, fungi, and plants	Heavy metals (lead, mercury, cadmium)	Gadd (2004), Ali et al. (2013)
Biotransformation	Microbial or plant-mediated conversion of toxic pollutants into less harmful or more manageable forms	Heavy metals, organic pollutants	Garbisu and Alkorta (2001), Harms et al. (2011)
Biostimulation	Enhancement of microbial activity by adding nutrients, oxygen, or electron acceptors	Petroleum hydrocarbons, polychlorinated biphenyls (PCBs)	Azubuike et al. (2016)

11.2.2.3 Mycoremediation

Employs fungi to degrade complex organic pollutants. White-rot fungi, such as *Phanerochaete chrysosporium*, produce ligninolytic enzymes capable of breaking down persistent organic pollutants like polycyclic aromatic hydrocarbons (Harms et al. 2011) (Table 11.2).

11.2.3 Key Microbial and Plant Species Involved

Several species are pivotal in bioremediation efforts (Table 11.3).

11.3 Soil and Water Contaminants in Agriculture

Soil and water pollution in agricultural systems are becoming an increasing concern worldwide because of the long-term buildup of harmful chemicals that are posing risks to the sustainability of the environment and food security. Different contaminants such as heavy metals, residues of pesticides, agrochemicals, industrial pollutants, and new emerging contaminants negatively impact soil fertility, microbial diversity, and crop yields. Integrated management of the contaminants using bioremediation measures to renew soil health and agricultural sustainability is needed (Ali et al. 2019; Kumar et al. 2022a, b).

Table 11.2 Types of bioremediations

Type of bioremediation	Description	Example pollutants	References
Microbial remediation	Utilizes bacteria, fungi, and archaea to degrade or detoxify pollutants through enzymatic activity	Hydrocarbons, pesticides, chlorinated solvents	Das and Chandran (2011), Azubuike et al. (2016)
Phytoremediation	Uses plants to extract, stabilize, degrade, or volatilize contaminants from soil, water, or air	Heavy metals, petroleum hydrocarbons, organic pollutants	Ali et al. (2013), Garbisu and Alkorta (2001)
Mycoremediation	Employs fungi to break down complex organic pollutants using extracellular enzymes like lignin peroxidases	Polycyclic aromatic hydrocarbons (PAHs), synthetic dyes, heavy metals	Harms et al. (2011), Pointing (2001)

Heavy metals like lead (Pb), cadmium (Cd), arsenic (As), mercury (Hg), and chromium (Cr) find their way into agricultural soils via wastewater irrigation, overuse of fertilizers, and industrial effluents. The metals are stable in the environment because they are non-biodegradable, resulting in bioaccumulation in plants and subsequent introduction into the food chain (Azhar et al. 2021). Chronic heavy metal exposure interferes with plant metabolism, leading to oxidative stress, lowered photosynthesis, and impaired nutrient uptake, finally lowering crop yield and quality (Mangal et al. 2023). Organophosphate, carbamate, and pyrethroid pesticide residues add to the problem of soil contamination. Even though they play a pest-controlling role, overuse of pesticides leads to soil degradation, microbial toxicity, and groundwater pollution (Zhao et al. 2021). The persistence of pesticide residues in crops results in bioaccumulation, which creates health hazards to consumers and promotes loss of biodiversity (Gulati et al. 2022).

Synthetic fertilizers, herbicides, and fungicides are pollutants of agrochemicals that change the composition of soil microorganisms and hinder nutrient cycling. Fertilizers' excess nitrogen and phosphorus cause eutrophication of aquatic ecosystems, resulting in fish kills and oxygen depletion (Liu et al. 2020). Correspondingly, industrial pollutants such as petroleum hydrocarbons, polychlorinated biphenyls (PCBs), and polycyclic aromatic hydrocarbons (PAHs) contaminate agricultural ecosystems via atmospheric deposition and effluent release. Such pollutants are highly persistent, bioaccumulate in plant tissues, and have toxic impacts on beneficial soil microorganisms, contaminating soil fertility and plant growth (Bhatt et al. 2022).

Table 11.3 Key microbial and plant species involved in bioremediation

Biological agent	Category	Role in bioremediation	Example pollutants	References
Pseudomonas putida	Bacteria	Degrades aromatic hydrocarbons and petroleum contaminants	Hydrocarbons, benzene, toluene	Das and Chandran (2011), Azubuike et al. (2016)
Deinococcus radiodurans	Bacteria	Tolerates extreme radiation and remediates radioactive waste	Radioactive metals (uranium, cesium)	Brim et al. (2000)
Trametes versicolor	Fungi	Decomposes synthetic dyes and endocrine-disrupting chemicals using ligninolytic enzymes	Synthetic dyes, pharmaceuticals	Pointing (2001)
Phanerochaete chrysosporium	Fungi	Produces lignin peroxidases that degrade persistent organic pollutants	Polycyclic aromatic hydrocarbons (PAHs), polychlorinated biphenyls (PCBs)	Harms et al. (2011)
Brassica juncea (Indian mustard)	Plant	Accumulates and removes heavy metals from contaminated soils	Lead, cadmium, arsenic	Ali et al. (2013), Garbisu and Alkorta (2001)
Populus deltoides (Eastern cottonwood)	Plant	Uptakes and metabolizes organic pollutants in groundwater	Trichloroethylene (TCE), heavy metals	Newman et al. (1997)

Emerging contaminants such as pharmaceuticals, microplastics, and endocrine-disrupting chemicals have only recently drawn interest as a result of their widespread presence in agricultural systems. Medicines like hormones and antibiotics find their way into soil and water pathways via livestock manure and wastewater irrigation, leading to antimicrobial resistance and soil microbial dysbiosis (Chen et al. 2023). Another serious issue is microplastic contamination, which changes soil texture, decreases water holding capacity, and inhibits root growth (Zhou et al. 2021). Moreover, endocrine-disrupting chemicals also influence plant hormonal equilibrium, causing growth abnormalities and lowered crop yields (Tudi et al. 2021).

To counteract the negative impacts of these contaminants, bioremediation methods like microbial degradation, phytoremediation, and mycoremediation provide holistic and environmentally friendly options. The use of bioremediation methods improves soil microbial diversity, enables pollutant breakdown, and redeems soil health for sustainable agricultural production (Wang et al. 2023). Integrated soil and water management practices together with sustainable agriculture are crucial to reducing contamination risks as well as food safety in the event of environmental problems.

Heavy metals and pesticide residues are the most persistent soil pollutants in agricultural soils, and they pose significant threats to soil health, crop yields, and food safety. Heavy metals persist in the soil as a result of industrial discharges, excessive use of fertilizers, and irrigation with wastewater (Ali et al. 2022a, b). In contrast to organic pollutants, heavy metals are non-biodegradable and remain in the environment, causing bioaccumulation and biomagnification along the food chain (Islam et al. 2022). Heavy metal toxicity in plants is expressed in several forms, such as oxidative stress, nutrient uptake disruption, inhibition of photosynthesis, and diminished enzymatic activity, which culminate in yield loss and deteriorated food quality (Mangal et al. 2023).

Pesticide residues of organophosphates, carbamates, neonicotinoids, and pyrethroids exacerbate water and soil pollution. Although pesticides are important for controlling pests and weeds, their overuse and non-regulatory application result in environmental persistence and non-target toxicity (Gulati et al. 2022). Most pesticides are slowly broken down and tend to concentrate in soil and groundwater, where they shift microbial community structure and decrease biodiversity (Zhao et al. 2021). Prolonged exposure to pesticides in agricultural ecosystems has been associated with the loss of pollinators, the development of pesticide-resistant insects, and human health hazards due to endocrine disruption and carcinogenicity (Tudi et al. 2021).

Heavy metal and pesticide residue bioremediation is a promising strategy for reducing their environmental footprint. Microbial remediation with metal-resistant microorganisms like *Pseudomonas*, *Bacillus*, and *Rhizobium* species allows the transformation and sequestration of toxic metals and lowers their bioavailability (Bhatt et al. 2022). Likewise, phytoremediation with hyperaccumulator plants like *Brassica juncea* (Indian mustard) and *Helianthus annuus* (sunflower) allows for the extraction and stabilization of heavy metals from polluted soils (Ali et al. 2019). Biodegradation of pesticides by fungal and bacterial genera like *Trichoderma harzianum* and *Pseudomonas putida* enhances the process of toxic residue decomposition, enhancing soil recovery (Chen et al. 2023). Combination of bioremediation practices with sustainable agriculture is necessary to help reduce the risk of contamination, provide food safety, and improve climate-resilient agricultural production systems.

11.4 Bioremediation Strategies for Sustainable Agriculture

Bioremediation is an eco-friendly strategy that uses biological agents like microorganisms, fungi, and plants to purify contaminated soil and water bodies in agricultural ecosystems. The mounting accumulation of heavy metals, pesticide residues, and organic pollutants makes it imperative to adopt bioremediation strategies to improve soil fertility, crop yields, and ecological safety (Bhatt et al. 2022). The three main bioremediation strategies—microbial bioremediation, phytoremediation, and mycoremediation provide effective solutions for sustainable agriculture systems.

Microbial bioremediation uses PGPR and fungi to decontaminate polluted soils by degrading, immobilizing, or transforming harmful compounds. PGPR like *Pseudomonas*, *Bacillus*, and *Rhizobium* induce metal tolerance in plants through siderophores, organic acids, and metallothioneins production that enhance sequestration of heavy metals and bioavailability decline (Vimal et al. 2022; Singh et al. 2024; Barwant et al. 2025c). Moreover, PGPR-mediated systemic plant resistance improves the tolerance of plants to pollutant-induced oxidative stress, enabling sustainable plant–microbe relationships in polluted ecosystems (Kour et al. 2023). Fungal genera like *Aspergillus niger* and *Trichoderma harzianum* are involved in pesticide, hydrocarbon, and industrial pollutant biodegradation using enzymatic processes that include laccases, peroxidases, and hydrolases (Kumar et al. 2022a, b). These microbial communities efficiently minimize environmental toxicity and can restore balance in soil microbiota.

Phytoremediation is a successful green technology where plants serve as natural remediators by taking up, stabilizing, or breaking down contaminants. Hyperaccumulator plant species like Brassica juncea (Indian mustard), *Helianthus annuus* (sunflower), and *Vetiveria zizanioides* (vetiver grass) are widely utilized for heavy metal phytoextraction (Ali et al. 2022a, b). The trees of Populus and Salix genera have the capacity for phytostabilization, which sequesters the contaminants in the soil from percolating leachate towards groundwater (Bhagwat et al. 2023). Phytodegradation under the catalytic action of the enzymes present within plants, for example, dehalogenases and oxidoreductases, degrades the organic pollutants polycyclic aromatic hydrocarbons (PAHs) and pesticides into nontoxic metabolites (Gupta et al. 2021). The combination of phytoremediation with biofertilizers and soil amendments increases plant tolerance and remediation efficiency.

Mycoremediation utilizes the potential of fungi to degrade and detoxify environmental toxins by enzymatic and biosorption processes. Fungi like *Pleurotus ostreatus* (oyster mushroom), Trametes versicolor, and *Phanerochaete chrysosporium* secrete ligninolytic enzymes, laccases, manganese peroxidases, and lignin peroxidases (Singh and Singh 2023), which are effective in degrading complex organic pollutants like pesticides, dyes, and pharmaceuticals (Singh et al. 2023a). The large surface area of fungal mycelia makes it possible to effectively biosorb heavy metals, immobilizing them and lowering their toxicity and mobility in soil (Tigini et al. 2022). The coupling of mycoremediation with biochar and organic amendments is also more efficient by enhancing the colonization of fungi and bioavailability of pollutants (Liu et al. 2023).

Combining these bioremediation techniques with agricultural practices provides a sustainable solution to mitigate pollution and restore soil health. The combined application of PGPR, hyperaccumulator plants, and mycoremediating fungi has the potential to improve the efficiency of remediation processes for long-term ecological gains. Improved omics technologies and microbial engineering have the potential to optimize the bioremediation pathways and enhance their applications in climate-resilient agriculture.

11.5 Technological Innovations in Bioremediation

Biotechnological advancements in bioremediation have significantly improved the efficacy and accuracy of pollutant degradation, soil detoxification, and ecosystem recovery. The use of genetic engineering, artificial intelligence (AI), the Internet of Things (IoT), and geospatial mapping has transformed bioremediation methods to allow real-time monitoring, optimization, and large-scale application. These technologies provide environmentally friendly solutions for the mitigation of environmental contamination in agricultural environments, ensuring long-term ecological stability (Jaiswal et al. 2023a, b).

Genetic engineering has become a promising means of accelerating microbial and plant-based bioremediation systems. Engineered microorganisms like recombinant Pseudomonas putida and Escherichia coli have enhanced capacity for the degradation of organic contaminants, including pesticides and hydrocarbons, through the expression of catabolic genes that belong to xenobiotic degradation pathways (Singh et al. 2022). The application of metal-chelating proteins like metallothioneins and phytochelatins into bacterial and fungal strains is found to improve their potential for immobilization and detoxification of heavy metals (Sharma et al. 2022a, b). Additionally, transgenic plants with stress-inducible genes like AtPCS1 (phytochelatin synthase) and MT1 (metallothionein) exhibit enhanced phytoremediation potential for heavy metal accumulation and detoxification (Ali et al. 2023a, b). Bioremediation has been further evolved by CRISPR-Cas9-based genome editing to produce targeted changes in microbial metabolic routes for improved efficiency and adaptability of pollutant degradation (Gul et al. 2023).

The combination of AI and IoT has revolutionized bioremediation monitoring and process optimization by enabling real-time monitoring, predictive modeling, and automation. Machine learning algorithms developed by AI are applied to large datasets for predicting microbial activity, pollutant degradation rates, and environmental factors affecting bioremediation efficiency (Xu et al. 2023a, b). IoT-enabled biosensors integrated with nanotechnology-based sensors allow for in-situ monitoring of heavy metal concentrations, microbial metabolic functions, and soil health factors to ensure effective and adaptive bioremediation measures (Huang et al. 2022a). Decision support systems empowered by AI support optimizing the choice of microbial consortia, supplementation of nutrients, and remediation site management to ensure enhanced effectiveness (Kumar et al. 2023a, b). The integration of AI and IoT supports remote observation of bioremediation areas with minimal human interventions, promoting consistent performance evaluation and risk prevention.

Geospatial mapping has proven to be an essential tool for accurate bioremediation through the combination of remote sensing, GIS (Geographic Information Systems), and satellite imagery to detect hotspots of contamination and monitor the progress of remediation. UAV surveys and hyperspectral imaging at high resolution deliver spatially explicit records of contamination levels in soil, vegetation status, and microbial activity on contaminated sites (Wang et al. 2022a, b). GIS spatial modeling supports the design of bioremediation strategies through forecasting pollutant dispersion patterns, microbial colonization behavior, and ecological restoration potential (Zhao et al. 2023a, b). Coupling geospatial technologies with AI analytics allows adaptive decision-making for maximizing resource allocation and remediation effectiveness in large-scale agricultural environments.

The intersection of genetic engineering, AI, IoT, and geospatial mapping in bioremediation has opened the door for a new era of high-precision environmental management. Such technologies improve the scalability, efficiency, and sustainability of bioremediation processes, which provide promising prospects for the mitigation of agricultural pollution while promoting ecosystem resilience. Future investigations should aim at combining multi-omics technologies, AI-based metabolic modeling, and blockchain-based data management to further improve bioremediation strategies for climate-resilient agricultural systems.

11.6 Integration of Bioremediation in Climate-Resilient Agriculture

The incorporation of bioremediation in climate-resilient agriculture is necessary to counteract the negative impacts of environmental pollutants as well as for improving the sustainability of agriculture. Climate change has widely impacted the effectiveness of bioremediation techniques by modifying microbial function, altering soil quality, and contaminant degradation dynamics. Sustainable agriculture management practices, with the support of policy structures, can maximize bioremediation process efficiency for long-term ecological stability (Ghosh et al. 2023).

Climate change impacts the effectiveness of bioremediation by altering temperature, precipitation patterns, and moisture levels in the soil, all of which directly impact microbial metabolism and pollutant breakdown. Higher temperatures increase microbial activity but can volatilize organic contaminants, decreasing availability for microbial breakdown (Singh et al. 2023b). Increased frequency of intense weather conditions, including droughts and floods, destabilizes microbial communities and leaches pollutants into groundwater, making bioremediation more challenging (Wang et al. 2022a, b).

Climate-related changes in soil pH and redox potential also influence the bioavailability and transformation of heavy metal and organic contaminants (Kumar et al. 2023a, b). Climate-resilient microbial consortia and bioengineered strains with improved adaptive traits are essential for maintaining bioremediation effectiveness under fluctuating climatic conditions (Huang et al. 2022b).

Sustainable farming practices are essential for pollutant management using integrated bioremediation strategies. Cover cropping, conservation tillage, and organic matter amendments increase soil microbial diversity, enhancing the natural attenuation of pollutants (Verma et al. 2023). Agroforestry systems with phytoremediation plants, including Populus spp. and Brassica spp., sequester heavy metals and enhance carbon sequestration and soil structure (Ali et al. 2023a, b). The addition of biochar and composted organic amendments increases microbial populations and pollutant breakdown through provision of nutrients and shelter for pollutant-degrading fungi and bacteria (Xu et al. 2023a, b). The inoculation with plant growth-promoting rhizobacteria (PGPR) and mycorrhizal fungi also enhances pollutant detoxification in addition to increasing plant resistance to environmental stresses (Sharma et al. 2022a, b). The integration of AI-based monitoring systems allows for the accuracy in the application of bioremediation approaches by evaluating levels of soil pollution, microbial functions, and environmental conditions in real-time (Jaiswal et al. 2023a, b).

A strong policy and regulatory system is essential to mainstream bioremediation in climate-resilient agricultural production systems. National and global policies must position bioremediation as a principal element of sustainable land management and environmental restoration initiatives (Kumar et al. 2023a, b). The Green Deal of the European Union and the Sustainable Development Goals (SDGs) of the United Nations focus on minimizing soil pollution through nature-based approaches, including bioremediation and phytoremediation (Wang et al. 2022a, b). Encouraging farmers to implement bioremediation measures by offering subsidies, carbon credits, and payments for ecosystem services can speed up the shift to sustainable agriculture (Huang et al. 2022a). Enhancing regulatory mechanisms for monitoring agrochemical use, industrial effluent, and land contamination is required to facilitate the successful application of bioremediation measures. The adoption of geospatial technologies and AI-driven risk assessment models can enhance policy enforcement by providing real-time data on pollution hotspots and remediation progress (Zhao et al. 2023a, b).

Effective integration of bioremediation in climate-resilient agriculture demands a multi-faceted strategy involving scientific research, sustainable management, and policy measures. The future agenda should include the design of climate-resilient microbial consortia, bioengineered crops with increased detoxification, and AI-driven decision support systems for maximizing bioremediation strategies under fluctuating environmental scenarios. Policy reinforcement and economic incentives will also further promote the widespread use of bioremediation, leading to sustainable agriculture and ecosystem restoration.

11.7 Delta Ecosystem Management

Delta ecosystems, with their intricate web of rivers, wetlands, estuaries, and floodplains, are agriculturally and ecologically important landscapes. Delta ecosystems promote biodiversity, supply water for irrigation, and contribute to food production, especially in high-density and climate-sensitive areas like the Ganges–Brahmaputra–Meghna Delta, Mekong Delta, and Nile Delta. Nonetheless, the enhancement of agriculture, agrochemical pollution, sedimentation, and sea-level rise due to climate change endanger the ecological stability and productivity of these systems (Syvitski and Saito 2007).

Successful delta ecosystem management means ensuring environmental harmony alongside facilitating sustainable agricultural production. One of the possible interventions is incorporating bioremediation technologies for avoiding pollution and reclaiming degraded land cover in delta areas. Native microbial consortia and phytoremediating plants can bioremediate heavy metal, pesticide, and excess nutrient contaminated soils, thus increasing the resilience of agricultural systems in deltas (Dotaniya et al. 2016).

11.7.1 Challenges in Delta Ecosystem Management

The primary issues with the management of delta ecosystems are intrusion of salinity, loading of nutrients, hydrological alterations, and destruction of wetlands through infrastructure construction. These are compounded by global climate change, which impacts precipitation patterns and raises the occurrence of high-intensity weather events. Saltwater intrusion in the Mekong and Sundarbans deltas drastically constrains crop diversity and yield potential, for instance (Anthony et al. 2015).

Conventional remediation methods like dredging and chemical treatment are unsustainable and environmentally harmful. Conversely, bioremediation is a natural solution that maintains nutrient cycling and soil quality without adding new pollutants. To be effective, however, bioremediation needs localized application that respects the specific hydrology, soil type, and pollutant profile of each delta region (Gupta et al. 2022).

11.7.2 Role of Geospatial Tools in Delta Management and Future

Geospatial technologies such as remote sensing, GIS, and drone-based surveys play a critical role in monitoring and managing delta ecosystems. These technologies facilitate real-time monitoring of land use dynamics, pollution hotspots, and vegetative cover, which are essential for planning and implementing bioremediation interventions (Kumar et al. 2021). Additionally, spatial models can be used to forecast degradation scenarios under varying land use and climate futures, supporting proactive ecosystem management.

Delta ecosystem management has to embrace integrated watershed strategies that integrate bioremediation, agroecology, and climate adaptation. Site-specific microbial formulations and salt-tolerant phytoremediators can be formulated specifically for deltaic environments that can make bioremediation much more effective. Policy support and community involvement are also crucial to make interventions sustainable in the long term. In the context of climate-resilient agriculture, delta ecosystem management is not only an environmental issue but a foundation for food and water security for millions. Leverage the synergistic potential of bioremediation and geospatial technologies provides an alternative route to reclaim delta ecosystems while enhancing sustainable agricultural productivity.

Delta regions are among the most vulnerable agricultural zones due to combined pressures of climate change (sea-level rise, flooding, salinization), population growth, pollution load, and loss of natural ecosystem buffering. In this context, bioremediation within sustainable agricultural systems offers multiple pathways to strengthen delta resilience and sustainability. Below are the key contributions, along with supporting evidence.

11.8 Limitations

Despite its promising prospects as a green technology in agriculture, there are various limitations that limit its large-scale suitability. The major restriction is the environmental condition variability and uncertainty—like pH, temperature, and soil texture—that can have critical impacts on the effectiveness of microbial and phytoremediation agents (Ghosh and Singh 2005). Additionally, bioavailability of contaminants tends to restrict the degradation process in soils with complicated pollutant mixtures or hydrophobic compounds (Adams et al. 2015).

Another limitation is the rate at which bioremediation is slower compared to physicochemical processes. Microorganisms take time to settle, grow, and activate enzymatic pathways that break down contaminants. This makes bioremediation less desirable for emergency contamination incidents or high-toxicity situations (Strong and Burgess 2008). An added concern are biosafety issues with the application of genetically engineered organisms (GMOs) or foreign microbial strains in open agricultural settings that are mostly unresolved, causing ecological and ethical issues (Sayler et al. 2012).

11.9 Research Gaps

Previous research has not fully elucidated the molecular processes and pathways of pollutant degradation under natural conditions. Most research is confined to laboratories or greenhouse environments, without validation in practical agricultural ecosystems (Kuppusamy et al. 2016). Additionally, there is a wide knowledge gap concerning synergistic interactions in plant-microbe-soil systems, particularly under stress caused by climate change.

There is also insufficient integration of geospatial technologies and remote sensing capabilities to monitor, predict, and optimize bioremediation operations in different agroecological zones. Such integration has the potential to provide real-time decision support, but is not effectively utilized in present agricultural operations (Paz et al. 2020). There is also a deficiency of long-term studies assessing the influence of bioremediation operations on soil health, crop yield, microbial diversity, and ecosystem resilience.

11.10 Future Scope

Future studies must target multi-omics strategies—such as genomics, proteomics, and metabolomics—to decode the functional capacities of bioremediation-associated microbial communities. This will enable the design of specific microbial consortia with increased throughput in the degradation of pollutants (Khan et al. 2019). Likewise, synthetic biology methods can be utilized to construct future-generation microbial agents that are efficient and safe for field deployment.

The use of geospatial analytics and machine learning algorithms can transform site evaluation, risk estimation, and bioremediation process optimization. These technologies together with IoT sensor real-time monitoring will improve the accuracy and scale of bioremediation in agriculture (Kumar et al. 2021). In addition, there is vast potential in policy-making and farmer training programs for promoting the use of bioremediation as a climate-resilient crop management practice. Interdisciplinary interactions among microbiologists, agronomists, environmental engineers, and data scientists will be critical for achieving the maximum potential of bioremediation in agroecological systems.

11.11 Conclusion

Bioremediation has become a key strategy for reducing soil and water pollution in agricultural ecosystems, leading to sustainable agricultural systems and ecosystem recovery. The combination of microbial, phytoremediation, and mycoremediation methods has shown great promise in decontaminating agroecosystems while enhancing soil quality and crop yields. The efficiency of bioremediation depends on environmental factors, pollutant properties, and the metabolic potential of bioremediating organisms. Newer biotechnological technologies, including genetic engineering, AI-based monitoring, and geospatial mapping, have further increased the effectiveness and accuracy of bioremediation operations, leading the way towards climate-resilient agriculture.

The rising trend of heavy metal pollution, pesticide residues, and emerging contaminants requires adopting integrated bioremediation approaches backed by sustainable agriculture practices. Conservation agriculture, biofertilizer, and organic amendments enhance soil microbial diversity and increase degradation of the pollutant. The application of plant growth-promoting rhizobacteria (PGPR) and mycorrhizal fungi has proved effective in improving the tolerance of plants to toxic contaminants while increasing nutrient acquisition and soil fertility. Phytoremediation involving the use of hyperaccumulator plants such as Brassica juncea and Populus spp. provides an inexpensive and sustainable method for the sequestration of heavy metals in polluted soils.

Even with extensive progress, various issues still need to be addressed in large-scale bioremediation in agriculture. Variations in temperature, precipitation, and soil moisture due to climate change influence microbial growth and the rates of pollutant degradation, prompting the establishment of climate-resistant microbial consortia. Furthermore, a lack of unified regulatory frameworks and economic incentives for farmers inhibits the use of bioremediation measures. Enhancing policy support, offering economic incentives, and incorporating AI-based decision support systems will be essential in maximizing bioremediation strategies for long-term sustainability.

Bioremediation is a critical methodology for mainstreaming sustainability within contemporary agricultural systems. Using microorganisms, fungi, and plants to break down or immobilize pollutants, bioremediation restores soil and water quality and enhances crop yields. In vulnerable deltaic areas, bioremediation reinforces resilience through reducing pollution, enhancing soil quality, and sustaining ecosystem equilibrium in the context of climate variability. Its affordability, flexibility, and compatibility with precision technologies like UAV surveys and geospatial monitoring complement its promise for widespread application. The results of this chapter emphasize that bioremediation is not merely a remediation technique but a pillar of climate-resilient agriculture, with important implications for agricultural development, environmental sustainability, and food security.

Future studies should aim to improve microbial and plant-based bioremediation efficiency using genetic engineering and adaptive breeding methods. The creation of bioengineered microbes and transgenic plants with improved pollutant degradation capacity will be important in solving persistent contaminants in agricultural environments. In addition, utilization of AI and IoT technologies for real-time monitoring and predictive modeling will facilitate precision bioremediation, ensuring efficient pollutant management in varied agro-climatic conditions.

References

Adams GO, Fufeyin PT, Okoro SE, Ehinomen I (2015) Bioremediation, biostimulation and bioaugmentation: a review. Int J Environ Bioremediation Biodegradation 3(1):28–39. https://doi.org/10.12691/ijebb-3-1-5

Ali H, Khan E, Ilahi I (2019) Environmental chemistry and ecotoxicology of hazardous heavy metals: Environmental persistence, toxicity, and bioaccumulation. J Chem 2019:6730305. https://doi.org/10.1155/2019/6730305

Ali H, Khan E, Ilahi I (2022a) Phytoremediation of heavy metals: mechanisms and applications. Environ Sci Pollut Res 29(16):24798–24812. https://doi.org/10.1007/s11356-022-19149-6

Ali H, Khan E, Ilahi I (2023a) Phytoremediation strategies for sustainable agriculture: advances and challenges. Environ Sci Pollut Res 30(14):19234–19248. https://doi.org/10.1007/s11356-023-25267-4

Ali H, Khan E, Ilahi I (2023b) Transgenic plants for phytoremediation: genetic modifications and field applications. Environ Sci Pollut Res 30(12):18456–18472. https://doi.org/10.1007/s11356-023-25267-4

Ali H, Khan E, Sajad MA (2013) Phytoremediation of heavy metals—concepts and applications. Chemosphere 91(7):869–881. https://doi.org/10.1016/j.chemosphere.2013.01.075

Ali S, Rizwan M, Noureen S, Anwar S, Rehman MZ (2022b) Phytoremediation of heavy metals: physiological and molecular mechanisms. Chemosphere 293:133642. https://doi.org/10.1016/j.chemosphere.2022.133642

Anthony EJ, Brunier G, Besset M, Goichot M, Dussouillez P, Nguyen VL (2015) Linking rapid erosion of the Mekong River delta to human activities. Sci Rep 5:14745. https://doi.org/10.1038/srep14745

Azhar MA, Basheer S, Abbas M, Akram MS, Waseem A (2021) Heavy metals in agricultural soils: sources, toxicological effects, and remediation strategies. Environ Monit Assess 193(12):750. https://doi.org/10.1007/s10661-021-09482-x

Azubuike CC, Chikere CB, Okpokwasili GC (2016) Bioremediation techniques–classification based on site of application: principles, advantages, limitations and prospects. World J Microbiol Biotechnol 32(11):180. https://doi.org/10.1007/s11274-016-2137-x

Barwant MM, Ogidi OI, Singh B, Patil VS, Sharma SAB (2025a) Climate change and pesticide dynamics. In: Babaniyi BR, Babaniyi EE (eds) The interplay of pesticides and climate change. Springer, Cham. https://doi.org/10.1007/978-3-031-81669-7_7

Barwant MM, Singh B, Rashid R, Rasool N, Lone R (2025b) Lignin as biofertilizer. In: Jawaid M, Ahmad A, Meraj A (eds) Handbook of lignin. Springer, Singapore. https://doi.org/10.1007/978-981-97-2664-6_55-1

Barwant MM, Singh B, Sharma S, Gore NT, Mohammad A (2025c) Role of mycorrhizal association in leguminous plant growth. In: Mathur P, Gupta A (eds) Recent trends and applications of leguminous microgreens as functional foods. Springer, Cham. https://doi.org/10.1007/978-3-031-75678-8_13

Barwant MM, Singh B et al (2025d) Proteome engineering: an evolutionary perspective of biochemical systems. In: Kaur S, Dhiman S, Tripathi M (eds) Microbial metabolomics. Springer, Singapore. https://doi.org/10.1007/978-981-96-4824-5_6

Bhagwat MA, Singh AR, Kumar R (2023) Sustainable phytoremediation strategies for heavy metal detoxification in contaminated soils. J Environ Manage 336:117561. https://doi.org/10.1016/j.jenvman.2023.117561

Bhatt P, Rene ER, Kumar AJ (2022) Bioremediation of agro-industrial waste pollutants: challenges and future perspectives. Sci Total Environ 819:152085. https://doi.org/10.1016/j.scitotenv.2022.152085

Brim H, Venkateswaran A, Kostandarithes HM, Fredrickson JK, Daly MJ (2000) Engineering *Deinococcus radiodurans* for metal remediation in radioactive mixed waste environments. Nat Biotechnol 18(1):85–90. https://doi.org/10.1038/71986

Chauhan P, Singh B, Gupta S, Jena NK, Kumar D (2024) Mushroom diversity and their biochemical composition. In: Kumar D, Gupta S, Ansari MJ, Sangeeta Hasan W, Kashyap S, Mehrotra S (eds) Mushroom magic: biochemistry and nutritional value of fungi, 1st edn. CRC Press. https://doi.org/10.1201/9781003570257-1

Chen W, Zhang L, Xu Z, Chen C (2023) Microbial bioremediation of pesticide residues: mechanisms and applications. Environ Pollut 319:120968. https://doi.org/10.1016/j.envpol.2022.120968
Cropway (2024) Leveraging geospatial technology for sustainable water management in agriculture: Solutions for a greener future. Retrieved from https://www.cropway.com/geospatial-technology-water-management/
Das N, Chandran P (2011) Microbial degradation of petroleum hydrocarbon contaminants: an overview. Biotechnol Res Int 2011:941810. https://doi.org/10.4061/2011/941810
Dotaniya ML, Meena VD, Latare AM, Rajendiran S, Coumar MV (2016) Role of bioremediation and biochar application in deltaic soil health management. Ecol Eng 97:268–275. https://doi.org/10.1016/j.ecoleng.2016.09.011
Gadd GM (2004) Microbial influence on metal mobility and application for bioremediation. Geoderma 122(2–4):109–119. https://doi.org/10.1016/j.geoderma.2004.01.002
Gadd GM (2010) Metals, minerals, and microbes: geomicrobiology and bioremediation. Microbiology 156(3):609–643. https://doi.org/10.1099/mic.0.037143-0
Garbisu C, Alkorta I (2001) Phytoextraction: a cost-effective plant-based technology for the removal of metals from the environment. Biores Technol 77(3):229–236. https://doi.org/10.1016/S0960-8524(00)00108-5
Ghosh M, Singh SP (2005) A review on phytoremediation of heavy metals and utilization of its byproducts. Appl Ecol Environ Res 3(1):1–18
Ghosh P, Singh R, Yadav M (2023) Bioremediation in climate-resilient agriculture: potential and challenges. Sci Total Environ 883:163247. https://doi.org/10.1016/j.scitotenv.2023.163247
Gul S, Mehmood T, Ahmed Z (2023) CRISPR-based genome editing for microbial bioremediation: prospects and challenges. Front Microbiol 14:1265039. https://doi.org/10.3389/fmicb.2023.1265039
Gulati M, Singh N, Verma R (2022) Impact of pesticide residues on soil microbial communities and environmental health. Ecotoxicol Environ Saf 236:113472. https://doi.org/10.1016/j.ecoenv.2022.113472
Gupta DK, Pandey A, Rai A (2021) Phytodegradation of organic pollutants: recent advances and future challenges. Environ Chem Lett 19:3089–3103. https://doi.org/10.1007/s10311-021-01275-2
Gupta S, Roy R, Das S (2022) Bioremediation strategies for sustainable management of agro-ecosystems in delta regions. Environ Sustain 5(1):81–91. https://doi.org/10.1007/s42398-022-00217-0
Harms H, Schlosser D, Wick LY (2011) Untapped potential: exploiting fungi in bioremediation of hazardous chemicals. Nat Rev Microbiol 9(3):177–192. https://doi.org/10.1038/nrmicro2519
Huang J, Chen Y, Zhang H (2022a) IoT-based biosensors for real-time monitoring of soil and water pollutants in bioremediation. J Environ Monit 24(8):3254–3271. https://doi.org/10.1039/D2EM00347H
Huang J, Chen Y, Zhang H (2022b) Role of AI and IoT in monitoring bioremediation processes under changing climates. Environ Monit Assess 194(7):5221–5238. https://doi.org/10.1007/s10661-022-10003-w
Islam MS, Ahmed MK, Habibullah-Al-Mamun M, Masunaga S (2022) Heavy metal contamination in agricultural soils: Sources, impacts, and sustainable remediation strategies. Environ Res 214:113950. https://doi.org/10.1016/j.envres.2022.113950
Jaiswal D, Yadav S, Tiwari R (2023a) Advances in bioremediation technologies: a futuristic approach for sustainable agriculture. Sci Total Environ 879:162502. https://doi.org/10.1016/j.scitotenv.2023.162502

Jaiswal D, Yadav S, Tiwari R (2023b) Advances in precision bioremediation: a futuristic approach for sustainable agriculture. J Environ Manage 345:119637. https://doi.org/10.1016/j.jenvman.2023.119637

Juwarkar AA, Singh SK, Mudhoo A (2010) A comprehensive overview of elements in bioremediation. Rev Environ Sci Biotechnol 9(3):215–288. https://doi.org/10.1007/s11157-010-9215-6

Khan N, Bano A, Rahman MA (2019) Effect of plant growth-promoting rhizobacteria and exogenous application of salicylic acid on phytoremediation potential of *Helianthus annuus* under chromium stress. Int J Phytorem 21(5):407–414. https://doi.org/10.1080/15226514.2018.1537241

Kour D, Rana KL, Yadav AN, Singh B (2023) Rhizosphere microbiome engineering for bioremediation and sustainable crop production. Appl Soil Ecol 184:104726. https://doi.org/10.1016/j.apsoil.2023.104726

Kumar A, Singh A, Pandey R (2021) Role of geospatial technologies in sustainable agriculture. In: Jat MK et al (eds) Geospatial technologies for sustainable agriculture. Springer, pp 35–60. https://doi.org/10.1007/978-3-030-73459-7_3

Kumar M, Singh E, Kumar P (2022a) Soil pollution and its impact on agricultural productivity: a review. J Environ Manage 310:114786. https://doi.org/10.1016/j.jenvman.2022.114786

Kumar P, Sharma R, Meena RS (2022b) Role of fungi in biodegradation of pesticides and heavy metals: an eco-friendly approach. J Hazard Mater 424:127546. https://doi.org/10.1016/j.jhazmat.2021.127546

Kumar P, Sharma R, Meena RS (2023a) AI-driven decision support systems for optimizing microbial bioremediation strategies. Environ Technol Innov 31:102458. https://doi.org/10.1016/j.eti.2023.102458

Kumar P, Sharma R, Meena RS (2023b) Climate change and bioremediation: adaptive strategies for sustainable agriculture. Environ Technol Innov 32:102782. https://doi.org/10.1016/j.eti.2023.102782

Kuppusamy S, Thavamani P, Venkateswarlu K, Lee YB, Naidu R, Megharaj M (2016) Remediation approaches for polycyclic aromatic hydrocarbons (PAHs) contaminated soils: technological constraints, emerging trends and future directions. Chemosphere 168:944–968. https://doi.org/10.1016/j.chemosphere.2016.10.115

Liu H, Zhang Y, Zhang W, Tang C (2023) Enhancing mycoremediation efficiency using biochar and organic amendments: mechanistic insights. Biores Technol 368:128303. https://doi.org/10.1016/j.biortech.2023.128303

Liu X, Zhang C, He Y (2020) Eutrophication in agricultural landscapes: causes, consequences, and control strategies. Environ Sci Technol 54(8):5030–5045. https://doi.org/10.1021/acs.est.9b07448

Mangal V, Sharma M, Kumar P, Singh J (2023) Heavy metal toxicity in plants: physiological, biochemical, and molecular mechanisms. Environ Res 221:115083. https://doi.org/10.1016/j.envres.2023.115083

Newman LA, Strand SE, Choe N, Duffy J, Ekuan G, Ruzinski S, Gordon MP et al (1997) Uptake and biotransformation of trichloroethylene by hybrid poplars. Environ Sci Technol 31(4):1062–1067. https://doi.org/10.1021/es960564i

Ngwenya BT, Kennedy L, Parsons I (2010) Bio-weathering of silicates: a chemical and microbiological perspective. Chem Geol 269(1–2):1–17. https://doi.org/10.1016/j.chemgeo.2009.11.003

Paz A, Martínez L, Alvarado S (2020) Applications of remote sensing for bioremediation monitoring. In: Puglisi A (ed) Remote sensing of soil salinization: impact on land management. CRC Press, pp 189–208

Pointing SB (2001) Feasibility of bioremediation by white-rot fungi. Appl Microbiol Biotechnol 57(1–2):20–33. https://doi.org/10.1007/s002530100745

Rawlings DE, Johnson DB (2007) The microbiology of biomining: development and optimization of mineral-oxidizing microbial consortia. Microbiology 153(2):315–324. https://doi.org/10.1099/mic.0.2006/001206-0

Reuters (2025) Comment: how empowering smallholder farmers with AI tools can bolster global food security. Retrieved from https://www.reuters.com/sustainability/land-use-biodiversity/comment-how-empowering-smallholder-farmers-with-ai-tools-can-bolster-global-food-2025-01-10/

Sample I (2025) UK soil breakthrough could cut farm fertiliser use and advance sustainable agriculture. The Guardian. Retrieved from https://www.theguardian.com/environment/2025/feb/22/uk-soil-breakthrough-could-cut-farm-fertiliser-use-and-advance-sustainable-agriculture

Saravanan A, Kumar PS, Duc PA, Rangasamy G (2023) Strategies for microbial bioremediation of environmental pollutants from industrial wastewater: a sustainable approach. Chemosphere 313:137323. https://doi.org/10.1016/j.chemosphere.2022.137323

Sayler GS, Ripp S, Simpson ML (2012) Biosensors and environmental monitoring. Environ Sci Technol 44(21):8359–8365. https://doi.org/10.1021/es102235t

Sharma S, Gupta R, Singh P (2022a) Genetic engineering of microbes for enhanced heavy metal bioremediation. Appl Microbiol Biotechnol 106(14):5127–5143. https://doi.org/10.1007/s00253-022-12003-7

Sharma S, Gupta R, Singh P (2022b) Microbial bioremediation in agroecosystems: current trends and future perspectives. Appl Microbiol Biotechnol 106(15):5187–5204. https://doi.org/10.1007/s00253-022-12003-7

Singh B, Singh VK (2023) Diversity of wood-inhabiting macrofungi from District Ayodhya, Uttar Pradesh, India. Kavak 59(3):1–11. https://doi.org/10.36460/Kavaka/59/3/2023/51-61

Singh B, Tripathi S, Tyagi S, Tripathi A, Dubey P, Kumar S (2024) Phytostimulation and induction of systemic resistance by plant growth promoting fungi. In: Mathur P, Roy S (eds) Plant microbiome and biological control. Sustainability in plant and crop protection, vol 20. Springer, Cham. https://doi.org/10.1007/978-3-031-75845-4_10

Singh H, Ward OP (2021) Biodegradation and bioremediation. Springer Science & Business Media

Singh P, Verma R, Sharma J (2022) Engineered microbial systems for the biodegradation of organic pollutants: advances and applications. Microb Biotechnol 15(9):2198–2214. https://doi.org/10.1111/1751-7915.14199

Singh P, Verma R, Sharma J (2023a) Climate change impacts on microbial bioremediation efficiency: a global perspective. Microb Biotechnol 16(8):2283–2295. https://doi.org/10.1111/1751-7915.14376

Singh P, Verma R, Sharma J (2023b) Mycoremediation of industrial pollutants: fungal diversity and their enzymatic potential. Fungal Biol Biotechnol 10(1):1–16. https://doi.org/10.1186/s40694-023-00159-2

Strong PJ, Burgess JE (2008) Treatment methods for wine-related and distillery wastewaters: a review. Bioremediat J 12(2):70–87. https://doi.org/10.1080/10889860802085383

Syvitski JPM, Saito Y (2007) Morphodynamics of deltas under the influence of humans. Global Planet Change 57(3–4):261–282. https://doi.org/10.1016/j.gloplacha.2006.12.001

Tigini V, Prigione V, Varese GC (2022) Fungal biosorption of heavy metals: mechanisms and applications. Environ Sci Technol 56(21):15120–15132. https://doi.org/10.1021/acs.est.2c05309

Tudi M, Ruan HD, Wang L, Lyu J (2021) Endocrine-disrupting chemicals in agriculture and their impact on plant growth and soil health. Agric Syst 190:103095. https://doi.org/10.1016/j.agsy.2021.103095

Verma N, Mishra P, Gupta S (2023) Conservation agriculture for mitigating pollution: synergistic role of bioremediation. Environ Pollut 315:120053. https://doi.org/10.1016/j.envpol.2023.120053

Vidali M (2001) Bioremediation. An overview. Pure Appl Chem 73(7):1163–1172. https://doi.org/10.1351/pac200173071163
Vijayaraghavan K, Yun YS (2008) Bacterial biosorbents and biosorption. Biotechnol Adv 26(3):266–291. https://doi.org/10.1016/j.biotechadv.2008.02.002
Vimal SR, Singh JS, Arora NK (2022) Role of plant growth-promoting rhizobacteria in environmental cleanup and sustainable agriculture. Front Microbiol 13:860195. https://doi.org/10.3389/fmicb.2022.860195
Wang X, Li Y, Zhang Z (2022a) Climate change-driven alterations in bioremediation efficacy: implications for soil health. Environ Res 309:119923. https://doi.org/10.1016/j.envres.2022.119923
Wang X, Li Y, Zhang Z (2022b) Remote sensing applications in precision bioremediation: hyperspectral imaging and UAV-based monitoring. Environ Pollut 309:119822. https://doi.org/10.1016/j.envpol.2022.119822
Wang X, Sun Z, Zhang H (2023) Bioremediation approaches for soil contamination: recent advances and future perspectives. J Hazard Mater 448:131715. https://doi.org/10.1016/j.jhazmat.2023.131715
Xu T, Luo W, Li X (2023a) Artificial intelligence in bioremediation: machine learning approaches for environmental monitoring. Environ Res 227:115872. https://doi.org/10.1016/j.envres.2023.115872
Xu T, Luo W, Li X (2023b) Sustainable agricultural practices for pollutant management: the role of bioremediation. Sci Total Environ 901:164059. https://doi.org/10.1016/j.scitotenv.2023.164059
Zhao L, Feng Y, Chen X (2023a) Geospatial mapping for precision bioremediation in climate-resilient agriculture. J Environ Monit 39(5):1835–1850. https://doi.org/10.1039/D3EM00182G
Zhao L, Feng Y, Chen X (2023b) GIS-based spatial modeling for precision bioremediation of contaminated farmlands. J Environ Manage 338:118622. https://doi.org/10.1016/j.jenvman.2023.118622
Zhao Q, Wang J, Li S (2021) Pesticide residues in agricultural soils and their effects on soil microbial diversity. Environ Sci Pollut Res 28(15):18742–18754. https://doi.org/10.1007/s11356-021-12614-6
Zhou Y, Wang J, Zhang X, Wang X (2021) Microplastic pollution in agroecosystems: sources, impact, and sustainable remediation strategies. Sci Total Environ 791:148298. https://doi.org/10.1016/j.scitotenv.2021.148298

Application of Remote Sensing and Geographic Information System in Precision Farming

12

Kirti Vardhan Pandey, Shachi Tiwari, Swarnima Tiwari, Mehrukh Iffat, Chandra Kant, and Anupriya Rana

Abstract

Remote Sensing (RS) and Geographic Information Systems (GIS) integration has transformed precision farming through heightened efficiency, sustainability, and decision-making. Remote sensing offers useful insights into crop health, soil moisture content, land use, and weather using satellite and UAV data, allowing for near real-time tracking of agricultural operations. GIS enables the management, analysis, and visualization of location-related data, thus optimizing resources and planning. Combined, RS and GIS enhance crop monitoring, allow for early detection of pests and disease, and optimize irrigation systems, eventually raising productivity while reducing environmental footprints. RS and GIS are equally important for soil mapping, yield prediction, and measuring climate change impacts on agriculture. Farmers can minimize chemical inputs, save water, and encourage sustainable practices by embracing RS- and GIS-based approaches. With ongoing development in these technologies, they possess great promise in solving world challenges concerning food security, climate change, and conservation of the environment.

K. V. Pandey
Department of Agricultural Meteorology, Acharya Narendra Deva University of Agriculture and Technology, Ayodhya, UP, India

S. Tiwari (✉) · S. Tiwari · M. Iffat · A. Rana
Department of Life Science, Keral Verma Subharti College of Science, Swami Vivekanand Subharti University, Meerut, UP, India
e-mail: tiwarishachi1998@gmail.com

C. Kant
School of Agricultural Sciences, Shri Venkateshwara University, Amroha, India

C. Singha et al. (eds.), *Sustainable Management of Delta Ecosystems Resilience*, Deltas of the World, https://doi.org/10.1007/978-3-032-16261-8_12

Abbreviations

AI	Artificial Intelligence
ANOVA	Analysis of Variance
ArcGIS	ArcGIS (ESRI GIS software)
AVHRR	Advanced Very High Resolution Radiometer
CCCI	Canopy Chlorophyll Content Index
CIGR	Commission Internationale du Génie Rural
DAS	Days After Sowing
DEM	Digital Elevation Model
ERTS	Earth Resources Technology Satellite
ENSO	El Niño–Southern Oscillation
EVI	Enhanced Vegetation Index
FAO	Food and Agriculture Organization
GEE	Google Earth Engine
GIS	Geographic Information System(s)
GI	Greenness Index
GYURI	General Yield Unified Reference Index
GSM	Global System for Mobile Communications
GPS	Global Positioning System
HNBs	Hyperspectral Narrowbands
HVIs	Hyperspectral Vegetation Indices
IDW	Inverse Distance Weighting
IoNT	Internet of Nano Things
IoT	Internet of Things
IRS	Indian Remote Sensing (satellite series)
IR	Infrared
IRRI	International Rice Research Institute
ISOXML	ISO XML (machine-compatible exchange format)
k-fold	K-fold cross-validation
LAI	Leaf Area Index
LULC	Land Use and Land Cover
MAE	Mean Absolute Error
MODIS	Moderate Resolution Imaging Spectroradiometer
N, P, K	Nitrogen, Phosphorus, Potassium
NASA	National Aeronautics and Space Administration
NDRE	Normalized Difference Red Edge Index
NDVI	Normalized Difference Vegetation Index
NDWI	Normalized Difference Water Index
NIR	Near-Infrared
NOAA	National Oceanic and Atmospheric Administration

OR (Orfeo)	OrfeoToolBox (tool name appearing in text)
PCA	Principal Component Analysis
PVI	Perpendicular Vegetation Index
QGIS	Quantum GIS
R^2	Coefficient of Determination (R-squared)
RGB	Red, Green, Blue
RF	Random Forest
RMSE	Root Mean Square Error
RVI	Ratio Vegetation Index
RS	Remote Sensing
SAM/ SNAP	Sentinel Application Platform (SNAP)
SAVI	Soil Adjusted Vegetation Index
SCADA/GSM etc.	(GSM appears as communication standard in the file)
SOM	Soil Organic Matter
SPAD	SPAD chlorophyll meter index (chlorophyll measurement)
SPOT	Satellite Pour l'Observation de la Terre
SRTM	Shuttle Radar Topography Mission
SVM	Support Vector Machine
TM/ETM +	Thematic Mapper/Enhanced Thematic Mapper Plus
TSF	Conventional single-rate (nitrogen) fertilization (TSF, as used in the doc)
UAV	Unmanned Aerial Vehicle
VCI	Vegetation Condition Index
VPMs	Vegetation Phenology Metrics
VRF	Variable-Rate Fertilization (variable-rate nitrogen fertilization)
TVI	Transformed Vegetation Index
VRI	(Variable-rate irrigation/related term appears contextually)
WiFS	Wide Field Sensor
XGBoost	EXtreme Gradient Boosting
USDA	United States Department of Agriculture

12.1 Introduction

Agriculture is a fundamental component of every nation's economy, contributing significantly to trade and economic strength. The primary objective of farmers, large-scale farm managers, and regional agricultural agencies is to produce food efficiently and cost-effectively. Remote sensing and Geographic Information Systems (GIS) have proven to be valuable tools for analyzing and visualizing agricultural landscapes, benefiting both the farming community and the agricultural industry. These technologies play a crucial role in enhancing agricultural productivity, reducing expenses, and improving land management (Fig. 12.2). GIS, in particular, is widely recognized as a powerful tool for monitoring land use and land cover changes (Kumar et al. 2015). The application of remote sensing and GIS in agriculture helps assess crop health, identify infestations, estimate yields, and analyze soil conditions. These technologies support various agricultural applications, including crop identification, area measurement, crop condition analysis, soil moisture assessment, yield prediction, and agricultural water management. Additionally, they are instrumental in agro-meteorology and other precision farming practices.

Remote sensing applications in agriculture cover several key areas (Table 12.1), such as biomass and yield estimation, monitoring vegetation health and drought stress, tracking crop phenological development, estimating crop acreage, and mapping croplands. They also help in assessing disturbances, analyzing land-use changes, and managing precision agriculture and irrigation (Atzberger 2013). GIS-based mapping facilitates the identification of crop locations, adaptation to environmental variables, monitoring of individual crop health, yield estimation, and overall optimization of agricultural production. By integrating land-use data with primary food crop statistics and mobile-based data collection, GIS plays a crucial role in identifying areas facing food insecurity and addressing its underlying causes. Moreover, it supports automated field operations, contributing to global efforts to combat hunger.

Farmers can leverage remote sensing data and sensor-equipped farm machinery to enhance decision-making for better planning and yield optimization. Factors such as historical crop yields, terrain characteristics, soil organic matter, pH levels, moisture, and nutrient content all play a role in precision farming. Modern harvesters equipped with GPS technology can measure crop yields in real time, while also assessing plant water content and chlorophyll levels at specific field locations. The rapid advancements in remote sensing and geospatial technologies are transforming agricultural practices, enabling efficient crop monitoring, stress management, regional yield estimation, and sustainable resource use to support long-term agricultural productivity (Kingra et al. 2016a, b).

Table 12.1 Data types, sources, and applications in precision agriculture

Data type	Source	Spatial resolution	Temporal resolution	Purpose
Optical imagery	Sentinel-2	10 m	5–10 days	Vegetation indices, phenology
SAR imagery	Sentinel-1	10–20 m	6–12 days	Soil moisture proxy
UAV multispectral	MicaSense (if used)	5–10 cm	Campaign based	High-resolution validation
Soil samples	Field campaign	Point	Once pre-season	Soil fertility mapping
Yield maps	Harvester	Varying (1–10 m)	Harvest	Model training/ validation

12.1.1 Research Gaps and Novelty of the Study

Despite considerable progress in applying Remote Sensing (RS) and Geographic Information Systems (GIS) to agriculture, several research gaps remain, particularly in delta regions where fragile ecosystems and intensive cropping coexist. First, limited long-term spatiotemporal analyses of cropping intensity and diversity restrict our understanding of agricultural dynamics in delta landscapes. Second, soil salinity—a critical issue caused by seawater intrusion—has not been comprehensively mapped using integrated RS, GIS, and ground-truth approaches. Third, multi-source data integration (satellite, UAV, ground surveys, and socio-economic datasets) is still underexplored, reducing the accuracy of yield prediction and resource management. Furthermore, the lack of delta-specific climate impact modeling hampers early-warning systems for floods, droughts, and salinity intrusion.

The novelty of this study lies in its delta-focused perspective on precision farming. By reviewing and synthesizing applications of RS and GIS across major deltas such as the Cauvery, Mahanadi, Nile, and Niger, this work identifies region-specific challenges while underlining the potential of advanced tools like hyperspectral sensors, UAV-based validation, and machine learning–driven yield modeling. Unlike generic reviews, the present study highlights how delta ecosystems demand tailored precision agriculture strategies that integrate environmental monitoring, land-use planning, and input optimization. This delta-oriented framework contributes to bridging global knowledge with local agricultural sustainability, providing a roadmap for future climate-resilient and resource-efficient farming practices.

12.2 Methodology

12.2.1 Study Design

The research emphasized mapping in-field variability in crop vigor and soil characteristics, creating zones of management for variable-rate input application, and building crop-yield models by merging RS data with ancillary datasets.

12.2.2 Data Acquisition

- Remote sensing data: Sentinel-2, Landsat 8/9, Planet Scope (optical multispectral), Sentinel-1 SAR for soil moisture, and UAV-based high-resolution imagery for validation.
- Auxiliary data: soil maps, DEM (SRTM), yield maps from the past, farm management records, and weather.
- Field-truth data: crop biomass, LAI, chlorophyll (SPAD), and soil samples (N, P, K, organic C, texture) obtained by stratified-random sampling based on initial NDVI variability.

12.2.3 Preprocessing of RS Data

Images were preprocessed by atmospheric and radiometric correction, geometric co-registration, cloud/shadow masking, and speckle filtering for SAR. UAV imagery was corrected with reflectance panels for accuracy.

12.2.4 Feature Extraction

Vegetation indexes (NDVI, SAVI, EVI, NDRE), biophysical variables (LAI, canopy chlorophyll), SAR backscatter coefficients, and temporal phenological indicators were obtained. Texture features from UAV imagery improved heterogeneity mapping.

12.2.5 Spatial Analysis and Zoning

Normalized input layers were diminished using PCA, then by clustering (K-means, Fuzzy C-means, or hierarchical). The obtained management zones were checked against field-truth yield and soil data by statistical tests (ANOVA/Kruskal–Wallis).

12.2.6 Yield Modeling

Spectral indices, SAR characteristics, DEM derivatives, soil attributes, and weather information were employed as predictors. Linear models, Random Forest, XGBoost, and deep learning were compared. Models were trained using k-fold cross-validation and tested using R^2, RMSE, MAE, and bias.

12.2.7 Accuracy Assessment

Statistical significance tests were used to test zoning accuracy. Accuracy of classification was quantified using confusion matrices and kappa statistics. Variable importance analysis was used to determine the top predictors of yield.

12.2.8 Operational Outputs

Confirmed areas were translated into prescription maps for variable-rate applications (irrigation, fertilization). Outputs were exported in machine-compatible formats like shapefiles or ISOXML.

12.2.9 Tools and Software

Processing was undertaken with QGIS, ArcGIS, SNAP, Google Earth Engine, Pix4D/Agisoft, Python (TensorFlow, scikit-learn) (Fig. 12.1).

12.2.9.1 Optical Imagery (Sentinel-2, 10 M, 5–10 days)

- Multispectral optical data with 10-m spatial resolution and a 5–10 day revisit interval are given by Sentinel-2.
- It is mainly utilized in order to estimate vegetation indices (e.g., NDVI, SAVI, EVI) and observe crop phenology (growth stages and seasonality).

12.2.9.2 SAR Imagery (Sentinel-1, 10–20 M, 6–12 days)

- SAR is cloud-independent and particularly suitable in rainy or cloudy periods. It can be used as a proxy for soil moisture and crop structure monitoring.
- Sentinel-1 offers Synthetic Aperture Radar (SAR) data with a spatial resolution of 10–20 m and a revisit interval of 6–12 days.

Fig. 12.1 Graphical representation of methodological framework for variable-rate crop management

12.2.9.3 UAV Multispectral Imagery (MicaSense, 5–10 cm, Campaign-Based)

Multispectral cameras mounted on UAVs such as MicaSense deliver highly detailed imagery (5–10 cm). Purchased on a campaign by campaign basis (specific field trips), such images are high resolution and offer detailed information on crop health and are utilized as ground-truth verification for satellite-based indices.

12.2.9.4 Soil Samples (Field Campaign, Point Data, Once Pre-Season)

- Soil samples are obtained from various field locations prior to the cropping season.
- They offer valuable information on soil fertility factors (N, P, K, organic matter, pH, texture).
- These point measurements are utilized to calibrate and verify satellite- or UAV-based soil fertility evaluations.

12.2.9.5 Yield Maps (Harvester, 1–10 M, Harvest Time)

- Yield monitors mounted on harvesters produce high-resolution spatial yield data (1–10 m according to machine calibration).
- These maps are the ground truth used for model training and evaluation, enabling the assessment of the accuracy with which remote sensing and GIS analyses estimate actual crop yields

12.3 Remote Sensing in Agriculture

Remote Sensing (RS) and Geographic Information Systems (GIS) play a vital role in crop inventory by identifying different crop types and detecting changes in cropping patterns. These technologies serve as valuable tools for conducting crop surveys and mapping agricultural areas (Profile 2016). For countries with agriculture-based economies, reliable and timely data on crop types, cultivation areas, and projected yields are essential for government planning and decision-making. Spectral information is a crucial component of remote sensing data used in crop modeling, as it is closely linked to canopy parameters that reflect crop health and growth stages. Crop-specific maps generated by integrating satellite imagery with survey data provide detailed land layouts and ownership information, benefiting agribusiness sectors such as seed and fertilizer companies.

Remote sensing technology significantly contributes to crop inventory by compiling databases on various crops. Numerous studies utilizing aerial photography and digital image processing techniques have demonstrated its effectiveness in agricultural monitoring. By reducing the need for extensive field data collection while improving estimation accuracy, remote sensing enhances the efficiency of agricultural assessments (Kingra et al. 2016a, b).

12.3.1 Fundamental Aspects of Remote Sensing in Agriculture

In the initial phases of satellite remote sensing, scientists were mostly interested in applying data to distinguish between various cover types, with specific interest in differentiating crop types for agricultural purposes. Agricultural remote sensing has recently evolved to understand the plant biophysical properties with more profound insight into agronomic conditions. Remote sensing has widely been used to monitor and analyze agricultural activities, especially in evaluating crop canopies. One of the strengths of this technology lies in the ability to offer continuous, non-destructive data acquisition, a feature that makes precision agriculture an indispensable tool to utilize (De Beurs and Townsend 2008). Remote sensing also has the advantage of offering a low-cost means of obtaining agricultural information over large geographical regions.

In India, remote sensing by satellite is utilized mainly to estimate crop area and production. It has the ability to transform agricultural productivity measurement by examining the biophysical properties of crops and soils (Liaghat and Balasundram 2010). Remotely sensed data acquired by satellites make important contributions to numerous agricultural applications, such as yield estimation (Doraiswamy et al. 2005; Bernerdes et al. 2012), monitoring crop phenology (Sakamoto et al. 2005), stress condition detection (Gu et al. 2007), and disturbance identification in agricultural landscapes. Combining remote sensing with GIS is of great use for the generation of spatio-temporal information layers, which can be used across a wide range of fields including floodplain mapping, hydrological modeling, surface energy flux estimation, urban planning, land use change detection, monitoring crop growth, and stress identification (Kingra et al. 2016a, b).

Technical progress has been prompted by the advent of hyperspectral or narrow-band sensors and the enhancement of spatial resolution of airborne and satellite-borne sensors. Hyperspectral remote sensing, in essence, has considerably improved crop classification accuracy. Thenkabail et al. (2004) made a detailed study of hyperspectral sensors (400–2500 nm) for crop discrimination utilizing data mining tools such as principal component analysis, lambda–lambda models, stepwise discriminant analysis, and derivative greenness vegetation indices. Different researchers have utilized different sensors having the ability to provide timely and reliable agricultural information at a small fraction of the cost of using the conventional methods of data collection.

12.3.2 Monitoring Vegetation Cover Using Remote Sensing

Remote sensing has become a vital tool for crop classification, acreage estimation, and yield estimation. Aerial photography and digital image processing methods have been used in various research studies to enhance agricultural monitoring. Nonetheless, remote sensing has immensely mitigated the necessity for wide field data collection while improving the accuracy of crop estimates (Kingra et al. 2016a, b). Hyperspectral data is very effective in enhancing the characterization, classification, modeling, and mapping of vegetation and crops over conventional broadband multispectral remote sensing (Thenkabail et al. 2011). This development has helped to determine 33 ideal Hyperspectral Narrow Bands (HNBs) and the same number of two-band normalized difference Hyperspectral Vegetation Indices (HVIs) utilized in classifying, modeling, and mapping agricultural crops worldwide, and their biophysical and biochemical characterizations (Thenkabail et al. 2013). In crop condition assessment, whereas some of the remote sensing methods target physical parameters like nutrient stress and water content to assess the health and yield of crops, others give a more general view at the regional level through remote sensing indices. One of the most popular vegetation assessment indices is the Normalized Difference Vegetation Index (NDVI) developed by Rouse et al. (1973a, b). NDVI is now the accepted vegetation index (Calvão and Palmeirim 2004; Wallace et al. 2004), and scientists have continued to establish extra indices for reducing the impact of soil background and atmospheric status upon spectral measurements. One such example is the Soil Adjusted Vegetation Index (SAVI) established by Huete (1988), which serves to reduce interference in vegetation data due to soil. A number of vegetation indices are employed in drought monitoring and evaluating crop productivity and health, such as NDVI, Vegetation Condition Index (VCI), Leaf Area Index (LAI), General Yield Unified Reference Index (GYURI), and Temperature Crop Index (TCI) (Doraiswamy et al. 2003; Ferencz et al. 2004; Prasad et al. 2006). For instance, Kogan et al. (2005) used vegetation indices based on Advanced Very High Resolution Radiometer (AVHRR) data to simulate corn yields and issue early drought warnings in China. Also, Hadria et al. (2006) showed the application of leaf area indices from four satellite systems to estimate wheat yield distribution and irrigation needs in semi-arid areas. Several vegetation indices that have been tailored for agricultural use are listed in the table below.

12.3.3 Crop Condition Assessment Using Remote Sensing

Remote sensing is important in agriculture as it offers timely spectral information for the monitoring of biophysical indicators of vegetation health. Plant physiological changes brought about by stress may cause their spectral reflectance and emission properties to change, so remote sensing can be an efficient means to detect and monitor crop stress

(Menon 2012). Regular crop monitoring throughout its growth cycle is essential for taking preventive measures and estimating potential yield losses due to stress factors. Crop development is influenced by multiple factors, including soil moisture availability, planting date, air temperature, day length, and soil condition, all of which directly impact plant health and productivity. For instance, corn yields may decline if temperatures are excessively high during pollination. Thus, temperature monitoring in this important stage can enhance yield estimates (Nellis et al. 2009). Likewise, drought can make land unfavorable for cultivation, impacting human beings, animals, biomass yields, and plant species (Siddiqui 2004). Over the last few years, satellite remote sensing-based monitoring of drought has found extensive use. Measurements like the Normalized Difference Vegetation Index (NDVI) and Vegetation Condition Index (VCI) have received international acclaim to detect agricultural drought over various ecosystems (Nicholson and Farrar 1994; Kogan 1995; Seiler et al. 2000; Wang et al. 2001; Anyamba et al. 2001; Ji and Peters 2003). Crop health and growth are normally measured through several vegetation indices such as ratio of reflectance, NDVI, Perpendicular Vegetation Index (PVI), Transformed Vegetation Index (TVI), and Greenness Index. In operational remote sensing, year-round NDVI profiles are employed to derive 12 Vegetation Phenology Metrics (VPMs), which assist in describing the reaction of farming crops to climatic fluctuations and land use practices (Reed et al. 1994).

12.3.4 Nutrient and Water Status Monitoring Using Remote Sensing

One of the key uses of remote sensing and GIS in precision agriculture is nutrient and water stress management. Identification of nutrient stress through remote sensing and GIS facilitates site-specific nutrient management, lowering cultivation expenses while enhancing the efficiency of fertilizers. In semi-arid and arid areas, precision farming technologies provide the optimal utilization of water resources. For instance, drip irrigation with remotely sensed data—like canopy vs. air temperature difference—can increase water use efficiency by reducing percolation and runoff losses (Das and Singh 1989). Water-stressed plants have greater spectral reflectance within the visible range than non-stressed ones. Crop water stress indices like NDVI, RVI, PVI, and Greenness Index (GI) are generally lower in stressed and greater in healthy, well-irrigated crops. The advent of microwave remote sensing has further facilitated precise estimation of soil moisture content in agricultural crops. Remote sensing information can also be employed to assess crop water requirement, soil moisture status, and growth stages of crops for the purposes of optimal irrigation management. Bandara (2003), for example, used NOAA satellite data to assess three large-scale irrigation schemes in Sri Lanka. The research contrasted remotely sensed estimates of the use of water by crops with available water to gauge the efficiency in irrigation. Improvements in hyperspectral bands in the thermal range have further improved remote sensing to analyze crop-soil interaction. With the incorporation of GPS technology, such data will deliver more precise and location-specific results to inform precision farming. Under wet tropical and subtropical conditions, nitrogen leaching is a high risk owing to spatial variability in soil characteristics, such as soil organic matter (SOM) content (Casa et al. 2011), water content (Delin and Berglund 2005), and yield variability (Blackmore et al. 2003; Bramley 2009). Such variability commonly results in inefficiency in conventional single-rate nitrogen fertilization (TSF), leading to over-fertilization on some areas and under-fertilization in other areas (Bredemeier and Schmidhalter 2005).

Variable-rate nitrogen fertilization (VRF), using real-time crop sensors, has been suggested as an alternative that can reduce the inefficiency. VRF provides for site-specific application of nitrogen, enhancing the efficiency of fertilization and maximizing crop growth (Singh et al. 2006; Li et al. 2010).

Time-consuming and labor-intensive, Remote sensing integrated with precision agriculture, however, presents an effective solution for site-specific weed management (Johnson et al. 1997; Moran et al. 1997; Lamb et al. 1999). Remote sensing depends on the spectral reflectance differences between weeds and crops in order to correctly detect weed infestations in the fields. This technology can produce weed distribution maps, enabling farmers to target herbicides where they are needed, minimizing chemical application and environmental damage.

Kaur et al. (2013) noted that radiance ratio and NDVI values were much greater in pure wheat plots than in solid weed areas. With NDVI and radiance ratio, wheat was discriminable from Rumex spinosus populations at more than 30 DAS. Various population levels of Rumex could be discriminable from one another at and above 60 DAS. Likewise, Kaur et al. (2014) illustrated that pure wheat fields were discriminable from Malva neglecta infestations at 30 DAS, with evident differences remaining for as long as 120 DAS. With GIS-based prescription maps for weeds, farmers are able to apply site-specific herbicides, resulting in cost-saving and environmentally friendly weed control measures. Not only does this eliminate labor and input expenses, but it also lowers the potential for herbicide resistance among weed populations.

12.3.5 Crop Evapotranspiration Monitoring Using Remote Sensing

Remote sensing is important in agriculture as it offers timely spectral information for the monitoring of biophysical indicators of vegetation health. Plant physiological changes brought about by stress may cause their spectral reflectance and emission properties to change, so remote sensing can be an efficient means to detect and monitor crop stress (Menon 2012). Regular crop monitoring throughout its growth cycle is essential for taking preventive measures and estimating potential yield losses due to stress factors. Crop development is influenced by multiple factors, including soil moisture availability, planting date, air temperature, day length, and soil condition, all of which directly impact plant health and productivity. For instance, corn yields may decline if temperatures are excessively high during pollination. Thus, temperature monitoring in this important stage can enhance yield estimates (Nellis et al. 2009). Likewise, drought can make land unfavorable for cultivation, impacting human beings, animals, biomass yields, and plant species (Siddiqui 2004).

Over the last few years, satellite remote sensing-based monitoring of drought has found extensive use. Measurements like the Normalized Difference Vegetation Index (NDVI) and Vegetation Condition Index (VCI) have received international acclaim to detect agricultural drought over various ecosystems (Nicholson and Farrar 1994; Kogan 1995; Seiler et al. 2000; Wang et al. 2001; Anyamba et al. 2001; Ji and Peters 2003). Crop health and growth are normally measured through several vegetation indices such as ratio of reflectance, NDVI, Perpendicular Vegetation Index (PVI), Transformed Vegetation Index (TVI), and Greenness Index. In operational remote sensing, year-round NDVI profiles are employed to derive 12 Vegetation Phenology Metrics (VPMs), which assist in describing the reaction of farming crops to climatic fluctuations and land use practices (Reed et al. 1994).

12.3.6 Weed Identification and Management Using Remote Sensing

Precision weed management is an advanced approach that enhances weed control strategies, improving overall crop health and yield. Traditional ground surveying methods for weed mapping are time-consuming and labor-intensive. However, remote sensing, combined with precision agriculture, offers an efficient alternative for site-specific weed management (Johnson et al. 1997; Moran et al. 1997; Lamb et al. 1999). Remote sensing relies on the differences in spectral reflectance between crops and weeds to accurately identify weed infestations in agricultural fields. This technology enables the creation of weed distribution maps, allowing farmers to apply herbicides precisely, reducing chemical usage and environmental impact.

Kaur et al. (2013) observed that radiance ratio and NDVI values were significantly higher in pure wheat fields compared to solid weed plots. Using NDVI and radiance ratio, wheat could be distinguished from Rumexspinosus populations beyond 30 days after sowing (DAS). Different Rumex population levels could be further discriminated from each other after 60 DAS. Similarly, Kaur et al. (2014) demonstrated that pure wheat fields could be differentiated from Malvaneglecta infestations after 30 DAS, with clear distinctions persisting up to 120 DAS. Using GIS-based weed prescription maps, farmers can implement site-specific herbicide applications, leading to cost-effective and environmentally sustainable weed control strategies. This approach not only reduces labor and input costs but also minimizes the risk of herbicide resistance in weed populations.

12.3.7 Pest and Disease Infestation Monitoring Using Remote Sensing

12.3.7.1 Some Site-Specific Management Strategies that Can Be Developed Using GIS Analysis of Pest Spatial Data Include (Table 12.2)

Targeted pesticide applications: GIS may be used to pinpoint areas of a field or orchard in which pest population levels exceed economic thresholds. Targeted, localized pesticide application is then possible, with less overall pesticide applied and at lower cost than through broadcast application.

Precision release of biological control agents: GIS maps of pest hotspots can inform the strategic release of natural enemies, e.g., predators or parasitoids, to control pest populations. Release of biocontrol agents at high-priority locations can maximize their effectiveness.

Table 12.2 Integrating remote sensing, GIS, and IoT for sustainable insect pest management

Category	Details
Technology & Applications	Remote Sensing—Early pest detection, monitoring of pest population dynamics, crop health assessment GIS—Spatial analysis of pest distribution, risk assessment, targeted interventions Wireless Sensor Networks—Real-time monitoring of pest populations and environmental conditions, automation of pest traps Decision Support Systems—Integration of data for optimizing control strategies, pest forecasting and risk assessment
Challenges & Future Directions	Challenges: High initial costs; Data management complexities; Capacity building among end-users Future Directions: Development of cost-effective solutions; Standardization of data protocols; Knowledge transfer through collaborative and participatory approaches
Benefits	Reduced insecticide usage—Targeted interventions based on pest population dynamics Minimized environmental impacts—Reduced non-target effects, decreased insecticide resistance risk Improved sustainability—Optimized strategies for long-term ecological and economic balance
Future Trends & Impacts	Precision agriculture + Big data analytics—Enhanced decision-making and predictive capabilities Autonomous robots—Improved efficiency and reduced labor costs IoT and nanotechnology adoption—Real-time, fine-scale monitoring and targeted delivery of pest control agents

Trap crop placement: Trap crops, which are more pest-attractive than the main crop, may be located strategically to attract pests away from the main crop. GIS analysis can determine where the best place is for trap crop strips, based on the spatial pattern of pest distribution.

Targeted cultural practices: GIS-based maps can inform site-specific cultural management practices. For instance, if it is observed that some areas are more susceptible to pest attacks, those locations may get additional attention in the form of sanitation practices, pruning, or rotation of crops.

Precision irrigation and fertilization: Irrigation and fertilization levels can be optimized using GIS-based pest risk maps to control crop stress and reduce conditions for pest development. Excessive watering or overfertilization may, at times, add to the pest issue.

Focused scouting and monitoring efforts: GIS-based risk maps can guide field scouting operations to where pest infestations are most likely to occur. This concentrated scouting enhances the effectiveness of pest monitoring and facilitates early detection and swift response to emerging outbreaks.

Wireless sensor networks present real-time information on pest infestation and environmental factors, allowing timely and precise interventions (Duan et al. 2017). Decision support systems synthesize multi-source data and models to provide useful recommendations for pest management, selecting the optimal time and location for applying control measures (Liao et al. 2019).

Combining precision agriculture methods with smart technologies has numerous advantages for controlling insect pests (Table 12.2). By giving precise and timely information regarding pest dynamics, these technologies allow farmers to make informed decisions and adopt targeted control strategies (Gómez-Candón et al. 2014). This strategy minimizes the excessive use of insecticides, averting environmental effects and the rise of insecticide resistance (Lofton et al. 2012). Moreover, precision agriculture methods can improve the efficacy and economic efficiency of pest control measures since treatments are only conducted when and where required (Fraser et al. 2009).

12.3.7.2 Some Practical Examples of Precision Agriculture Methods for Insect Pest Management, Along with Support from International Organizations and Their Potential Profitability and Impact (Table 12.2)

Remote sensing for locust outbreak monitoring: Satellite data is utilized by the United Nations Food and Agriculture Organization (FAO) to track and forecast desert locust outbreaks across Africa and Asia. It identifies likely breeding areas through the analysis of vegetation health and weather patterns, enabling targeted control activities. The early warning system assists in safeguarding crops and livelihoods, with a presumed benefit–cost ratio of 1:50 for preventing locusts.

Unmanned Aerial Vehicles (UAVs) for precision insecticide spraying: In China, UAVs equipped with high-resolution cameras and precision sprayers are used to target cotton bollworm infestations. The drones identify pest hotspots and apply insecticides only where needed, reducing pesticide use by up to 50% compared to traditional methods. This technology, supported by government subsidies, improves both profitability and environmental sustainability for cotton farmers.

Wireless sensor networks for real-time pest monitoring: In Spain, the European Union (EU) has sponsored a project that has set up wireless sensors in olive groves to track populations of olive fruit fly. The sensors report pest infestation and send data to a central station, which notifies farmers when threshold levels are reached (Fig. 12.2).

This real-time monitoring allows for the application of insecticides in a timely and focused manner, lowering costs and reducing environmental impacts. The project proved to reduce the use of pesticides by 30% and yield by 20%.

GIS-based decision support for integrated pest management: The International Rice Research Institute (IRRI) in Indonesia created a GIS-based decision support system for managing rice pests. The system combines data on pest population, weather, and crop status to create risk maps and provide targeted interventions. Farmers using the system indicated a 15–20% increase in yield and a 30–40% decrease in the cost of pesticides. The Asian Development Bank funds scaling up of this technology.

Hyperspectral imaging for early pest detection: The US Department of Agriculture (USDA) in the United States is exploring the application of hyperspectral imaging to

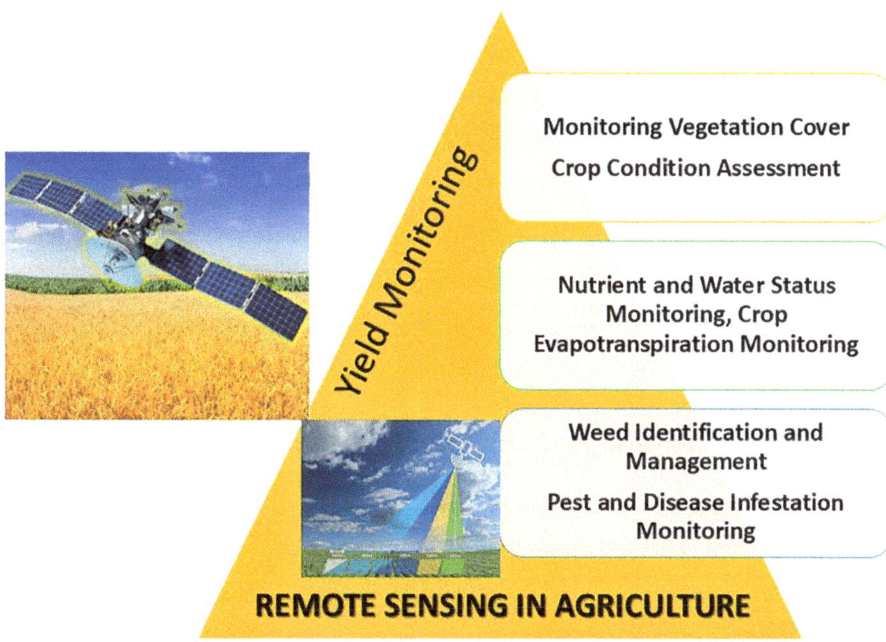

Fig. 12.2 Applications of remote sensing in agriculture

identify early warnings of insect pest infestations in crops. Hyperspectral imaging has the ability to detect stress in plants prior to the appearance of symptoms, enabling predictive pest management. Preliminary estimates indicate this could cut pesticide use by 20–30% and raise crop yields by 10–15%, thereby improving farmers' profitability.

Though there is the potential for benefits, precision agriculture methods of insect pest control are subject to some challenges. The high costs of initial investment in procuring and sustaining sophisticated technologies have the potential to act as a limiter for small farmers (De Luca et al. 2019). The technical sophistication of these technologies necessitates specialized training and expertise, which could impede their adoption by a wider group of users (Peña-Barragán et al. 2011). In addition, the sheer volume of data produced by smart technology raises data management and interpretation issues, calling for strong data analytics and decision support capabilities (Liakos et al. 2018). To fully realize the potential of precision agriculture and smart technologies for insect pest control, future efforts should be directed toward the development of low-cost, easy-to-use, and interoperable instruments that are easily adoptable by farmers in varied agroecosystems (Weiss and Baret 2017).

Partnership between researchers, technology vendors, and end-users is important to facilitate the effective integration of these technologies into current pest control practices (Huang et al. 2018). Capacity building and knowledge transfer programs also play a significant role to enhance farmers' and extension workers' awareness and adoption of precision agriculture methods (Singh et al. 2016). Here, we discuss the use of precision agriculture methods and smart technologies in the detection, tracking, and control of insect pests. We mention important strategies including remote sensing of the early detection of pests, GIS-based spatial analysis for site-specific interventions, wireless sensor networks for real-time monitoring of pests, and decision support systems for decision-making in pest control activities.

We also emphasize challenges and future perspectives for their integration into sustainable insect pest control practices. In presenting a complete understanding of present knowledge and research gaps, this review will inform future avenues for harnessing precision agriculture and smart technologies for sustainable and effective insect pest management.

12.3.8 Remote Sensing for Early Pest Detection Satellite Imagery

Satellite remote sensing has become a useful technique for massive insect pest surveillance and detection (Zhang et al. 2019).

Satellite high-resolution imagery offers spatially explicit data concerning vegetation vigor and crop stress, which can be used as indicators of pest infestation (Patel et al. 2013).

Vegetation indices, like the Normalized Difference Vegetation Index (NDVI) and Enhanced Vegetation Index (EVI), extracted from satellite imagery, have been extensively applied to evaluate crop health and identify abnormalities related to pest infestations (Huang et al. 2012).

For instance, MODIS and Landsat satellite imagery has been effectively used to monitor and map the spatial extent of Helicoverpa armigera (cotton bollworm) infestations within cotton fields (Joyce et al. 2008). By observing temporal changes in vegetation indices, scientists revealed areas with compromised crop vigor, reflecting possible bollworm damage. Satellite remote sensing has also been used to track other locust swarms, such as Locusta migratoria (migratory locust), in order to allow timely control interventions (Luo et al. 2013).

12.3.9 Aerial Photography and UAVs

UAVs and aerial photography provide greater spatial resolution and versatility than satellite data and can be used to conduct detailed pest monitoring at the field and regional levels (Liaghat et al. 2010). Detailed high-resolution aerial images collected by manned aircraft or UAVs have the ability to observe slight changes in crop health and delineate pest hotspots within fields (Santoso et al. 2016). Near-infrared (NIR) and visible cameras attached to UAVs allow crop stress assessment and pest damage detection (Paap et al. 2008).

UAVs mounted with multispectral and hyperspectral sensors have been effectively used to identify and map Diaphorina citri (Asian citrus psyllid) infestations in citrus plantations (Lan et al. 2009). By comparing spectral signatures related to psyllid-induced stress, scientists detected infected trees and created site-specific management techniques. Likewise, UAV-based remote sensing has also been utilized to identify and monitor Rhopalosiphum padi (bird cherry-oat aphid) infestations in wheat fields, facilitating precise insecticide application (Jiang et al. 2008).

12.3.10 Spectral Analysis Techniques

Advancements in the methods of spectral analysis further improved the ability of remote sensing in detecting insect pests (Sankaran et al. 2010). Hyperspectral imaging, which acquires data in many narrow bands of the spectrum, offers high-resolution information on plant biochemical and physiological characteristics (Hillnhütter et al. 2011). From the hyperspectral signatures, it is possible to identify minor variations in plant health that relate to infestation by pests even before symptomatic expression (Stenberg 2017). Machine learning models like SVM and RF have also been used on hyperspectral data to identify and map infested regions (Bajwa et al. 2017). SVM classifiers, for example, trained on hyperspectral data correctly identified Spodoptera frugiperda (fall armyworm) infestations in maize crops and exhibited good classification accuracies (Lowe et al. 2017).

In the same way, RF algorithms have also been applied to classify hyperspectral data and map infestations of Aphis gossypii (cotton aphid) in cotton crops, allowing for targeted control actions (Tremblay et al. 2012).

12.4 GIS-Based Spatial Analysis for Targeted Interventions

12.4.1 Spatial Pattern Analysis

Geographic information systems (GIS) are important to precision agriculture because they enable the integration, analysis, and visualization of spatial data concerning insect pest populations (Tremblay et al. 2012). GIS-based spatial pattern analysis methods, including spatial autocorrelation and hotspot analysis, allow the detection of pest distribution patterns and infestation hotspot delineation in fields (Yang et al. 2009). Such analysis offers useful information on the spatial dynamics of pest populations and informs targeted interventions and resource allocation optimization (Fitzgerald et al. 2010). Spatial autocorrelation techniques, including Moran's I and Getis-Ord Gi* statistics, were utilized to determine the spatial clustering of Bemisia tabaci (whitefly) populations in tomato plantations (Mahlein et al. 2016). Researchers used identification of high-density areas with high levels of spatial clusters to create site-specific management strategies, where control was targeted at high-risk areas. Similarly, hotspot analysis using the Getis-Ord Gi* statistic has been employed to map the spatial distribution of Tuta absoluta (tomato leafminer) infestations in tomato greenhouses, enabling targeted insecticide.

12.4.2 Geostatistical Modeling

Geostatistical modeling methods, including kriging and inverse distance weighting (IDW), are useful tools for interpolating data on pest populations and creating continuous maps of pest distribution (Banu 2015). These tools use spatially referenced data from pest samples to make estimates pest densities at unsampled points and gaining an overall picture of pest distribution in fields (Kamilaris et al. 2017). Geostatistical models facilitate the detection of pest gradients and the creation of management zones for interventions (Tzounis et al. 2017). Applied to detect and monitor Rhopalosiphum padi (bird cherry-oat aphid) populations in wheat fields to allow for insecticide applications targeting specific locations (Jiang et al. 2008).

12.4.3 Spectral Analysis Techniques

Improvement in the spectral analysis methods has also improved the potential of remote sensing in detecting insect pests (Sankaran et al. 2010). Hyperspectral imaging, which records data in many narrow spectral bands, offers precise information on biochemical and physiological characteristics of plants (Hillnhütter et al. 2011). Subtle changes in plant condition related to infestations by pests can be detected by hyperspectral signature analysis even before the appearance of any visible symptoms (Stenberg 2017). Machine learning models like support vector machines (SVM) and random forests (RF) have been used to classify and map areas of pest infestation from hyperspectral data (Bajwa et al. 2017).

For example, SVM classifiers learned from hyperspectral data accurately identified infestations of Spodoptera frugiperda (fall armyworm) in maize, with high classification accuracies (Lowe et al. 2017). Likewise, RF models have been employed to study hyperspectral images and chart Aphis gossypii infestations in cotton crops and facilitate targeted control measures (Tremblay et al. 2012).

12.4.4 Optimization of Control Interventions

Optimization of the choice and use of control interventions can be done by DSS taking multiple elements into account, including the dynamics of pest population, the effectiveness of various control methods, as well as the economic and environmental implications (Oberti and Shapiro 2016). These systems rely on mathematical algorithms for optimization to find the most efficient and environmentally friendly control measures for a specific problem (Gonzalez-de-Santos et al. 2017). Optimization of control interventions by DSS reduces insecticide use by farmers, saves them money, and makes pest control strategies more efficient (Bechar and Vigneault 2017). For instance, a DSS has been designed to maximize the deployment of pheromone traps and mating disruption methods in the control of Cydia pomonella (codling moth) in apple production (Miranda et al. 2019). The system combines information on moth population dynamics, orchard properties, and performance of various pheromone formulations to decide on the optimal trap and dispenser density and placement. By maximizing the application of these control strategies, the DSS allows growers to control codling moth populations while reducing pheromone-based management costs. Crop Yield and Production Forecasting Remote sensing contributes to crop yield forecasting through setting up statistical-empirical relationships between vegetation indices and yield (Thenkabail et al. 2002; Casa and Jones 2005).

Correct yield estimation is a critical component for national food policy decision-making, providing adequate resources and informed choices. Crop yield is subject to several factors, such as crop type, water and nutrient availability, weed competition, pest and disease attacks, and weather conditions. These factors affect the spectral response curve, with its growth and degradation characteristic indicating crop health and performance. High-resolution remote sensing technologies, including IRS P3 WiFS (Wide Field Sensor), IRS-1C WiFS, and LISS3, provide high-frequency observations, enabling the building of growth profiles and yield-related parameters at regional scales (Menon 2012).

12.5 Crop Yield and Production Forecasting

Remote sensing plays a crucial role in forecasting crop yields by establishing statistical-empirical relationships between vegetation indices and yield (Thenkabail et al. 2002; Casa and Jones 2005). Accurate yield predictions are essential for national food policy planning, enabling better resource allocation and decision-making. Crop yield is influenced by multiple factors, including crop variety, water and nutrient availability, weed competition, pest and disease infestations, and weather conditions. These variables impact the spectral response curve, with its growth and decay patterns reflecting crop health and performance.

Advanced remote sensing technologies, such as IRS P3 WiFS (Wide Field Sensor), IRS-1C WiFS, and LISS3, offer high-frequency observations, making it possible to construct growth profiles and extract yield-related parameters at regional levels (Menon 2012). These advancements improve the accuracy and efficiency of crop monitoring and forecasting, ultimately benefiting agricultural planning and food security.

12.5.1 Implications for Agriculture in Delta Regions

Enhanced Crop Monitoring: The integration of RS and GIS allows for precise tracking of crop phenology, aiding in timely agricultural interventions.

Land Use Planning: Understanding LULC dynamics assists policymakers in making informed decisions to balance development and agricultural sustainability.

Environmental Conservation: Monitoring environmental changes helps in identifying areas at risk, enabling the implementation of conservation strategies.

12.6 Limitations, Future Scope in Delta-Related Agricultural Studies: The Role of RS and GIS

12.6.1 Limitations

- Data Availability and Resolution: High-resolution satellite data are often costly or unavailable for continuous monitoring, limiting the frequency and precision of observations in delta regions.

- Cloud Cover and Weather Dependency: Delta areas frequently experience cloud cover, particularly during monsoon seasons, which hampers optical remote sensing applications.
- Limited Ground Truth Data: Accurate validation of RS and GIS outputs requires field data, which can be sparse or difficult to obtain in remote or inundated deltaic zones.
- Technical Expertise and Infrastructure: The effective implementation of RS and GIS in agricultural monitoring in delta regions requires trained personnel and computational infrastructure, which may be lacking in developing regions.

12.6.2 Research Gaps

- Temporal Dynamics of Cropping Patterns: There is insufficient research on the spatiotemporal analysis of cropping intensity and diversity in delta areas using long-term RS datasets.
- Soil Salinity Mapping: While salinity is a major concern in deltas, particularly due to sea water intrusion, integrated studies using RS, GIS, and ground surveys remain limited.
- Integration of Multisource Data: More studies are needed that combine satellite imagery, UAV data, and socio-economic variables for a holistic understanding of agricultural systems in deltaic landscapes.
- Climate Impact Assessment: There is a lack of predictive modeling integrating RS and GIS with climate scenarios to forecast the impact of sea-level rise and extreme weather events on agricultural productivity.

12.6.3 Advances in Spaceborne Remote Sensing

The evolution of spaceborne remote sensing over the past 50 years, from Sputnik 1 to WorldView-1, has been remarkable. Present trends (Fig. 12.3) indicate several key innovations shaping the future of this field:

Diverse Data Sources—The availability of data from multiple sensors with varying spatial, spectral, and radiometric characteristics is expanding. These datasets will be sourced from a variety of platforms, enhancing data accessibility and application.

Data Harmonization—Significant advancements have been made in integrating and synthesizing data from diverse sensors. This enables the seamless use of datasets with varying characteristics, allowing vendors to harmonize and add value to multi-source data products.

Constellations of Satellites—Missions such as Rapid Eye, with its 6.5-m resolution, five spectral bands (including a red-edge band), and global coverage every 13 days, highlight the increasing importance of high-resolution and high-frequency data collection.

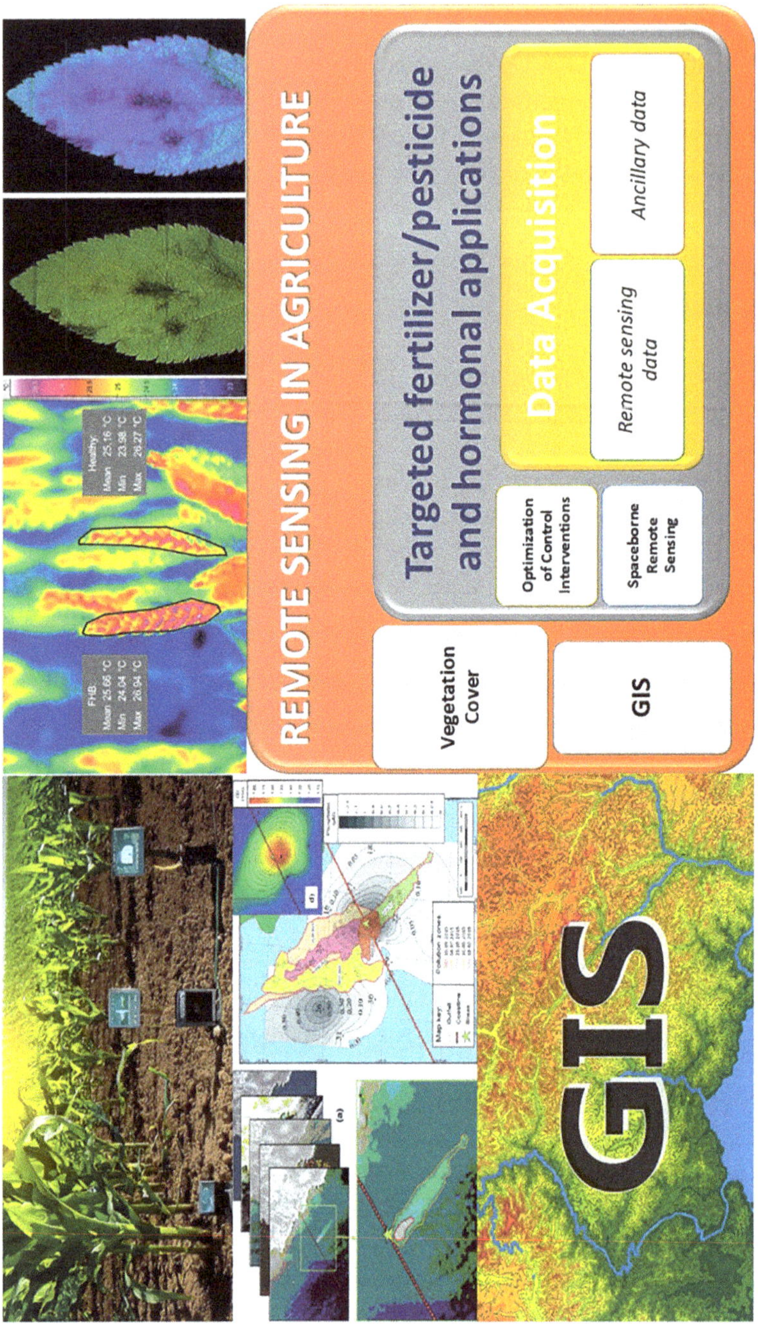

Fig. 12.3 Remote sensing applications in precision agriculture

Managing such vast amounts of data will require innovations in storage, processing, and retrieval.

Micro-Satellites—These specialized satellites, designed for specific geographic regions or targeted applications, are gaining traction due to their cost-effectiveness and tailored data collection capabilities.

Global Wall-to-Wall Coverage—For environmental and natural resource applications, comprehensive global coverage remains essential. Satellites like Landsat will continue to play a crucial role in long-term Earth observation efforts.

Hyperspectral and Hyperspatial Sensors—The increasing availability of hyperspectral and high-resolution data presents new challenges in data mining, processing, storage, and retrieval. These challenges necessitate advanced computational solutions and efficient data management strategies.

Enhanced Data Access and Visualization—Advances in data presentation and accessibility—through platforms like Google Earth—are expanding the reach of remote sensing applications. These innovations will attract new users and drive broader adoption in environmental science and natural resource management. With continued advancements in sensor technology, data integration, and accessibility, remote sensing will play an increasingly vital role in monitoring, analyzing, and managing Earth's resources.

12.6.4 Future Scope in Delta Regions

Future research and development in Remote Sensing (RS) and Geographic Information Systems (GIS) applications for precision farming should proceed towards more integrated, adaptive, and farmer-oriented strategies. Although current applications have shown remarkable improvements in crop monitoring, soil mapping, and input optimization, the future phases of innovation should focus on scalability, accessibility, and adaptability in relation to environmental and socio-economic conditions.

Variable-rate input systems continue to be a priority, especially for deltaic ecosystems and areas where soil fertility, salinity, and water availability are highly variable within short distances. Generation of sophisticated prescription maps for irrigation, fertilization, and pesticide use will enable the farmer to implement site-specific strategies, preventing wastage of resources and avoiding environmental impacts such as nutrient leaching and depletion of the groundwater. This system will be helpful in tackling immediate concerns of agricultural sustainability in sensitive landscapes.

Early warning systems must be enhanced by combining high-resolution satellite data and real-time weather forecasts, hydrological models, and pest networks. These systems would allow prompt detection of floods, droughts, pest outbreaks, and disease outbreaks and hence offer farmers and policymakers proactive management options. The integration of remote sensing and wireless sensor networks, as exemplified in pest management case studies, points towards the possibility of operational early warning systems that reduce crop losses and ensure food security.

The idea of digital twin modeling holds a promising horizon. By developing dynamic, data-driven models of farm fields, farmers and planners are able to simulate various management and climate scenarios virtually prior to their application in real life. This shall not only enhance decision-making but also enable policymakers to analyze long-term effects of climate change, land-use changes, and resource allocation policies in susceptible ecosystems like the Nile, Cauvery, and Mahanadi deltas (Table 12.3).

Community-based GIS platforms will be key to democratizing precision agriculture. By engaging local farmers in data collection, validation, and interpretation, participatory GIS strategies can close the distance between high-tech applications and farm-level practices. This assimilation of local knowledge with high-quality geospatial datasets maximizes the chances that solutions are locally meaningful, socially inclusive, and adopted on the ground.

Table 12.3 Summary of recent studies on delta agriculture using remote sensing and GIS

Study	Region	Focus area	Key findings	RS/GIS role
SICKLE Dataset (2023)	Cauvery Delta, India	Paddy cultivation	Dataset of 2,370 samples from 388 plots developed, annotated with crop parameters including sowing, transplanting, harvesting dates, and yield	Multi-sensor satellite imagery (Landsat-8, Sentinel-1, Sentinel-2) applied for crop monitoring and yield prediction
Mahanadi Delta Study (2023)	Mahanadi Delta, India	Land Use/Land Cover (LULC) changes	Significant growth in forests and barren lands observed, accompanied by a decline in agricultural fields between 2010 and 2020	Landsat imagery and GIS tools employed for LULC classification and change detection
Niger Delta Assessment (2006)	Niger Delta, Nigeria	Coastal environmental changes	Development-induced alterations in land cover identified, exerting substantial impacts on the coastal environment	Landsat TM and ETM + images integrated with GIS for temporal environmental change analysis
Egypt Delta Monitoring (2014)	Nile Delta, Egypt	Agricultural land degradation	Progressive urban encroachment documented, resulting in marked reduction of agricultural lands over the past decade	Multi-temporal Landsat data combined with GIS techniques used for monitoring land cover changes

Last but not least, the integration of cutting-edge technologies like the Internet of Things (IoT), Artificial Intelligence (AI), machine learning, and hyperspectral remote sensing will reshape the analytical functions of precision agriculture. Field sensors with IoT capability can monitor soil moisture, nutrient levels, and microclimate factors continuously, while AI and machine learning can convert enormous data into actionable information for yield estimation, stress identification, and allocation of resources. Hyperspectral sensors, with their capacity to collect high-resolution spectral information, will also improve the accuracy of crop discrimination, nutrient determination, and disease detection.

Together, these guidelines will enhance the contributions of RS and GIS to the development of precision farming as a foundation of climate-resilient, resource-conserving, and food-secure agri-food systems. Coupling technological progress with participatory strategies and region-based tailoring, research in the future can guarantee that precision farming is not just a scientific breakthrough but also a workable and fair solution for sustainable agriculture globally.

12.7 Conclusion

The combination of Remote Sensing (RS) and Geographic Information Systems (GIS) is a paradigm shift in precision agriculture, providing new solutions for sustainable agricultural production. Compared to conventional farming practices that are based primarily on visual observation and generalized input application, RS and GIS supply spatially explicit, timely, and precise information about the dynamics of farmland. Such technologies enable comprehensive evaluation of crop health, fertility levels of the soil, moisture levels, infestation by pests and diseases, and yield heterogeneity. This enables farmers and policymakers to make well-informed choices that increase productivity and reduce wastage of resources.

The most powerful support provided by RS and GIS is their capacity for monitoring growth of crops and stress indicators on large areas with the least labor input. Vegetation indices like NDVI, SAVI, and EVI from satellite and UAV imagery provide uninterrupted, non-destructive monitoring of crop health, whereas SAR data provides improved analysis of soil moisture even in the presence of clouds. Likewise, GIS allows spatial datasets like soil maps, topography, and weather to be combined, thus providing an option for the demarcation of zones for variable-rate application of inputs like fertilizers, irrigation, and pesticides. This site-specific management enhances efficiency, lowers costs of production, and reduces environmental degradation.

Aside from input optimization, RS and GIS help in forecasting yield, climate change adaptation, and long-term planning in agriculture. Yield models that combine spectral indices, weather, and soil parameters generate precise predictions that aid in market planning and food security policy. In deltaic areas and other fragile ecosystems, RS and

GIS also contribute significantly to monitoring land use change, mapping soil salinity, and flood/drought hazard assessments. These applications enhance resilience in highly vulnerable situations subject to environmental change and human impacts.

Despite their huge potential, a number of challenges still constrain the universal adoption of RS- and GIS-based precision agriculture. Satellite imagery of high resolution and UAV data can be costly and need advanced technical skills in processing and interpretation. Cloud cover is an ongoing drawback for optical remote sensing, while combination of multi-source datasets into unified management systems is computationally challenging. Moreover, the unavailability of ground-truth data in many developing areas limits validation and accuracy. Tackling these challenges demands capacity development, infrastructure investment, and cross-disciplinary research that links technological innovation with concrete application in the field.

Despite this, the development path of remote sensing and GIS holds a bright future. Hyperspectral sensors, micro-satellites, IoT-based farm sensors, and AI-based data analytics are anticipated to improve the accuracy, availability, and pricing of these technologies. When coupled with decision-support systems, digital twin modeling, and participatory GIS platforms, RS and GIS can offer integrated solutions to manage crop production, save resources, and mitigate climate threats.

Finally, RS and GIS are no longer supplementary tools but integral parts of contemporary precision agriculture. Their ability to rework unprocessed spatial information into usable knowledge guarantees more intelligent farming methods, enhanced yields, and better resource durability. By breaching current obstacles and embracing technical advancements, precision agriculture driven by RS and GIS can play a significant role in achieving global food security, climatic resilience, and sustainable agricultural landscape management. In (Table 12.3) delta regions such as the Cauvery, Mahanadi, Nile, and Niger, RS and GIS have effectively tracked paddy crop phenology, land-use/land-cover changes, soil salinity risks, and urban encroachment, thereby revealing critical threats to agricultural sustainability. Integration of multisource datasets (optical, SAR, UAV, and ground truth) significantly improves the accuracy of yield modeling and resource management. Vegetation indices (NDVI, SAVI, EVI) and SAR backscatter provide reliable indicators for crop health, moisture stress, and input zoning, reducing chemical and water wastage.

Delta-specific applications of RS and GIS not only enhance crop monitoring and early warning systems but also guide policymakers in land-use planning and environmental conservation.

References

Anyamba A, Tucker CJ, Eastman JR (2001) NDVI anomaly patterns over Africa during the 1997/98 ENSO warm event. Int J Remote Sens 22:1847–1859

Atzberger C (2013) Advances in remote sensing of agriculture: context description, existing operational monitoring systems and major information needs. Remote Sens 5(2):949–981

Bajwa SG, Rupe JC, Mason J (2017) Soybean disease monitoring with leaf reflectance. Remote Sens 9(2):127

Bandara KMPS (2003) Monitoring irrigation performance in Sri Lanka with high frequency satellite measurements during the dry season. Agric Water Manag 58(2):159–170

Banu S (2015) Precision agriculture: tomorrow's technology for today's farmer. J Food Process Technol 6(8):1

Batra N, Islam S, Venturini V, Bisht G, Jiang L (2006) Estimation and comparison of evapotranspiration from MODIS and AVHRR sensors for clear sky days over the Southern Great Plains. Remote Sens Environ 103:1–15

Bechar A, Vigneault C (2017) Agricultural robots for field operations. Part 2: operations and systems. Biosys Eng 153:110–128

Bernardes T, Meriera MA, Adami M, Giarolle A, Rudorff BFT (2012) Monitoring biennial bearing effect on coffee yield using MODIS remote sensing imagery. Remote Sens 4:2492–2509

Blackmore S, Godwin RJ, Fountas S (2003) The analysis of spatial and temporal trends in yield map data over six years. Biosci Eng 84:455–466

Bramley RGV (2009) Lessons from nearly 20 years of precision agriculture research, development, and adoption as a guide to its appropriate application. Crop Past Sci 60:197–217

Bredemeier C, Schmidhalter U (2005) Laser-induced chlorophyll fluorescence sensing to determine biomass and nitrogen uptake of winter wheat under controlled environment and field conditions. In: Stafford JV (ed) Precision agriculture, proceedings. Academic Publishers, Wageningen, pp 273–280

Calvao T, Palmeirim JM (2004) Mapping Mediterranean scrub with satellite imagery: biomass estimation and spectral behaviour. Int J Remote Sens 25:3113–3126

Casa R, Jones HG (2005) LAI retrieval from multiangular image classification and inversion of a ray tracing model. Remote Sens Environ 98:414–428

Casa R, Cavalieri A, Locascio B (2011) Nitrogen fertilization management in precision agriculture: a preliminary application example on maize. Italian J Agron 6:23–27

De beurs K, Townsend P (2008) Estimating the effect of gypsy moth defoliation using MODIS. Remote Sens Environ 112: 3983–3990

De Luca G, Silva JM, Cerasoli S, Araújo J, Campos J, Di Fazio S, Modica G (2019) Object-based land cover classification of cork oak woodlands using UAV imagery and OrfeoToolBox. Remote Sens 11(10):1238

Delin S, Berglund K (2005) Management zones classified with respect to drought and waterlogging. Prec Agric 6:321–340

Doraiswamy PC, Sinclair TR, Hollinger S, Akhmedov B, Stern A, Prueger J (2005) Application of MODIS derived parameters for regional crop yield assessment. Remote Sens Environ 97:191–202

Dutta D, Kundu A, Patel NR, Saha SK, Siddiqui AR (2015) Egypt J Remote Sens Space Sci 18:53–63

Ferencz CSP, Bognar J, Lichtenberger D, Hamar GY, Tarcsai G, Timar G, Molnar SZ, Pasztor P, Steinbach B, Szekely OE, Ferencz, FerenczArkos I (2004) Crop yield estimation by satellite remote sensing. Int J Remote Sens 25(20):4113–4149

Fitzgerald GJ, Rodriguez D, O'Leary G (2010) Measuring and predicting canopy nitrogen nutrition in wheat using a spectral index—the canopy chlorophyll content index (CCCI). Field Crop Res 116(3):318–324

Fountas S, Mylonas N, Malounas I, Rodias E, Hellmann Santos C, Pekkeriet E (2020) Agricultural robotics for field operations. Sensors 20(9):2672

Franke J, Menz G (2007) Multi temporal wheat disease detection by multi spectral remote sensing. Precis Agric 8:161–172

Franklin S (2001) Remote sensing for sustainable forest management. Lewis Publisher, Boca Raton, p 407

Gonzalez-de-Santos P, Ribeiro A, Fernandez-Quintanilla C, Lopez-Granados F, Brandstoetter M, Tomic S, Kaplanis G et al (2017) Fleets of robots for environmentally-safe pest control in agriculture. Precis Agric 18(4):574–614

Gu Y, Brown JF, Verdin JP, Wardlow B (2007) A five year analysis of MODIS NDVI and NDWI for grassland drought assessment over the central great plains of the United States. Geophys Res Lett 34

Hadria R, Duchenin B, Lahrouni A, Khabba S, Er-Raki S, Dedieu G, Chehbouni AG, Olioso A (2006) Monitoring of irrigated wheat in a semiarid climate using crop modeling and remote sensing data: impact of satellite revisit time frequency. Int J Remote Sens 27(6):1093–1117

Hillnhütter C, Mahlein AK, Sikora RA, Oerke EC (2011) Remote sensing to detect plant stress induced by Heteroderaschachtiiand Rhizoctonia solani in sugar beet fields. Field Crop Res 122(1):70–77

Huang J, Liao H, Zhu Y, Sun J, Sun Q, Liu X (2012) Hyperspectral detection of rice damaged by rice leaf folder (Cnaphalocrocismedinalis). Comput Electron Agric 82:100–107

Huang Y, Chen ZX, Yu T, Huang XZ, Gu XF (2018) Agricultural remote sensing big data: management and applications. J Integr Agric 17(9):1915–1931

Huete AR (1988) A soil-adjusted vegetation index (SAVI). Remote Sens Environ 25:295–309

Jackson RD (1986) Remote sensing of biotic and abiotic plant stress. Annu Rev Phytopathol 24:265–286

Jackson RD, Idso SB, Reginato RJ, Pinter PJ (1981) Canopy temperature as a crop water stress indicator. Water Resour Res 17:1133–1138

Ji L, Peters AJ (2003) Assessing vegetation response to drought in the northern Great Plains using vegetation and drought indices. Remote Sens Environ 87:85–98

Jiang JA, Tseng CL, Lu FM, Yang EC, Wu ZS, Chen CP, Liao CS et al (2008) A GSM-based remote wireless automatic monitoring system for field information: a case study for ecological monitoring of the oriental fruit fly, *Bactrocera dorsalis* (Hendel). Comput Electron Agric 62(2):243–259

Johnson GA, Cardina J, Mortensen DA (1997) Site–specific weed management: current and future direction. In: The state of site–specific management for agriculture, pp 131–147

Joyce KA, Dingle JG, Lauer TE, Olson BD (2008) Remote sensing of cottony cushion scale (Iceryapurchasi) and its biological control agent (Rodoliacardinalis) on Pittosporum tobira using hyperspectral imaging. J Appl Remote Sens 2(1):023509

Kamilaris A, Prenafeta-Boldú FX (2018) Deep learning in agriculture: a survey. Comput Electron Agric 147:70–90

Kaur R, Jaidka M (2014) Spectral reflectance characteristics to distinguish Malvaneglecta in wheat (Triticumaestivum). Indian J Agric Sci 84(10):1243–1249

Kaur R, Jaidka M, Kingra PK (2013) Study of optimum time span for distinguishing Rumexspinosus in wheat crop through spectral reflectance characteristics. Proc Natl Acad Sci India Sect B Biol Sci

Kingra PK, Majumder D, Chandra B, Viswavidyalaya K, Singh SP (2016a) Application of remote sensing and GIS in agriculture and natural resource management under changing climatic conditions. Agric Res J 53(3):295–302

Kingra PK, Majumder D, Singh SP (2016b) Application of remote sensing and GIS in agriculture and natural resource management under changing climatic conditions. Agric Res J 53(3):295–302

Kogan F, Yang B, Guo W, Pei Z, Jiao X (2005) Modelling corn production in China using AVHRR-based vegetation health indices. Int J Remote Sens 26(11):2325–2336

Kumar N, Yamaç SS, Velmurugan A (2015) Applications of remote sensing and GIS in natural resource management 20(1):1–6

Lamb DW, Weedon MM, Rew LJ (1999) Evaluating the accuracy of mapping weeds in seeding crops using airborne digital imaging: Avena spp. in seeding triticale. Weed Res 39(6): 481–492

Lan Y, Zhang H, Lacey R, Hoffmann WC, Wu W (2009) Development of an integrated sensor and instrumentation system for measuring crop conditions. Agric Eng Int CIGR J 11:1–16

Lee W, Alchanatis V, Yang C, Hirafuji M, Moshou D, Li C (2010) Sensing technologies for precision specialty crop production. Comput Electronic Agric 74:2–33

Li Y, Chen D, Walker CN, Angus JF (2010) Estimatingthe nitrogen status of crops using a digital camera. FieldCrops Res 118:221–227

Liaghat S, Balasundram SK (2010) A review: the role of remote sensing in precision agriculture. Am J Agric Biol Sci 5(1):50–55

Liaghat S, Balasundram SK, Mohammadi J, Meon S (2010) Early detection of basal stem rot in oil palm plantations using airborne hyperspectral sensing: a preliminary study. Pertanika J Sci Technol 18(S):47–53

Liakos KG, Busato P, Moshou D, Pearson S, Bochtis D (2018) Machine learning in agriculture: a review. Sensors 18(8):2674

Lowe A, Harrison N, French AP (2017) Hyperspectral image analysis techniques for the detection and classification of the early onset of plant disease and stress. Plant Methods 13(1):80

Luo J, Huang W, Zhao J, Zhang J, Zhao C, Ma R (2013) Detecting aphid density of winter wheat leaf using hyperspectral measurements. IEEE J Selected Topics Appl Earth Observations Remote Sens 6(2):690–698

Mahlein AK (2016) Plant disease detection by imaging sensors—parallels and specific demands for precision agriculture and plant phenotyping. Plant Dis 100(2):241–251

Mahlein AK, Rumpf T, Welke P, Dehne HW, Plümer L, Steiner U, Oerke EC (2013) Development of spectral indices for detecting and identifying plant diseases. Remote Sens Environ 128:21–30

Menon ARR (2012) Remote sensing applications in agriculture and forestry. A paper from the proceedings of the Kerala environment congress, pp 222–235

Miranda J, Ponce P, Molina A, Wright P (2019) Sensing, smart and sustainable technologies for Agri-Food 4.0. Comput Ind 108:21–36

Mirik et al (2013) Remote monitoring of wheat streak mosaic progression using sub-pixel classification of Landsat 5 TM imagery for site specific disease management in Winter Wheat. Adv Remote Sens 2:16–28

Moran MS, Inoue Y, Barnes EM (1997) Opportunities and limitations for image–based remote sensing in precision crop management. Remote Sens Environ 61(3):319–246

Moran MS, Clarke TR, Inoue Y, Vidal A (1994) Estimating crop water deficit using the relation between surface-air temperature and spectral vegetation index. Remote Sens Environ 49:246–263

Nayyar A, Puri V, Le DN (2017) Internet of nano things (IoNT): next evolutionary step in nanotechnology. Nanosci Nanotechnol 7(1):4–8

Neale CMU, Jayanthi H, Wright JL (2005) Irrigation water management using high resolution airborne remote sensing. Irrig Drain Syst 19(3–4):321–336

Nellis MD, Pricey KP, Rundquist D (2009) Remote sensing of cropland agriculture. In: The SAGE handbook of remote sensing, vol 26
Nicholson SE, Farrar TJ (1994) The influence of soil type on the relationships between NDVI, rainfall, and soil moisture in semiarid Botswana: I. NDVI response to rainfall. Remote Sens Environ 50:107–120
Oberti R, Shapiro A (2016) Advances in robotic agriculture for crops. Biosys Eng 146:1–2
Paap A, Askraba S, Alameh K, Rowe J (2008) Photonic-based spectral reflectance sensor for ground-based plant detection and weed discrimination. Opt Express 16(2):1051–1055
Patel N, Yadav K, Singh A (2013) Monitoring spatio-temporal dynamics of beetles (Coleoptera) diversity in different agroecosystems using light trap: a case study. J Environ Biol 34(3):483–489
Peña-Barragán JM, Ngugi MK, Plant RE, Six J (2011) Object-based crop identification using multiple vegetation indices, textural features and crop phenology. Remote Sens Environ 115(6):1301–1316
Prasad AK, Chai L, Singh RP, Kafatos M (2006) Crop yield estimation model for Iowa using remote sensing and surface parameters. Int J Appl Earth Obs Geoinf 8(1):26–33
Profile SEE (2016) Remote sensing and GIS Applications in agriculture some of the authors of this publication are also working on these related projects: assessment of Climate change effect on mustard yield in western Haryana View project Computation of agro-meteorological indices in Barley cultivars under different growing environments at Hisar View project
Reed BC, Brown JHF, Vander Zee D, Loveland TR, Merchant JW, Ohnlen DO (1994) Measuring phenological variability from satellite imagery. J Veg Sci 5:703–714
Riedell WE, Osborne SL, Hesler LS (2004) Insect pest and disease detection using remote sensing techniques. In: Proceedings of 7th international conference on precision agriculture. Minneapolis, MN, USA
Rouse JW, Haas RH, Schell JA, Deering DW (1973a) Monitoring vegetation systems in the Great Plains with ERTS. Third ERTS Symp NASA 351(1):309–317
Rouse JW Jr, Haas RH, Schell JA, Deering DW (1973b) Monitoring the vernal advancement and retrogradation (green wave effect) of natural vegetation (No. NASA-CR-132982)
Sakamoto T, Yokozawa M, Toritani H, Shibayama M, Ishitsuka N, Ohno H (2005) A crop phenology detection method using time series MODIS data. Remote Sens Environ 96:366–374
Sankaran S, Mishra A, Ehsani R, Davis C (2010) A review of advanced techniques for detecting plant diseases. Comput Electron Agric 72(1):1–13
Santoso H, Tani H, Wang X (2016) A simple method for detection and counting of oil palm trees using high-resolution multispectral satellite imagery. Int J Remote Sens 37(21):5122–5134
Seiler RA, Kogan F, Wei G (2000) Monitoring weather impact and crop yield from NOAA AVHRR data in Argentina. Adv Space Res 26:1177–1185
Shamshiri RR, Weltzien C, Hameed IA, Yule IJ, Grift TE, Balasundram SK, Chowdhary G et al (2018) Research and development in agricultural robotics: a perspective of digital farming. Int J Agric Biol Eng 11(4):1–14
Siddiqui AR (2004) Regional evaluation of desertification hazards in the Aridlands of Western Rajasthan (an unpublished Ph. D.thesis). AMU, Aligarh, Uttar Pradesh, India, p 221
Singh A, Ganapathysubramanian B, Singh AK, Sarkar S (2016) Machine learning for high-throughput stress phenotyping in plants. Trends Plant Sci 21(2):110–124
Singh I, Srivastava IA, Chandna P, Gupta R (2006) Crop sensors for efficient nitrogen management in sugarcane. Potential and constraints. Sugar Technol 8:299–302
Sruthi S, Mohammed Aslam MA (2015) Agricultural drought analysis using the NDVI and land surface temperature data; a case study of Raichur District. In: International conference on water resources, coastal and ocean engineering, vol 4, pp 1252–1264

Stenberg JA (2017) A conceptual framework for integrated pest management. Trends Plant Sci 22(9):759–769

Thenkabail PS, Lyon GJ, Huete A (2011) Advances in hyperspectralremote sensing of vegetation and agricultural crops. In: Hyperspectral remote sensing of vegetation. CRC Press/Taylor and Francis Group, Boca Raton, London, New York, vol 1, pp 3–29

Thenkabail PS, Enclona EA, Ashton MS, Van Der Meer B (2004) Accuracy assessments of hyperspectral waveband performance for vegetation analysis applications. Remote Sens Environ 91(3–4):354–376

Thenkabail PS, Mariotto I, Gumma MK, Middleton EM, Landis DR, Huemmrich KF (2013) Selection of Hyperspectral Narrowbands (HNBs) and Composition of Hyperspectral Twoband Vegetation Indices (HVIs) for biophysical characterization and discrimination of crop types using field reflectance and Hyperion/EO-1 data. IEEE J Selected Topics Appl Earth Observations Remote Sens 6(2):427–439

Thenkabail PS, Smith RB, De-Pauw E (2002) Evaluation of narrowband and broadband vegetation indices for determining optimal hyperspectral wavebands for agricultural crop characterization. Photogramm Eng 68:607–621

Tremblay N, Wang Z, Cerovic ZG (2012) Sensing crop nitrogen status with fluorescence indicators. A review. Agron Sustain Dev 32(2):451–464

Tremblay N, Wang Z, Ma BL, Belec C, Vigneault P (2009) A comparison of crop data measured by two commercial sensors for variable-rate nitrogen application. Precis Agric 10(2):145–161

Tzounis A, Katsoulas N, Bartzanas T, Kittas C (2017) Internet of Things in agriculture, recent advances and future challenges. Biosys Eng 164:31–48

Vidal A, Pinglo F, Durand H, Devaux-Ros C, Maillet A (1994) Evaluation of a temporal fire risk index in Mediterranean forests from NOAA thermal IR. Remote Sens Environ 49:296–303

Wallace JF, Caccetta PA, Kiiveri HT (2004) Recent developments in analysis of spatial and temporal data for landscape qualities and monitoring. Australian Ecol 29:100–107

Wang J, Price KP, Rich PM (2001) Spatial patterns of NDVI in response to precipitation and temperature in the central GreatPlains. Int J Remote Sens 22:3827–3844

Weiss M, Baret F (2017) Using 3D point clouds derived from UAV RGB imagery to describe vineyard 3D macro-structure. Remote Sens 9(2):111

Wilhite DA, Glantz MH (1985) Understanding the drought phenomenon: the role of definitions. Water Int 10:111–120

Williams D, Stauffer M, Leung K (1979) A forester's look at the application of image manipulation techniques to Landsat data. In: Symposium on remote sensing for vegtation damage assessment, February 14–16, Washington, The Society, Falls Church, VA, pp 221–229

Yang C, Everitt JH, Bradford JM (2009) Evaluating high resolution SPOT 5 satellite imagery for crop identification. Comput Electron Agric 65(2):347–354

Zhang J, Huang Y, Pu R, Gonzalez-Moreno P, Yuan L, Wu K, Huang W (2019) Monitoring plant diseases and pests through remote sensing technology: a review. Comput Electron Agric 165:104943

Spatiotemporal Variation in Land Use Land Cover in Piyali River Sub-Basin in the Response to Local Climate Change Using Multispectral Remote Sensing Data

13

Suvendu Halder, Satiprasad Sahoo, Tumpa Hazra, and Anupam Debsarkar

Abstract

Climate change poses significant social, economic, and environmental challenges for farmers whose livelihoods depend on natural resources. Analyzing Land Use Land Cover (LULC) changes has proven to be an essential method for monitoring transformations across various temporal and spatial scales. The Normalized Difference Vegetation Index (NDVI) serves as a valuable remote sensing tool, effectively capturing vegetation patterns across diverse ecological regions and aiding in the study of phonological cycles. This research utilized remote sensing and Geographic Information System (GIS) techniques to analyze LULC changes over 34 years in Piyali sub-basin. Additionally, some questionnaires were conducted with local farmers to explore the relationship between climate change and NDVI. LULC maps based on NDVI for 1990, 2000, 2010 and 2024 were created using Arc GIS. Regression analysis (R^2) was employed to examine the correlation between temperature and vegetation cover (NDVI). Findings revealed an increase in built-up areas from 2.25% to 9.70%, coupled with a decline in vegetation cover by 12% between 1991 and 2024. Mean

S. Halder
Faculty of Interdisciplinary Studies, Law & Management, Jadavpur University, Kolkata, India
e-mail: suvenduhalder52@gmail.com

S. Sahoo
International Center for Agricultural Research in the Dry Areas (ICARDA), Cairo, Egypt
e-mail: satiprasad.iitk@gmail.com

T. Hazra · A. Debsarkar (✉)
Department of Civil Engineering, Jadavpur University, Kolkata, India
e-mail: anupamju1972@gmail.com

T. Hazra
e-mail: tumpa.hazra@jadavpuruniversity.in

C. Singha et al. (eds.), *Sustainable Management of Delta Ecosystems Resilience*,
Deltas of the World, https://doi.org/10.1007/978-3-032-16261-8_13

NDVI values declined from 0.24 to 0.20 during this period. Survey responses indicated that 85% of farmers observed changes in climate, 70% noted agriculture affected by climate change and 81% reported climate changing due to deforestation. Regression analysis demonstrated a negative relationship between temperature and NDVI. This study highlights practical approaches and tools for local adaptation and sustainable governance of agricultural systems under changing climatic conditions, offering valuable insights for policymakers and urban planners in managing LULC at the local scale.

13.1 Introduction

Climatic change is the long-term modification of Earth's climatic patterns, mostly brought on by human activity that causes extreme weather occurrences and global warming. This change has largely impacted Land Use Land Cover (LULC) globally due to shifting temperatures and precipitation concentrations (Abdullahi et al. 2018; Pradhan et al. 2020). Land Use Land Cover (LULC) alteration refers to modifications to the Earth's natural or man-made surface features. The LULC modifications are helpful in illustrating and evaluating the impact of different development paths, as well as in assisting decision-makers in developing scientific plans (Li 2020) and putting into practice workable policies that preserve natural resources (Shah et al. 2021; Tariq & Shu., 2020). The LULC changes were analyzed as a crucial tool for evaluating global changes at different time and space scales (Hoffmann et al. 2012). It is a rapid and widespread process that is driven by human activity, and these changes frequently have an impact on people (Zahoor et al. 2019). According to Ahmed et al. (2013), anthropogenic activity also appears to be causing drastic changes to the Earth's surface. Through a reduction in ecological facilities, the LULC modifications intensified the interplay between resources, governmental uncertainties, and societies in response to socioeconomic calamities and global warming (Omran et al. 2012).

The Normalized Difference Vegetation Index (NDVI), a popular remote sensing metric, compares the reflectance of red and near-infrared (NIR) light from the Earth's surface to determine the health, density, and vigor of vegetation (Weng et al. 2004). According to Sabagh et al. (2020), the NDVI as a parameter used to measure the quantity, quality, and development of vegetation by relating red and near-infrared spectral information.

Numerous investigations have examined the connection between Land Surface Temperature (LST) and NDVI. The climate is a significant element that has an impact on the structure of vegetation dynamics (Hussain et al. 2020). A number of data sets and methodologies have been employed in the classification and mapping of LULC changes by remote sensing (Din et al. 2022). According to Pal and Ziaul (2017), Landsat images have greatly aided in the classification of different land features at a larger scale. Debnath et al. (2022) used RS/GIS to study the effects of LULC change and climate change in the

Brahmaputra basin. Kogo et al. (2019) reviewed the effects of climate change and changes in land use and cover in Africa using geospatial methods. The study examined changes in land use and cover in 1995, 2001, 2010, and 2017 using remote sensing and GIS tools. The maximum likelihood technique in ENVI version 5.4 was used to pre-process and classify multispectral Landsat (TM, ETM+, and OLI) images. All of the photos had overall categorization accuracy of >80%. The findings showed significant conversions in a range of trends and sizes for every form of land use and land cover. Remote sensing evaluation of climate-driven vegetation change in South Asian watersheds was presented by Hassan et al., 2024. To ensure accuracy in classifying various land types, image processing and classification were done using the Support Vector Machine (SVM) classifier and ArcGIS software. Fu and Weng (2016) demonstrated the NDVI were evaluated using Landsat imagery such as an Operation Land Imager (OLI), Enhanced Thematic Mapper (ETM+), and Thematic Mapper (TM). According to Zhang et al. (2017), remote sensing (RS) has become crucial for monitoring changes in vegetation because it provides effective satellite information regarding changes in plant cover and LST.

In India, the primary threat to food security, agriculture, and rural livelihoods for billions of people by climate change. The region of land where rainfall and surface runoff flow into a particular tributary or segment of the main river system defines a river sub-basin, which is a smaller hydrological unit inside a larger river basin. Unfortunately, West Bengal's ecosystems as well as river sub-basin are deteriorating as a result of climate change, which is considered a serious issue that has a substantial negative economic impact. Currently, considerable human influence in the Piyali sub-basin has resulted from unpredictable weather patterns. Groundwater is depleted during the dry season due to excessive consumption in home and agricultural settings. The growing population having a severe impact on agriculture as more land is needed to meet basic human needs. Studying how climate change affects agriculture and other land cover in the Piyali River Sub-Basin of the Sundarbans is therefore essential. The primary focus of this study was the geographical pattern analysis of LULC changes, which relates to climate variability.

Few studies have explicitly addressed the Piyali sub-basin of the Indian Sundarbans, which is extremely susceptible to anthropogenic pressures and climate variability. Instead, several have employed NDVI and remote sensing to study vegetation dynamics and LULC variations. In order to capture the socio-ecological aspects of climate change at the sub-basin level, previous research has primarily focused on broad-scale vegetation cover change or climatic impacts without integrating spatiotemporal LULC analysis, NDVI trend evaluation, and farmer perceptions. This study has unique in that it combines statistical regression techniques, local farmer surveys, and multi-decadal satellite-based NDVI analysis to provide a more comprehensive understanding of the connections between vegetation and climate. This two-pronged strategy enables both qualitative validation (through farmers' observations) and quantitative evaluation (via NDVI and regression with climate variables), which has been frequently ignored in previous research.

Therefore, the following were the main goals of this study: (i) to employ NDVI-based classification to examine changes in Land Use Land Cover (LULC) in the Piyali sub-basin over a 34-year period (1990–2024). (ii) Regression analysis will be used to evaluate the connection between vegetation dynamics (NDVI) and climate variables (rainfall and temperature). (iii) to combine the perspectives of farmers with the findings of remote sensing to gain a deeper understanding of the socio-environmental effects of land cover changes and climate change.

13.2 Material and Method

13.2.1 Study Area

The Sundarban region in West Bengal, India, is home to the tidal estuary Piyali River. The Piyali flows south after splitting off from the Bidyadhari River 14 km below Bamanghata. The Kultala Gang connects it to the Matla River, which in turn connects it to the Thakuran. This river was shortened from 51.254 km to 43.64 km in total length. Piyali, Marapai, Piruldah, Nabagram, Tilpi, Badukhola, Merigaunj, Kirtankhola, Kaksha, Andhari, Kundhankhali, Garankhati, and other localities are traversed by this river as it runs from north to south. Over the course of time, LULC eventually gave way to alternative land cover on 60,584 hectares of the Piyali sub-basin. This sub-basin regime extends both longitudinally and latitudinally from 88°25′ E to 88°37′30″ E and from 22°05′ N to 22°24′50″ N (Fig. 13.1). Over the research area, a geographical range in rainfall has been recorded for 1207–1301 mm.

13.2.2 Field Survey

The survey was carried out in 50 different places over the entire study area. Purposive sampling was used to choose respondents, with an emphasis on senior and experienced farmers having a history of farming. A questionnaire was created in order to use a survey method to gather data from the investigated area. In addition to talking with farmers and gathering local data, a questionnaire-based field survey approach was used. The survey had two main parts: perceptions of climate change (temperature, precipitation) and factors that contribute to LULC change (urbanization, soil degradation, and deforestation). The latitude and longitude points were gathered and stored using the Global Positioning System (GPS) software. In the Study area, interviews were conducted with fifty knowledgeable and senior farmers. A pre-planned questionnaire that was presented to farmers during a face-to-face meeting helped to record the data concerning the change in vegetation cover area. The survey data was analyzed and additional graphics were created using

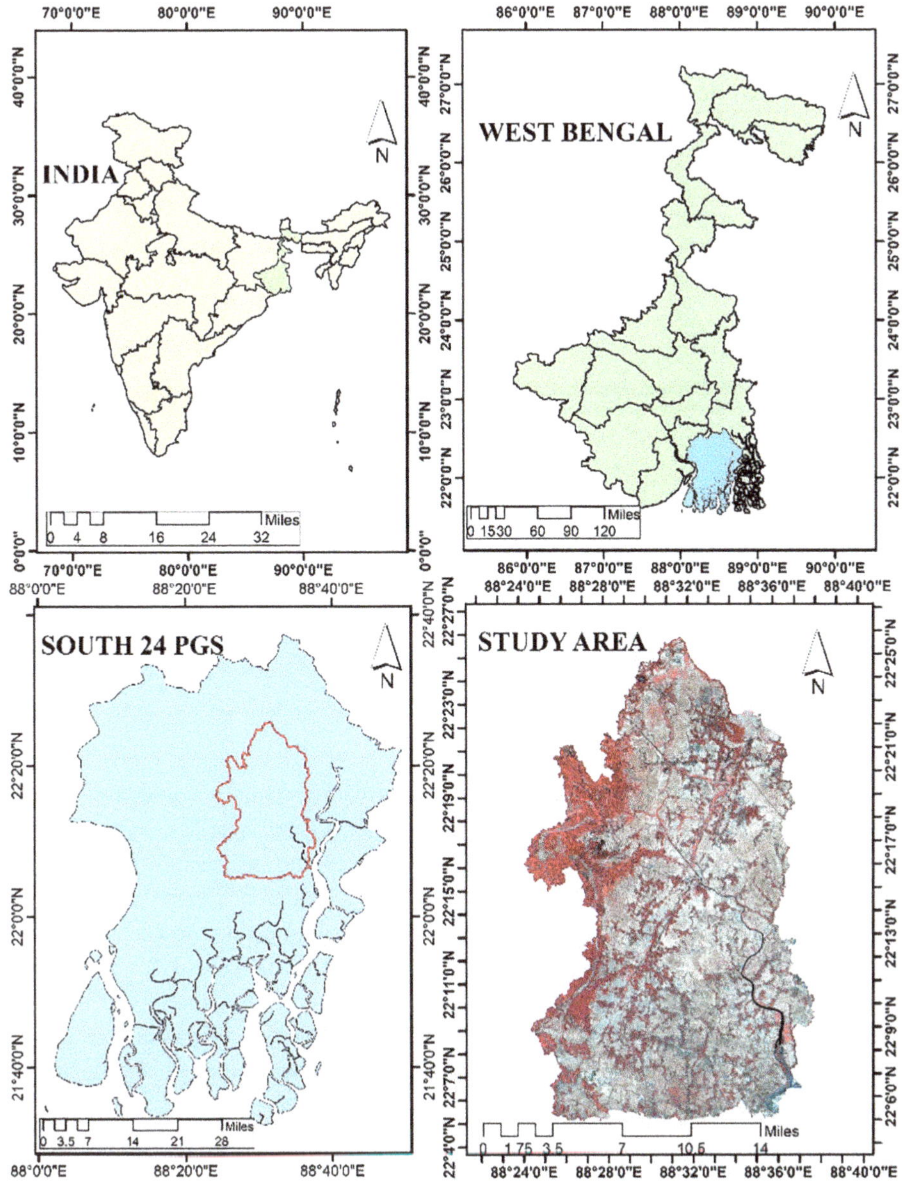

Fig. 13.1 Location of research area (Piyali river sub-basin)

Table 13.1 Spectral details of the satellite imageries

Landsat TM and Landsat ETM+		Landsat OLI/TRIS	
Band	Spectral resolution (μm)	Band	Spectral resolution (μm)
1	0.45–0.52	1	0.433–0.453
2	0.52–0.60	2	0.450–0515
3	0.63–0.69	3	0.525–0.600
4	0.76–0.90	4	0.630–0.680
5	1.55–1.75	5	0.845–0.885
7	2.08–2.35	6	1.560–1.660
		7	2.100–2.300
		8	0.500–0.680
		9	1.360–1.390
		10	10.30–11.30
		11	11.50–12.50

Geographical Information System (GIS) and Microsoft Excel. For categorical replies, percentages and frequency distributions were computed. Responses that were inconsistent or contradictory (such as missing information) were examined and eliminated. ArcGIS 10.8 was used to cross-validate the summarized results with the LULC/NDVI results.

13.2.3 Satellite Data

The study utilized freely available Landsat images with a spatial resolution of 30 m × 30 m, acquired from the USGS Earth Explorer platform (https://earthexplorer.usgs.gov). The Landsat 5 TM (1990, 2000), Landsat 7 ETM+ (2010), and Landsat 8 OLI (2024) were utilized. To ensure temporal consistency for each reference year, one cloud-free image taken during the January–March peak vegetation season was chosen. Every image were pre-processed using ArcGIS 10.8 and ERDAS IMAGINE 2014 for atmospheric correction, radiometric calibration, and layer stacking. Thus, the analysis is based on temporal snapshots rather than mean annual composites, ensuring comparability across the four decades. Table 13.1 illustrated the Spectral details of the selected satellite imageries.

13.2.4 Climate Data

The high-resolution gridded dataset Climatic Research Unit (CRU) Time-Series version (TS) 4.05 (0.5 × 0.5, NET CDF) provided the climatic data like temperature and rainfall data (https://crudata.uea.ac.uk/cru/data/hrg/). Arc GIS software was used to analyze this climate dataset for the years 1990, 2000, 2010, and 2020.

13.2.5 NDVI Calculation

The different LULC classes were identified with the use of the NDVI data. NDVI classes along with vegetation-based change evaluations were determined using Arc GIS 10.8 (Hussain et al. 2020).

$$\mathrm{NDVI} = \frac{NIR - RED}{NIR + RED} \tag{13.1}$$

In a spectrum, RED denotes reflectance in the red band while NIR denotes reflectance in the near-infrared band. A thorough explanation of the method was depicted in Fig. 13.2.

13.3 Results

13.3.1 Approaches of Cultivators About LULC and Climate Changing

A survey was created to learn more about farmers' perceptions of climate change, their replies about how it affects their farming methods, and the mitigation techniques they employ (Siddiqui et al. 2020; Weng 2009). The questionnaire was split up into several sections for this reason, and the findings from each section are explained in the research area one at a time. A large number of farmers who responded to questions about climate change components said that agriculture has been affected by the changing climate. As shown in Table 13.2, the majority of respondents (85%) believed that the climate has changing, and 70% of farmers saw how it affected their farming methods. Almost 81% of respondents said that the primary cause of climate change such as rising temperature is tree chopping. Furthermore, 61% of farmers in the research period believed that climate change would pose a serious threat to the agricultural industry in the future.

According to the Fig. 13.3, 32% of farmers in the research area rely on tub-well water, and 45% farmer utilize canal water for irrigation. Rainwater was used about 15% by farmers for irrigation during the monsoon season. There has been little utilization of pond water (8%) for agriculture in this research region for irrigation purpose because there are few ponds available and water is limited during the summer.

13.3.2 LULC Changes Based on NDVI

Satellite images are used to analyze vegetation metrics (NDVI) based on the spectral characteristics of vegetation (Usaman et al. 2013; Campo-Bescós et al. 2013). Based on the vegetative index, the amount of vegetation in LULC was estimated using several Landsat bands. Vegetation cover (NDVI) for 1990, 2000, 2010, and 2024 is determined in

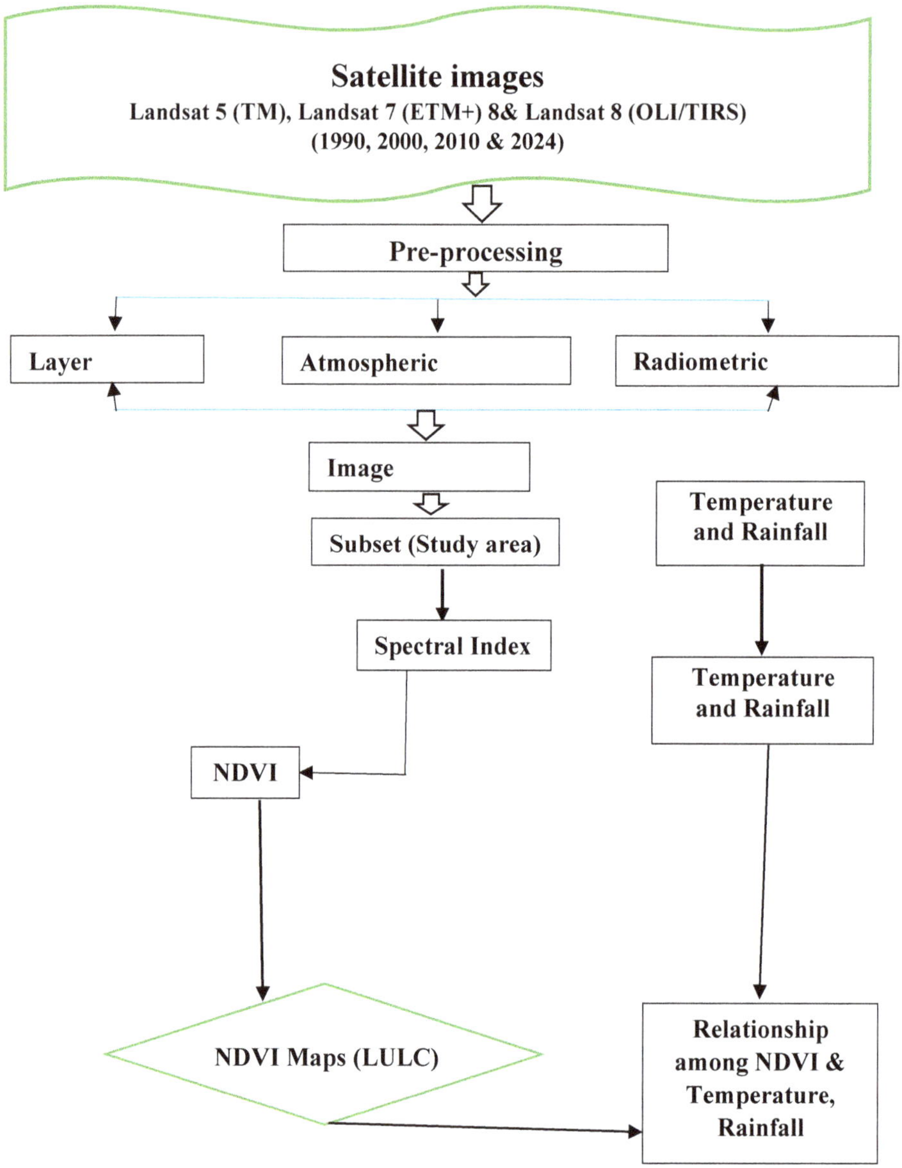

Fig. 13.2 Methodology of current research

Table 13.2 Different statements of respondents in Piyali river sub-basin region

Statement	Agree (%)	Disagree (%)	Not sure (%)	Do not know (%)
Changing of climate	85	5	6	4
Agriculture effected by climate change	70	2	20	8
Climate change due to deforestation	81	12	5	2
Climate change threats in future	61	6	23	10

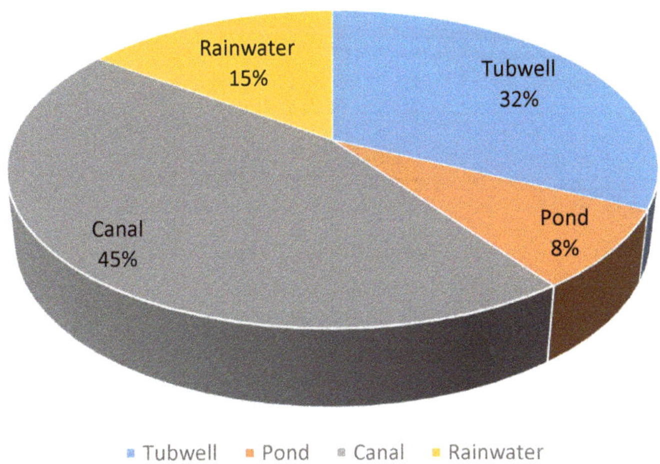

Fig. 13.3 Different sources of irrigation with concentration in study sub-basin

the study area using ERDAS Imagine 2015. Arc GIS 10.8 was utilized to generate spatial distribution map of NDVI values for the purpose of identification of LULC features.

According to NDVI classification, the Piyali River sub-basin's LULC category was analyzed in 1990, 2000, 2010, and 2024 (Fig. 13.4). The region was surrounded by a variety of land features, including vegetation, agriculture, build-up, and water bodies.

The "build-up area" expanded 55.01 km^2 (9.13%) to 154.16 km^2 (25.29%) during 1990 to 2024 (Table 13.3). Two categories of vegetation cover, such as dense and sparse vegetation, were recognized.

In 1990, 2000, 2010 and 2024, thick vegetation encompassed 182.10, 88.43, 86.17, and 79.28 km^2 of the overall land area, respectively. The vegetation area has shrunk between 1990 and 2024 (Fig. 13.5).

The research region's "vegetation" tends to have declined negatively (−17%) depicted in Fig. 13.6. The largest amount of agricultural land was found 304.30 km^2 in the decade 2000. The amount of agriculture land was increased by 71.58 km^2 between 1990 and

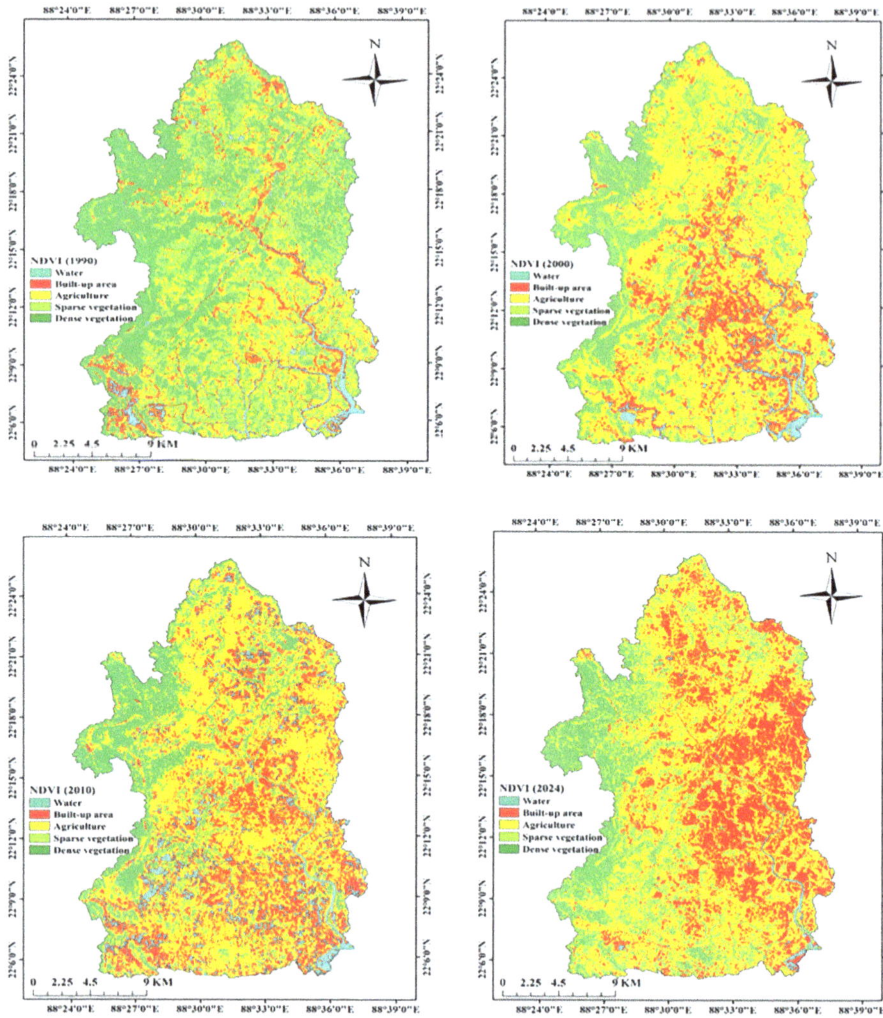

Fig. 13.4 LULC category according to NDVI classification in the years 1990, 2000, 2010 and 2024

2024. The water bodies were measured between 1990 and 2024 and ranged from 16.54 km^2 (2.75%) to 3.93 sq. km (0.56%). As a result water bodies area were dropped by − 2.19% during 1990 to 2024. Water bodies and vegetative regions have been replaced by built-up areas over the past 34 years.

On average, the NDVI low values dropped from −0.904 to −0.085 and the NDVI peak values dropped from 0.817 to 0.381 during 1990 to 2024. Vegetation cover has declined throughout the entire study area. The highest NDVI score was 0.817, indicating the greatest proportion of vegetation cover (forests, bushes, and crops), while the lowest

Table 13.3 Different types of LULC categories with cover in area for the year 1990–2024

LULC classes	1990		2000		2010		2024		1990–2024 (Changing LULC)	
	km^2	%	km^2	%	km^2	%	km^2	%	km^2	%
Water bodies	16.54	2.75	16.53	2.74	33.89	5.63	3.93	0.56	−12.61	−2.19
Built-up area	55.01	9.13	97.86	16.25	144.28	23.95	154.16	25.59	99.15	16.46
Agriculture	147.37	24.47	304.30	50.52	302.96	50.29	218.95	36.35	71.58	11.88
Sparse vegetation	201.35	33.43	95.21	15.80	35.03	5.81	146.52	24.32	−54.83	−9.11
Dense vegetation	182.10	30.23	88.43	14.68	86.17	14.30	79.28	13.16	−102.8	−17

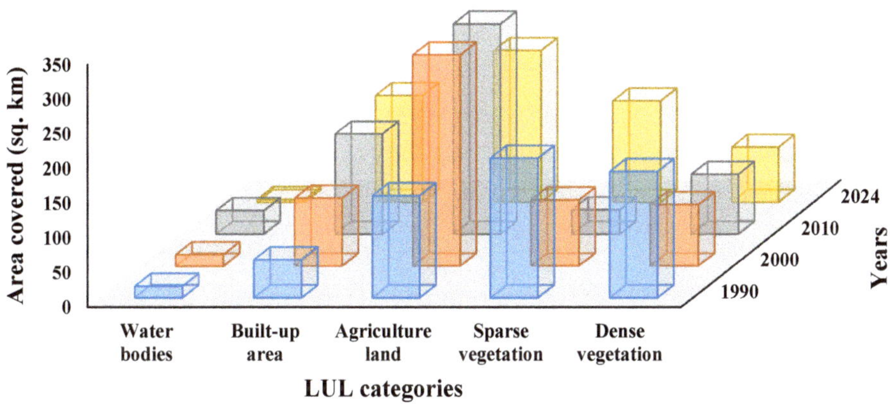

Fig. 13.5 LULC categories concentration in sq.km for the year 1990–2024

NDVI value was -0.904, indicating a greater amount of water bodies in 1990. The greatest NDVI score was 0.464 in 2000, indicating much less plant cover than in 1990. There was enough water bodies, as shown by the lowest NDVI value of −0.346 in 2000 (Table 13.4). The vegetation cover was found higher than the preceding decade in 2010, as evidenced by the maximum NDVI value of 0.555. The study area was more deforested, as indicated by the maximum NDVI value of 0.381 in 2024. Since this year's vegetation is much less abundant and the region receives less rainfall, it contains less water, as indicated by the minimal NDVI value of −0.085. Overall, the study area's NDVI value showed that plant cover continued to decline over time as a result of an increase in built-up and agricultural

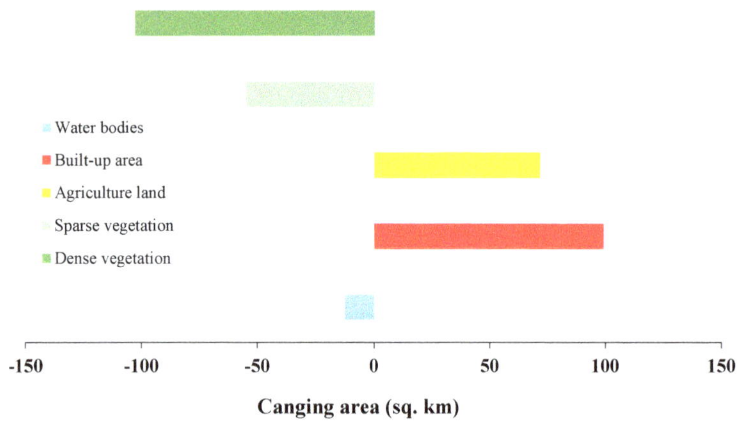

Fig. 13.6 LULC classes changing over the years in Piyali river sub-basin (1990–2024)

Table 13.4 NDVI and temperature values over the years in Piyali river sub-basin

Year	NDVI				Year	Temperature (Max.)			
	Min	Max	Mean	SD		Min (°C)	Max (^{0}c)	Mean (°C)	SD
1990	−0.904	0.817	0.13	0.47	1990	24.29	24.57	24.43	0.09
2000	−0.346	0.464	0.09	0.25	2000	24.37	24.57	24.47	0.06
2010	−0.368	0.555	0.12	0.27	2010	24.40	24.62	24.50	0.07
2024	−0.085	0.381	0.15	0.14	2020	24.66	24.87	24.76	0.07

areas. It was observed that the NDVI value fluctuated significantly from the past to the present.

13.3.3 Climate Factors

The high-resolution gridded dataset CRU TS 4.05 (0.5 × 0.5, NET CDF) was used to gather the temperature and precipitation data. The spatial map of rainfall and temperature was interpolated using an Inverse Distance Weighting (IDW) tool in ArcGIS. The temperature and rainfall maps of the Piyali River sub-basin was shown in the Fig. 7a. For the years 1990–2024, the low temperature was recorded at 24.29 °C, while the high temperature ranged to 24.87 °C. Maximum temperature has been observed in the north-eastern portion of the research region, whereas the southern portion of the sub-basin area experienced the lowest temperatures. In the Piyali River Sub-Basin, the mean temperature was recorded at 24.43 to 24.76 °C between 1990 and 2024. We found that over the past 34 years, the temperature has risen by 0.58 °C as a result of growing build-up areas and

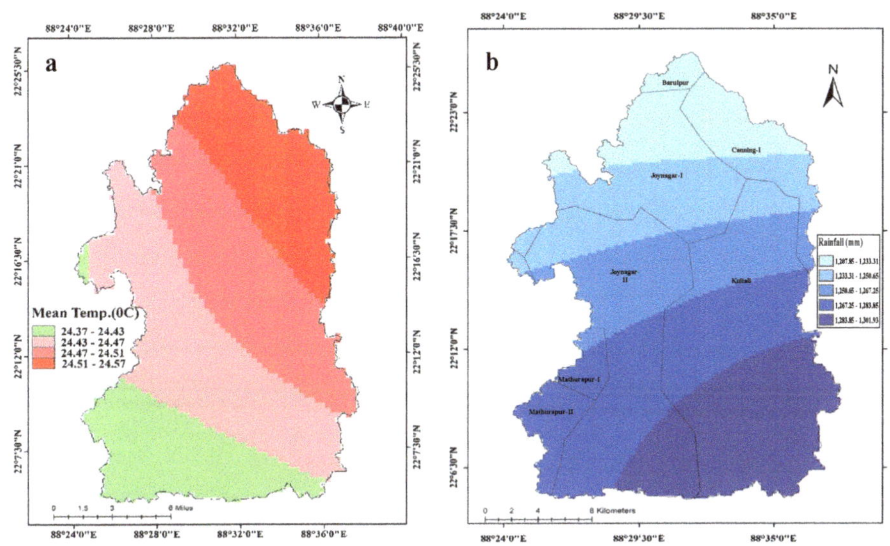

Fig. 13.7 Mean temperature (**a**) and rainfall (**b**) distribution map in Piyali river sub-bsin

declining vegetation cover. The build-up areas exhibited the highest temperature, while the water bodies and vegetation-covered areas experienced the lowest temperature according to our study result.

The Piyali sub-basin's rainfall distribution map has been shown in Fig. 7b. It is evident that the sub-basin regime's northern region received the least amount of rainfall. Conversely, the southern portion of the research sub-basin received the majority of rainfall.

The graph indicates the spatiotemporal association between mean temperature (°C) and rainfall (in cm) at five different locations within the basin regime (Fig. 13.8). There is a noticeable upward trend in rainfall from site 1 (about 120.8 cm) to site 5 (approximately 130.4 cm). Precipitation increases gradually throughout the basin regime, with site 1 recording the lowest rainfall and site 5 recording the greatest. This spatial variation raises the possibility of local hydroclimatic causes causing heavier precipitation in the basin's terminal or downstream regions. The mean temperature, on the other hand, shows a progressive downward trend from site 1 ($\approx$24.57 °C) to site 5 ($\approx$24.37 °C). Although there is not much of a temperature difference across sites ($\approx$0.2 °C), the steady reduction indicates a negative correlation with rising rainfall. This inverse relationship is common in humid regimes, where increased rainfall causes local temperatures to be moderated by evapotranspiration and cloud cover.

These patterns are important for comprehending the interplay between climate and basin-scale hydrology. The observed increase in rainfall points to the possibility of

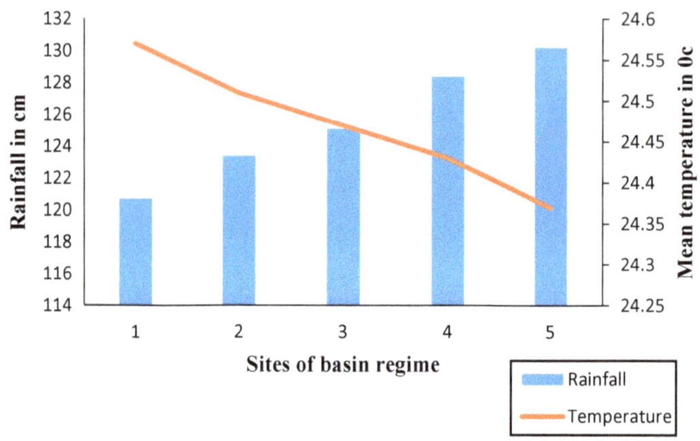

Fig. 13.8 Temperature and rainfall graph in Piyali river sub-basin

increased surface runoff and groundwater recharge at downstream locations. In the meantime, the temperature gradient shows the basin's microclimatic variability, which can affect evapotranspiration rates, vegetation distribution, and water resource management.

13.3.4 Correlation Between NDVI and Climate Factors

Vegetation cover has been impacted by global atmospheric change in the past (Tariq & Shu., 2020). Temperature and rainfall, two aspects of the climate, were linked to LULC more and more. Regression analysis (R^2) was used to determine the association between temperature and vegetation cover (NDVI) in the study area. There was a negative correlation between temperature and NDVI, according to the regression (R^2) tendency.

The current study found that the regression coefficients (R2) between temperature and NDVI (vegetation) were −0.795 for the year 1990 to 2024 (Fig. 13.8). The relation between rainfall and NDVI exhibited a positive correlation of 0.958 (Fig. 13.9).

According to regression analysis, the research area's NDVI must drop as the temperature rises. The results of the study indicated that the temperature in the built-up region higher than the areas covered by vegetation and other land. Additionally, our results indicated that the vegetation dynamics mostly influenced by the cumulative temperature during the growing season.

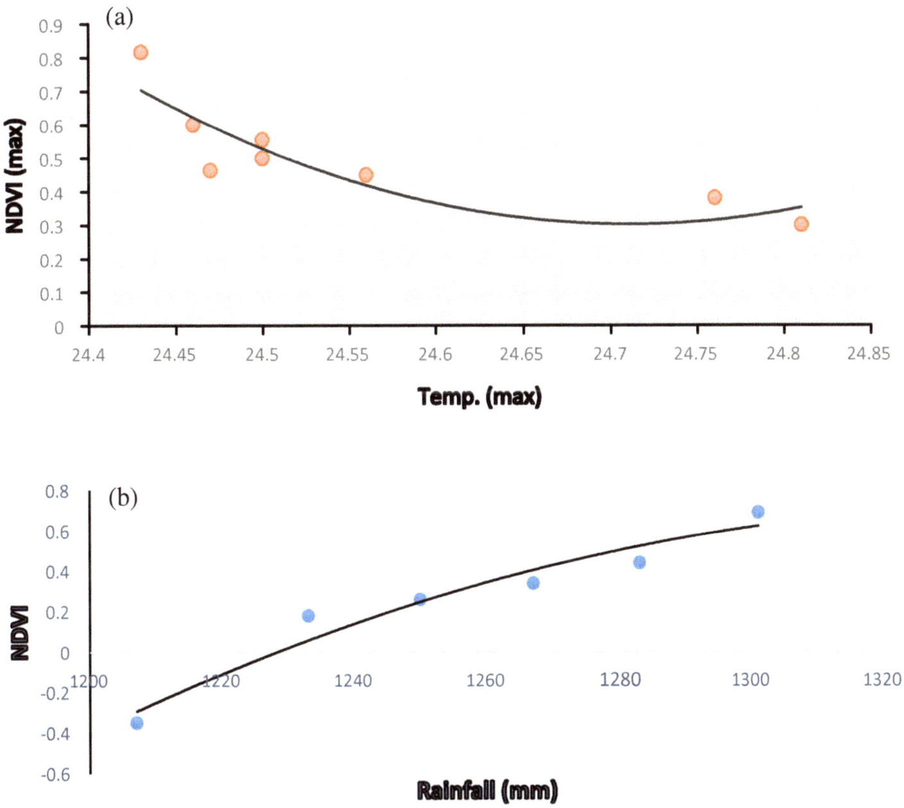

Fig. 13.9 **a** The relation between temperature and NDVI of study area, **b** The relation between Rainfall and NDVI of research area

13.4 Discussion

This study examined that the Piyali River sub-basin's natural resources were affected by climate change between 1990 and 2024. To identify which NDVI classes in the research area are most susceptible to changes in the environment, we employed multi-temporal remote sensing data. This approach is crucial for improving comprehension of the dynamics of the LULC pattern and environmental changes. The research area's rapidly growing built-up structure undoubtedly had the biggest impact on LULC pattern. The results of this study were demonstrated that notable shifts in vegetation patterns and other land use in the Piyali sub-basin over the last 34 years. Rising temperatures have a negative impact on vegetation health and cover, according to the regression study showing a negative association between temperature and NDVI. This is consistent with earlier research that demonstrated that climate factors, especially temperature, are important determinants of vegetation dynamics (Hussain et al. 2020; Zhang et al. 2017). The region's green cover appears to have deteriorated over time, as seen by the NDVI values falling from 0.24 in 1990 to 0.20 in 2024 and a 12% decrease in vegetation cover. The delicate ecosystems of West Bengal, which are already under stress from numerous climatic events, saline water intrusion, and unpredictable monsoon patterns, should be especially concerned about this trend. The built-up areas were increased from 2.25% to 9.70% highlights the strain urbanization, which causes habitat fragmentation and deforestation, both of which degrade ecosystem resilience. Local knowledge was incorporated into survey data to enhance the quantitative analysis of current research.

The research highlights the everyday experiences of rural communities facing environmental instability, with 70% reporting agricultural effects and 85% of farmers acknowledging climate change. Scientific evidence supports the strong belief held by 81% of farmers that deforestation is a contributing factor to climate change. Deforestation decreases the ability of forests to sequester carbon, which raises CO2 concentrations in the atmosphere and speeds up global warming. Furthermore, a reduction in tree cover increases heat stress and modifies local rainfall patterns by lowering evapotranspiration and soil moisture retention. Forest decline and climate variability in the Piyali sub-basin are directly linked by these processes.

The NDVI has a strong association with green plant, which was determined mathematically from the spectral bands in the satellite image (Weng et al. 2004). According to Sabagh et al. (2020), the NDVI as a parameter used to measure the quantity, quality, and development of vegetation. In this research, the use of NDVI as a stand-in for vegetation change has proven to be a reliable method, confirming its applicability for regional vegetation analysis. Anthropogenic factors have significantly changed the LULC structure of the Piyali sub-basin, especially uncontrolled built-up expansion, intensified agriculture, and excessive groundwater extraction. In addition to causing habitat fragmentation, the conversion of vegetated land into croplands and communities also compromises watershed resilience by interfering with the management of surface runoff and soil infiltration capacity. These changes brought about by humans exacerbate the consequences of climate change by generating feedback loops that further deteriorate ecosystem services.

These LULC alterations have long-term effects on watershed hydrology in addition to vegetation dynamics. When there is less plant cover, infiltration is reduced, surface runoff is increased, and flash floods are more likely to occur during periods of heavy rainfall. On the other hand, dwindling groundwater recharge makes extended dry spells worse. A destabilized hydrological cycle results from the cumulative effect, increasing agriculture's susceptibility to floods and droughts. Such changes could worsen saline intrusion in the Sundarbans and put food security at much greater risk.

Outcomes from this study have significant policy ramifications. First, sustainable urban planning, reforestation projects, and immediate land use zoning are required due to data-driven proof of vegetation loss and growing built-up areas. The second requirement is that climate adaptation methods be localized, combining traditional ecological knowledge with GIS-based monitoring technologies to make them community-responsive and data-rich.

13.5 Policy Recommendations for Climate-Resilient Watershed Management

This study suggests a number of actions to maintain the climatic resilience and hydrological sustainability of the Piyali sub-basin in light of the observed vegetation reduction, the fast urbanization, and the negative vegetation–temperature correlation:

Sustainable Land Use Planning—Enforce stringent zoning laws to limit uncontrolled built-up growth. Eco-sensitive infrastructure development can be guided by incorporating LULC maps into planning frameworks.

Reforestation and Afforestation Programs—large-scale native species planting along riparian zones and degraded areas to regulate microclimates, increase carbon sequestration, and restore vegetation cover.

Watershed-Based Agricultural Management—Encourage crop diversity, climate-smart agriculture (CSA), and effective irrigation to lessen soil deterioration and water stress.

Groundwater Regulation—To ensure aquifer recharge and avoid over-extraction during dry seasons, monitoring and community-led governance procedures should be established.

Flood and Salinity Management—To balance runoff, stop salt intrusion, and enhance water storage during extreme events, build small-scale check dams, recharge ponds, and embankment reinforcements.

Integration of Local Knowledge—Enhance farmer involvement in watershed governance by integrating GIS/RS-based monitoring tools with traditional ecological knowledge.

Policy Synergy—Local watershed management plans should be consistent with national initiatives for sustainable agriculture, climate adaptation, and the Sustainable Development Goals (SDGs) of the UN.

In addition to reducing the negative impacts of urbanization, deforestation, and climate stress on the Piyali sub-basin, these actions taken together would stabilize the hydrological and ecological systems of the area, increasing its long-term resilience.

13.6 Conclusions

This work uses remote sensing, NDVI analysis, and GIS techniques to effectively emphasize the effects of anthropogenic activities and climate change on the dynamics of Land Use Land Cover (LULC) in the Piyali sub-basin over a 34-year period. The results show that built-up areas have significantly increased while vegetation cover has decreased, indicating that human strain on land resources continues increasing.

A deterioration in vegetative health, probably caused by both land conversion and warming temperatures, can be seen in the observed drop in mean NDVI values from 0.24 to 0.20.

A negative correlation between temperature and vegetation cover was validated by the regression analysis, supporting previous research on the detrimental impacts of climate change on plant ecosystems. Moreover, community understanding of climate-related adverse effects was highlighted by local farmer surveys, where most respondents linked climate change and deforestation to decreased agricultural production and changed weather patterns. This study not only described the significance of NDVI and geospatial toolsQuery in detect LULC alteration but also underscores the need for integrated, climate-resilient land management policies. It demands that local stakeholders be included in sustainable land use planning, afforestation initiatives, and climate adaption plans that are based on both empirical evidence and firsthand knowledge from the community. A more thorough picture of landscape changes could be provided by future studies that integrate socioeconomic factors, examine hydrological implications, and use satellite data with higher temporal resolution. Thus, our study promotes evidence-based policymaking in climate-vulnerable areas like the Sundarbans and offers important insights into sustainable land governance.

References

Abdullahi S, Pradhan B (2018) Land use change modeling and the effect of compact city paradigms: Integration of GIS-based cellularautomata and weights-of-evidence techniques. Environ Earth Sci 77:251

Ahmed B, Kamruzzaman MD, Zhu X, Shahinoor RMD, Choi K (2013) Simulating land cover changes and their impacts on land surface temperature in Dhaka, Bangladesh. Remote Sens 5:5969–5998

Campo-Bescós MA, Muñoz-Carpena R, Southworth J, Zhu L, Waylen PR, Bunting E (2013) Combined spatial and temporal effects of environmental controls on long-term monthly NDVI in the Southern Africa Savanna. Remote Sens 5:6513–6538

Debnath J, Sahariah D, Lahon D (2022) Geospatial modeling to assess the past and future land use-land cover changes in the Brahmaputra Valley, NE India, for sustainable land resource management. Environ Sci Pollut Res 30:106997–107020. https://doi.org/10.1007/s11356-022-24248-2

Din MSU, Mubeen M, Hussain S, Ahmad A, Hussain N, Ali M A, Nasim W (2022) World Nations priorities on climate change and food Security. In: Building climate resilience in agriculture. Springer, Cham, pp 365–384

Fu P, Weng Q (2016) A time series analysis of urbanization induced land use and land cover change and its impact on land surface temperature with Landsat imagery. Remote Sens Environ 175:205–214

Hoffmann P, Krueger O, Schlünzen KHA (2012) Statistical model for the urban heat island and its application to a climate change scenario. Int J Climatol 32:1238–1248
Hussain S, Ahmad A, Wajid A, Khaliq T, Hussain N, Mubeen M, Farid HU, Imran M, Hammad HM, Awais M (2020) Irrigation scheduling for cotton cultivation. In: Cotton production uses. Springer, Singapore, pp 59–80
Kogo BK, Kumar L, Koech R (2019) Analysis of spatio-temporal dynamics of land use and cover changes in Western Kenya. Geocarto Int 36(4):376–391. https://doi.org/10.1080/10106049.2019.1608594
Li W (2020) Mapping urban impervious surfaces by using spectral mixture analysis and spectral indices. Remote Sens 2020(12):94
Omran ESE (2012) Detection of land-use and surface temperature change at different resolutions. J Geogr Inf Syst 4:189–203
Pal S, Ziaul S (2017) Detection of land use and land cover change and land surface temperature in English Bazar urban centre. Egypt J Remote Sens Space Sci 20:125–145
Pradhan B, Al-Najjar HAH, Sameen MI, Tsang I, Alamri AM (2020) Unseen land cover classification fromhigh-resolution orthophotos using integration of zero-shot learning and convolutional neural networks. Remote Sens 12:1676
Sabagh AE, Hossain A, Islam MS, Iqbal MA, Fahad S, Ratnasekera D, Llanes A (2020) Consequences and mitigation strategies of heat stress for sustainability of soybean (*Glycine max* L. Merr.) Production under the changing climate. In: Plant stress physiology. IntechOpen, London, UK
Shah SHIA, Yan J, Ullah I, Aslam B, Tariq A, Zhang L, Mumtaz F (2021) Classification of aquifer vulnerability by using the drastic index and geo-electrical techniques. Water 13:2144
Siddiqui S, Ali Safi MW, Rehman NU, Tariq A (2020) Impact of climate change on land use/land cover of Chakwal District. Int J Econ Environ Geol 11:65–68
Tariq A, Shu H (2020) CA-Markov chain analysis of seasonal land surface temperature and land use landcover change using optical multi-temporal satellite data of Faisalabad, Pakistan. Remote Sens 12:3402
Usman U, Yelwa SA, Gulumbe SU, Danbaba A, Nir R (2013) Modelling relationship between NDVI and climatic variables using geographically weighted regression. J Math Sci Appl 1:24–28
Weng Q (2009) Thermal infrared remote sensing for urban climate and environmental studies: methods, applications, and trends. ISPRS J Photogramm Remote Sens 64:335–344
Zahoor SA, Ahmad S, Ahmad A, Wajid A, Khaliq T, Mubeen M, Hussain S, Din MSU, Amin A, Awais M (2019) Improving water use efficiency in agronomic crop production. In: Agronomic crops. Springer, Singapore, pp 13–29
Zhang X, Wang D, Hao H, Zhang F, Hu Y (2017) Effects of land use/cover changes and urban forest configuration on urban heat islands in a Loess Hilly Region: case study based on Yan'an City, China. Int J Environ Res Public Health 14:840

www.ingramcontent.com/pod-product-compliance
Ingram Content Group UK Ltd.
Pitfield, Milton Keynes, MK11 3LW, UK
UKHW051838180926
475692UK00005B/85
* 9 7 8 3 0 3 2 1 6 2 6 0 1 *